Organometallic Catalysts and Olefin Polymerization

Springer

Berlin
Heidelberg
New York
Barcelona
Hong Kong
London
Milan
Paris
Singapore
Tokyo

R. Blom, A. Follestad, E. Rytter, M. Tilset, M. Ystenes (Eds.)

Organometallic Catalysts and Olefin Polymerization

Catalysts for a New Millennium

 Springer

Dr. Richard Blom
Hydrocarbon Process Chemistry
Sintef Applied Chemistry
P.O. Box 124 Blindern
N-0314 Oslo, Norway

Arild Follestad
Borealis AS
3960 Stathelle
Norway

Prof. Erling Rytter
Institute of Chemical Engineering
NTNU
7491 Trondheim
Norway

Prof. Mats Tilset
Department of Chemistry
University of Oslo
P.O. Box 1033 Blindern
0315 Oslo
Norway

Prof. Martin Ystenes
Department of Chemistry
NTNU
N-7491 Trondheim
Norway

ISBN 3-540-41402-9 Springer-Verlag Berlin Heidelberg New York

Library of Congress Cataloging-in-Publication Data
Die Deutsche Bibliothek – CIP-Einheitsaufnahme

Organometallic catalysts and olefin polymerization : catalysts for a
new millenium / R. Blom ... (ed.). – Berlin ; Heidelberg ; New York ;
Barcelona ; Hong Kong ; London ; Milan ; Paris ; Singapore ; Tokyo :
Springer, 2001
    ISBN 3-540-41402-9

Springer-Verlag Berlin Heidelberg New York
a member of BertelsmannSpringer Science+Business Media GmbH

© Springer-Verlag Berlin Heidelberg 2001
Printed in Germany

Cover design: E. Kirchner, Heidelberg
Typesetting: Camera-ready by authors

Printed on acid-free paper   SPIN 10771409   02/3020 hu – 5 4 3 2 1 0 –

Transition State – A Preface

"Catalysis is more art than science", probably all of you have heard and even used this expression. Whether it is true or not, it alludes to the experience that new catalysts are hard to find, and near impossible to predict. Hard work and a lifetime of experience is invaluable. However, a keen mind might give insight into where to search, but not necessarily about where to find the answers.

Historically, "quantum leaps" have often arisen from serendipity - we all know the story about the nickel-contaminated reactor that triggered further research towards the first coordination catalyst for ethene polymerization. Taking advantage of this event, Karl Ziegler became the first chemist to earn both a Nobel prize and a fortune for the same invention. A broken NMR tube helped Walter Kaminsky discover the effect of high concentrations of methylaluminoxanes as cocatalysts for metallocenes. When air reacted with the concentrated trimethyl aluminum solution, sufficient amounts of methylaluminoxanes were formed, and the lazy catalyst dormant in the NMR tube suddenly became sensationally active. Ziegler and Kaminsky were lucky and had the genius needed to take advantage of their luck.

The age of olefin polymerization catalysis is approaching half a century. Within the third millennium, we may assume that most of the questions we ask today will be answered, however, scientists are driven to obtaining answers within their lifetimes. What is science, but a hunger to understand, to understand why experiments do not work and get answers that make sense.

The annual production of polyolefins is a hundred million tons, with a value of approximately half a trillion Norwegian kroner. Although other technologies take a major share of this, still roughly a third is made with the help of catalysts described in the Ziegler patent of 1953. Hence there should be more than enough motivation to ensure a sufficient driving force to find better catalysts and processes. However, economics lives its own life, for an economist, a hen is just the tool that an egg needs to produce another egg. In the same way, science can be viewed as a tool to create profit. The ultimate questions of how and why the catalysts work may not necessarily create profit at all, and even if they did, the shareholders might not understand – or we might not be able to convince them. It is also a part of the problem that science can be frustratingly difficult. A scientific investigation leads to a number of answers, but also generates a high number of new questions that need to be answered. In this manner science will never obtain an exact solution, but will converge towards an increasingly better understanding of Nature in terms of descriptive and predictive power. Several ideas on how heterogeneous Ziegler-Natta and chromium catalysts work have been tested, some with success, but none reaching universal acceptance. The models can, in principle, not be proven, so no model study can be taken as proof for the function of the real catalysts. The complexity of the issue has led to an enormous amount of data, a lot of which is of high quality, but also data of lower quality that may lead to erroneous conclusions. The pool of knowledge, although rich and useful, is too complex for an authoritative answer to crystallize. The complexity is a marvel for a scientist seeking the ultimate challenge, but not always for those who pay for solving problems.

With the event of the new well-defined single site catalysts and with the help of advanced quantum calculations we now have the potential of understanding what is going on. It has been stated that scientists working on traditional Ziegler-Natta catalysts know more than they assume, and that those working on metallocenes know less than they claim. We may all soon experience that theoretical calculations will be required to verify experimental results before they are accepted. For some dedicated experimentalists this may be unbearable, but for much of science today this is already standard.

When will the power of quantum chemical calculations be sufficient to test our models and find out which data are correct? Give me enough computer resources and we can do it now, said one of the OCOP2000 participants. If this were to be the case, the experimentalists among us may be well advised to publish all our ideas before the theoreticians do. Fortunately, shareholders still seem to have limited confidence in basic science, so this may take some years, but not many – the art may be about to yield to science!

Two years have gone since the related conference in Hamburg. Only two years, and yet exciting, unexpected findings were presented in this conference. These proceedings include 39 selected papers from the 42 lectures and 121 posters presented at the conference. We hope that their combined impact will give an impression of transition in this field, as well as an overview of some of the most significant new developments in this dynamically influential field of science.

We would like to thank all the authors and all the participants at the conference for their scientific contributions and for bearing the major share of the costs of the conference. Generous financial support from the Norwegian Research Council, the Vista foundation and Borealis is gratefully acknowledged. Our coordinator Vesla Haegh is thanked for being the solid rock everyone of us could rely on when confusion was at its worst. Students from the University of Oslo and from NTNU, Trondheim, are thanked for their enthusiastic assistance during the conference, and finally, the University of Oslo is acknowledged for offering the use of the Sophus Lie lecture hall.

October 2000   Richard Blom
                                                                 Arild Follestad
                                                                 Erling Rytter
                                                                 Mats Tilset
                                                                 Martin Ystenes

Contributors

Table of Contents

1. Group IV Catalysts
and Cocatalysts

UV/VIS Studies on the Activation of Zirconocene-Based Olefin-Polymerization Catalysts

Ulrich Wieser and Hans-Herbert Brintzinger

Fachbereich Chemie, Universität Konstanz, D-78457 Konstanz, Germany
E-mail: hans.brintzinger@uni-konstanz.de

Abstract. Equilibria leading to the formation of active zirconocene catalysts were studied by observing changes in the positions of ligand-to-metal charge transfer bands of these complexes. UV/VIS-spectra of $Me_2Si(Ind)_2ZrX_2$ (X = Cl, Me) and $Me_2Si(2\text{-}Me\text{-}Benzind)_2ZrX_2$ (X = Cl, Me) treated with MAO or, in the case of X = Me, with $PhNMe_2H^+$ $(F_5C_6)_4B^-$ indicate that binuclear species of the type $(Cp^x_2ZrMe)_2(\mu\text{-}Me)^+$ are not formed in the MAO-activated reaction systems. Spectra obtained for MAO-activated catalysts are identical irrespective of whether they are derived from dichloride, dimethyl or biphenolate zirconocene derivatives. The abstracted ligands thus appear to be without coordinative contact to the cationic Zr center. Spectra of the species generated with large MAO excess (>1000:1) indicate that cationic trimethylaluminium adducts are formed, as judged by their similarity to spectra of heterometallic dinuclear cations of the type $Cp^x_2Zr(\mu\text{-}Me)_2AlMe_2^+$ $(F_5C_6)_4B^-$. These TMA adducts thus appear to be the catalytically most active entities in MAO-activated reaction systems.

Introduction

The reaction steps which lead to the activation of zirconocene catalysts by methylalumoxane (MAO) are still not adequately understood today, although this reagent is the most frequently used activator for zirconocene-based olefin-polymerization catalysts.[1] Previous studies on reaction systems of the type $Cp^x_2ZrCl_2$/MAO, mainly by NMR and UV/VIS methods, [2-5] uniformly indicate that Cl/Me exchange leads, at relatively low [Al]/[Zr] ratios of 10-20:1, to a monomethyl monochloride species $Cp^x_2ZrClMe$. At higher [Al]:[Zr] ratios, subsequent equilibria appear to lead to complex species with reduced electron densities at their zirconium centers. Discussed in this regard are contact-ion pairs of the type $Cp^x_2ZrMe^{+\cdots}XMAO^-$ (X=Cl, Me, O) [5-11] and homo- and hetero-binuclear cationic species such as $(Cp^x_2ZrMe)_2(\mu\text{-}Me)^+$ or $Cp^x_2Zr(\mu\text{-}Me)_2AlMe_2^+$, presumably associated with their $XMAO^-$ counteranions.[5,9]

UV/VIS methods are particularly suitable to study these reaction sequences, since the characteristic ligand-to-metal charge-transfer bands of zirconocene complexes [10] are not obscured even by a large excess of MAO (which shows no absorption in this spectral region) and, at the same time, quite sensitive to changes in the coordination sphere of the Zr center.[12-14] Deffieux and coworkers have observed that the absorption bands of $C_2H_4Ind_2ZrCl_2$ in toluene undergo, at [Al]/[Zr] ~ 20, first a hypsochromic shift due to formation of $C_2H_4Ind_2ZrClMe$, and then, at [Al]/[Zr] ~150 and ~5000, respectively, two successive bathochromic shifts which indicate the formation of cationic species with more positively charged Zr centers. [4,14] Of these species, the latter appears to be the active catalyst for the polymerization of propene.

To clarify the nature of these active species, we have conducted extended studies on changes in the UV/VIS-spectra of complexes such as $Me_2Si(Ind)_2ZrCl_2$ (**1A**) or $Me_2Si(2\text{-Me-Benzind})_2ZrCl_2$ (**2A**), [15,16] and of some of their ligand substitution derivatives in the presence of MAO [17] or - for comparison - in the presence of a borate activator. [18]

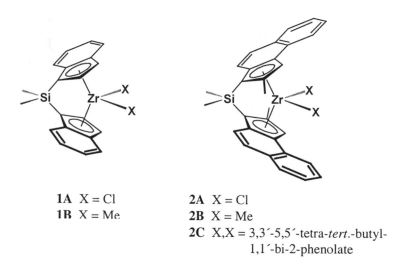

1A X = Cl
1B X = Me

2A X = Cl
2B X = Me
2C X,X = 3,3´-5,5´-tetra-*tert.*-butyl-
           1,1´-bi-2-phenolate

2. Experimental Part

Manipulations were performed under argon by the use of Schlenk techniques or in a nitrogen-filled glovebox with vacuum antechamber. Glassware was dried by heating to 150°C in a dynamic vacuum before being brought in contact with zirconocene solutions.

2.1 Materials

$Me_2SiInd_2ZrCl_2$ was synthesized according to literature reports.[15] $Me_2Si(2\text{-Me-Benz[e]Ind})_2ZrCl_2$ and N,N-dimethylanilinium tetra(pentafluorophenyl)borate were obtained as gifts from BASF AG. Zirconocene dimethyl derivatives were prepared by methods described by Samuel and Rausch.[19] Biphenolate complex **2C** was donated by Dr. Robert Damrau (Targor GmbH). Methylalumoxane (MAO) in toluene solution (total Al content 1.8 M, Al as $AlMe_3$ 30%) was obtained as a gift from Witco (Bergkamen) and used without further purification. Toluene was dried over molecular sieves, purified by refluxing over sodium metal and distilled under dry argon.

2.2 Measurements

Reaction mixtures were prepared by adding to 15 ml of a 2.0-6.0A10^{14} M solution of the appropriate zirconocene complex in toluene small volume increments of 1.8 M MAO or of 1.5*10^{-3} M borate in toluene solutions. UV/VIS spectra of the reaction mixtures were recorded on a Cary 50 Varian spectrometer. For these measurements, an immersion probe with fibre-optical light guide was placed in a specially designed Schlenk vessel. Absorbance values were corrected for dilution by the activator solutions added.

3. Results and Discussion

3.1 Activation of Me$_2$Si(Ind)$_2$ZrCl$_2$ (1A) and Me$_2$Si(2-Me-Benz[e]Ind)$_2$ZrCl$_2$ (2A) with MAO

Addition of increasing amounts of MAO to a toluene solution of Me$_2$Si(Ind)$_2$ZrCl$_2$ causes the spectral changes represented in Figure. As observed by Deffieux and coworkers for the system C$_2$H$_4$Ind$_2$ZrCl$_2$/MAO [4,14], three transformations occur successively: Generation of the monomethyl species Me$_2$Si(Ind)$_2$ZrClMe gives rise to an absorption band at 409 nm, while increasing [Al]/[Zr] ratios cause a first bathochromic shift to 456 nm and, at higher [Al]/[Zr] ratios, a second bathochromic shift to 496 nm. In each case, isosbestic points indicate a clean conversion reaction. An apparent half-way conversion for each of these reaction steps is reached at the [Al]/[Zr] ratios indicated in Table 1.

Similar observations pertain to the reaction system Me$_2$Si(2-Me-Benz[e]Ind)$_2$ZrCl$_2$/MAO (Figure 1): The absorption band for the monomethyl species Me$_2$Si(2-Me-Benz[e]Ind)$_2$ZrClMe appears at 406 nm, while the two bathochromic shifts observed at increasing [Al]/[Zr] ratios generate first a maximum at 448 nm and then two maxima at 430 nm and 484 nm, respectively.

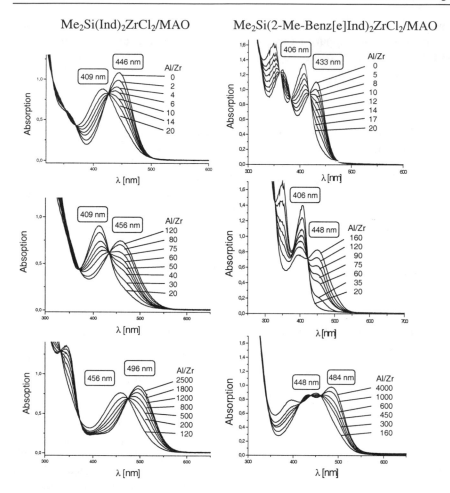

Me₂Si(Ind)₂ZrCl₂/MAO   Me₂Si(2-Me-Benz[e]Ind)₂ZrCl₂/MAO

Fig. 1. UV/VIS-spectra of Me₂Si(Ind)₂ZrCl₂ (**1A, left**) and Me₂Si(2-Me-Benz[e]Ind)₂ZrCl₂ (**2A, right**) with increasing amounts of MAO; **top:** monometylation; **middle:** 1. bathochromic shift; **bottom:** 2. bathochromic shift; [Zr(**1A**)] = 6,0*10⁻⁴ mol/l; [Zr(**2A**)] = 2,4*10⁻⁴ mol/l; 20°C; toluene

The first two conversions, i.e. the formation of the monomethyl complex and the first bathochromic shift are completed within a time span of ca. 2 min after addition of MAO, required to take the first spectrum. The completion of the second bathochromic shift shows some time lag, however. After addition of MAO to Me₂Si(2-Me-Benz[e]Ind)₂ZrCl₂ in a ratio of [Al]/[Zr] = 1000:1, the intensity of the high-wavelength band at 484 nm reaches about 85% of its final value within less than 2 min, but continues to increase slightly even after several hours.

While this time dependence will in principle affect the half-completion values of the [Al]/[Zr] ratio given for the second bathochromic shift in Table 1, this interference is rather limited, since most of the intensity increase occurs relatively fast: When the [Al]/[Zr] ratio is augmented e. g. from 1000:1 to 4000:1, generation of the band at 484 nm is practically complete within a few minutes. Since the exchange of equatorial ligands in a zirconocene cation is quite unlikely to require reaction times of several hours, we assume that some rearrangement of the MAO framework limits the rate at which this reaction step goes to completion.

Table 1. [Al]/[Zr] ratios at half-way conversion for each of the MAO activation steps

complex	[Al]/[Zr] required for monomethylation	[Al]/[Zr] required for 1. bathochromic shift	[Al]/[Zr] required for 2. bathochromic shift
1A	6	40	500
2A	10	60	500
1B	-	25	500
2B	-	45	500

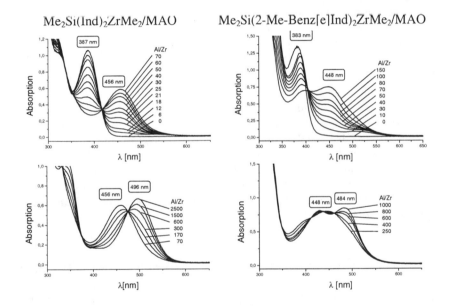

Fig. 2. UV/VIS-spectra of $Me_2SiInd_2ZrMe_2$ (**1B, left**) and $Me_2Si(2-Me-Benz[e]Ind)_2ZrMe_2$ (**2B, right**), respectively, with increasing amounts of MAO; **top:** 1. bathochromic shift ; **bottom:** 2. bathochromic shift; $[Zr(\mathbf{1B})] = 6{,}0*10^{-4}$ mol/l; $[Zr(\mathbf{2B})] = 2{,}4*10^{-4}$ mol/l; 20°C; toluene.

3.2 Activation of the dimethyl complexes 1B and 2B with MAO

Addition of MAO to one of the zirconocene dimethyl complexes **1B** or **2B** results in the immediate appearance of the absorption bands at 456 and 448 nm, respectively, which were generated also by adding excess MAO to the respective zirconocene dichlorides. The same holds for the species arising from the second bathochromic shift at still higher [Al]/[Zr] ratios: Their absorption maxima occur at the same positions, 496 and 484 nm, respectively, as in the dichloride reaction systems (see Figure 2).

Even with the biphenolate zirconocene complex Me$_2$Si(2-Me-BenzInd)$_2$Zr-3,3´-5,5´-tetra-*tert.*-butyl-1,1´-bi-2-phenolate **2C** as starting material, addition of MAO generates again the same bands at 430 and 484 nm as in the cases described above. This MAO-induced transformation is initially slower by at least one order of magnitude than that in an analogous reaction system derived from the dichloride **2A**; it goes to completion noticeably faster, however (see Figure 3).

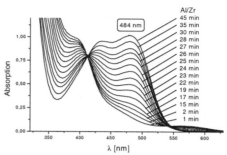

Fig. 3. Time-dependent UV/VIS-spectra of Me$_2$Si(2-Me-BenzInd)$_2$Zr-(3,3´-5,5´-tetra-*tert.*-Bu-1,1´-bi-2-phenolate) + MAO; [Al]/[Zr] ratio = 1000:1; [Zr] = 2,4*10^{-4} mol/l; 20°C; toluene

Whether Cl$^-$, Me$^-$ or a biphenolate dianion is abstracted by excess MAO is thus without influence on the spectra of the resulting cationic species. Since UV/VIS spectra of these cations depend very sensitively on the nature of their counteranions (vide infra), we can conclude that these abstracted anionic ligands are no longer in coordinative contact with the Zr center in these cationic species.

Nevertheless, some influence of these ligands on the MAO-induced anion abstraction is apparent, in that the [Al]/[Zr] ratios required to reach half-way conversions for these transformations are somewhat smaller for **1B** and **2B** than for **1A** and **2A** (Table 1).

3.3 Activation of the dimethyl complexes 1B and 2B with PhNMe$_2$H$^+$(F$_5$C$_6$)$_4$B$^-$

Homometallic binuclear cations of the type (Cp$^x_2$ZrMe)$_2$(μ-Me)$^+$ have been found to arise from cationic zirconocene complexes in the presence of excess dimethyl complex. [20,21] Thus, reaction of Me$_2$Si(Ind)$_2$ZrMe$_2$ (**1B**) with one half equivalent of dimethyl anilinium perfluorotetraphenyl borate has been shown by Bochmann and Lancaster [20] by NMR methods to generate a methyl-bridged homobinuclear species (Me$_2$Si(Ind)$_2$ZrMe)$_2$(μ-Me)$^+$ (F$_5$C$_6$)$_4$B$^-$. This reacts with further dimethylanilinium borate to form the contact ion pair Me$_2$Si(Ind)$_2$ZrMe$^+$ (F$_5$C$_6$)$_4$B $^-$ (see Scheme 1). Apparently, excess zirconocene dimethyl complex can displace the perfluorotetraphenyl borate anion from its contact with the zirconocene methyl cation. (Me$_2$Si(Ind)$_2$ZrMe)$_2$(μ-Me)$^+$ is thus found to be the dominant species in solutions of Me$_2$Si(Ind)$_2$ZrMe$_2$ in the presence of ½ equivalent of a borane or borate activator,[20] while the more highly substituted dimeric cation $(Me_2Si(2\text{-Me-Benz}[e]Ind)_2ZrMe)_2(\mu\text{-Me})^+$ is not formed under analogous conditions.[22] A crystal structure of such a dimeric cation with a sterically crowded borate counter anion has recently been reported by Marks and coworkers. [23] These dinuclear species appear to participate in catalytic polymerization reactions as a stabilizing depository for reactive zirconocene alkyl cations.[22]

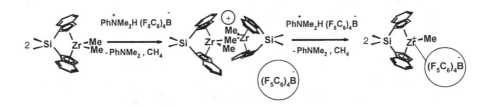

Scheme 1. Reactions of $Me_2Si(Ind)_2ZrMe_2$ (**1B**) with $PhNMe_2H^+$ $(F_5C_6)_4B^-$

Addition of about 0.5 equivalents of $PhNMe_2H^+$ $(F_5C_6)_4B^-$ to $Me_2Si(Ind)_2ZrMe_2$, dissolved in toluene at $-60°C$, causes the UV/VIS band of the dimethyl complex at 387 nm to disappear while giving rise to a new band at 438 nm, which we assign to the dimeric cation $(Me_2Si(Ind)_2ZrMe)_2(\mu\text{-Me})^+$. Upon addition of further dimethylanilinium borate this band disappears again while a new, broad band appears with a maximum at ~ 560 nm, which is to be assigned to the contact ion pair $Me_2Si(Ind)_2ZrMe^+$ $(F_5C_6)_4B^-$ (see Figure 4).

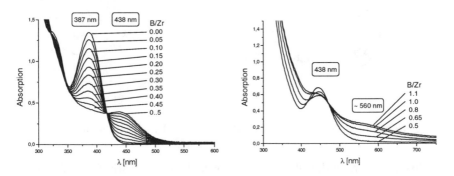

Fig. 4. UV/VIS-spectra of $Me_2Si(Ind)_2ZrMe_2$ **1B** + $PhNMe_2H^+$ $(F_5C_6)_4B^-$; **left:** generation of $(Me_2Si(Ind)_2ZrMe)_2(\mu\text{-Me})^+$ $(F_5C_6)_4B^-$; **right:** generation of $Me_2Si(Ind)_2ZrMe^+$ $(F_5C_6)_4B^-$; $[Zr] = 4,0*10^{-4}\,mol/l$; $-60°C$; toluene

With $Me_2Si(2\text{-Me-BzInd})_2ZrMe_2$ **2B**, for which formation of a dimeric species has been shown to be prevented by steric hindrance of the ligands,[22] reaction with 0.5 equiv. of $PhNMe_2H^+$ $(F_5C_6)_4B^-$ in toluene at $-60°C$ does indeed generate directly, i. e. even in the presence of unreacted dimethyl complex, the contact ion pair $Me_2Si(2\text{-}Me\text{-}BzInd)_2ZrMe^+$ $(F_5C_6)_4B^-$, which shows very broad bands at 420 nm and 510 nm.

For MAO-activated reaction systems we can conclude, from the closely similar behaviour of both reaction systems, **1B/MAO** and **2B/MAO**, and from the absence of the characteristic absorption band of $(Me_2Si(Ind)Me)_2(\mu\text{-}Me)^+$ at 438 nm, that binuclear cations are not present to any detectable extent. It remains to be clarified to which degree different ligand structures and/or lower zirconocene concentrations used in our UV/VIS study contribute to the divergence of this result from those of earlier NMR studies. [5,9]

The cationic species generated by excess MAO have their absorption maxima at substantially shorter wave lengths than the perfluorotetraphenyl borate contact-ion pairs. These differences indicate a higher electron density at the zirconium center and thus a stronger coordination of the MAO-generated anion than of $(F_5C_6)_4B^-$. They document, in any case, that the charge-transfer absorptions of zirconocene cations are highly sensitive with respect to the nature of their counter anions.

3.4 Formation of heterometallic binuclear cations with Al_2Me_6

Trimethylaluminum can displace the borate anions from zirconocene methyl cations to form Me-bridged heterometallic binuclear cations of the type $(Cp^x_2Zr(\mu\text{-}Me)_2AlMe_2^+$ (see Scheme 2).[9,20] These complexes appear to be active in propene polymerization unless an excess of trimethyl aluminum suppresses displacement of $AlMe_3$ from the Zr center. [20]

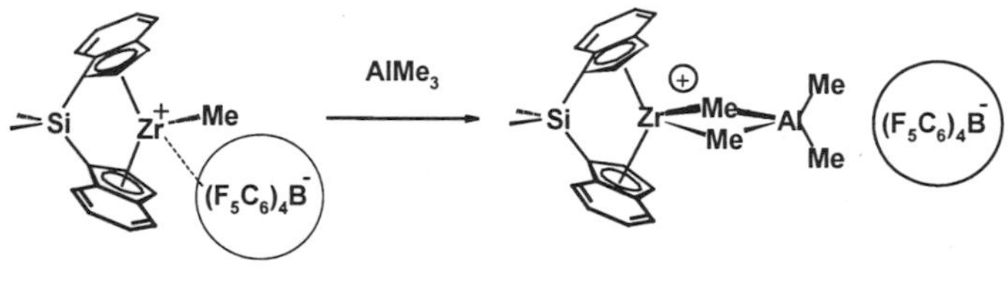

Scheme 2. Reaction of $Me_2Si(Ind)_2ZrMe^+$ $(F_5C_6)_4$ B$^-$ with TMA.

When a mixture of dimethyl complex **1B** and Al_2Me_6 ([Al]:[Zr] =1.2:1, $-60°C$) in toluene solution is slowly treated with a slight excess of the dimethylanilinium borate activator, the initial absorption band at 387 nm is replaced by a new band at 488 nm. Analogous experiments with the dimethyl complex **2B** lead to the formation of two absorption bands at 415 and 485 nm (see Figure 5). The positions of all these absorption bands are remarkably similar to those observed after activation of **1B** and **2B**, or of their dichloride analogs with a large excess of MAO.

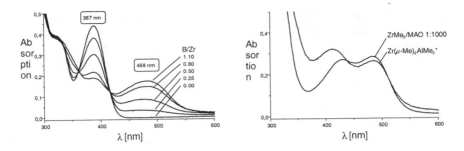

Fig. 5. UV/VIS-spectra for the generation of the heterometallic dinuclear cation $(Me_2Si(Ind)_2Zr\,(\mu\text{-}Me)_2AlMe_2^+$ by reaction of $Me_2Si(Ind)_2ZrMe_2$ (**1B**) + TMA + $PhNMe_2H^+\,(F_5C_6)_4B^-$ (**left**); and comparison of the spectrum of $Me_2Si(2\text{-}Me\text{-}Benz[e]Ind)_2ZrCl_2/MAO$ (**2A**, $[Al]:[Zr] = 4000:1$) with that of $(Me_2Si(2\text{-}Me\text{-}Benz[e]Ind)_2Zr(\mu\text{-}Me)_2AlMe_2^+$ (**right**); $[Zr] = 2,0*10^{-4}$ mol/l; $-60°C$; toluene

4. Conclusions

The absorption bands of the cationic species generated by MAO from various starting complexes are independent of the nature of the abstracted anion. These anions thus appear to be precluded from contact with the Zr center, possibly by exchange of a primarily formed $MAOX^-$ anion with excess MAO under formation of a neutral MAOX cluster and a $MAOMe^-$ anion.[24] The cation arising from the first bathochromic shift would then presumably be in contact with AlOAl, AlMe or $AlMe_2$ groups at the periphery of such a $MAOMe^-$ anion.

In any case, the anion present after the first bathochromic shift is apparently still too strongly coordinated to the Zr center to be displaced by $AlMe_3$. This is indicated also by the relatively short wavelength of the absorption band characteristic for the contact ion present at this stage. Even addition of excess trimethyl aluminum to reaction mixtures at this stage of MAO activation does not induce formation of the species present after the second bathochromic shift. Excess MAO, on the other hand, is apparently capable of transforming this anion into another species of substantially reduced coordinating power, which is then displaced from the Zr center by $AlMe_3$ present in the reaction mixture. The observation that this transformation occurs with some time lag, which depends on the nature of the abstracted anion and on the concentration of excess MAO, is in line with the view that a restructuring of the anionic MAO cluster is involved in this reaction step.

With regard to catalytic activities, data reported by Deffieux and coworkers [14a] suggest that the species present after the first bathochromic shift is rather inefficient as a catalyst for the polymerization of α-olefins. Apparently, olefins cannot displace the $MAOX^-$ or $MAOMe^-$ anions present at this stage to a degree sufficient to maintain substantial rates of olefin insertion.

The species generated by the largest excess of MAO, on the other hand, are likely to be the AlMe$_3$ adducts of the respective zirconocene cation, i. e. hetero-dinuclear cations of the type Cp$^x_2$Zr(μ-Me)$_2$AlMe$_2^+$. This is strongly indicated by the close coincidence of the bands arising from the second MAO-induced bathochromic shift with that of an authentic AlMe$_3$ adduct of the respective zirconocene methyl cation.

The most active species in reaction systems formed with high MAO excess are then probably the AlMe$_3$ adducts of the zirconocene methyl cations, Cp$^x_2$Zr(μ-Me)$_2$AlMe$_2^+$ MAOX$^-$ or closely related species, which arise by displacement of a more weakly coordinating, restructured anion by AlMe$_3$ contained or formed in the reaction system. Addition of excess AlMe$_3$ - e. g. together with higher amounts of MAO - would then be expected, in accord with observations [20,25], to decrease activities again by disfavoring the displacement of AlMe$_3$ from the Zr center by an olefin substrate.

Similar views have recently been proposed by Babushkin et al. [5] on the basis of NMR studies on related reaction systems. Attempts to substantiate this view and to clarify its details are presently under way in our laboratory.

5. Acknowledgements

This work was supported by BMBF and by BASF AG. We thank C.K. Witco GmbH for gifts of MAO solutions, Dr. Robert Damrau for a gift of complex **2C** and Dr. Dmitrii Babushkin for valuable suggestions.

References

1   H. Sinn; W. Kaminsky; H. Hoker, Eds. Alumoxanes; Macromolecular Symposia *97*; Hüthig & Wepf: Heidelberg, **1995** Germany

2   W. Kaminsky; R. Steiger, *Polyhedron* **1988**, *7*, 2375

3   D. Cam; U. Giannini, *Macromol. Chem.* **1992**, *193*, 1049

4   D. Coevoet; H. Cramail; A. Deffieux, *Macromol. Chem. Phys.* **1998**, *199*, 1451

5   D. E. Babushkin; N. V. Semikolenova; V. A. Zakharov; E. P. Talsi, *Macromol. Chem. Phy.*, **2000**, *201*, 558

6   P. G. Gassman; M. R. Callstrom, *J. Am. Chem. Soc.* **1987**, *109*, 7875

7   A. R. Siedle; W. M. Lamanna, A. R. Newmark; J. N. Schroepfer, *J. Mol. Cat.* **1998**, *128*, 257

8   C. Sishta; R. M. Hathorn; T. J. Marks, *J. Am. Chem. Soc.* **1992**, *114*, 1112

9   (a) I. Tritto, R. Donetti, M. C. Sacchi, P. Locatelli, G. Zannoni, *Macromolecules*, **1997**, *30*, 1247; (b) I. Tritto, R. Donetti, M. C. Sacchi, P. Locatelli, G. Zannoni, *Macromolecules* , **1999**, *32*, 264

10 E. Giannetti, G. M. Nicoletti, R. Mazzocchi, *J. Polym. Sci., Polym. Chem. Ed.*, **1985**, *23*, 2117

11   C. J. Harlan; S. G. Bott; A. R. Barron, *J. Am. Chem. Soc.* **1995**, *117*, 6465

12   P. J. J. Pieters, J. A. M. Van Beek, M. F. H. Van Tol, *Macromol. Rapid Commun.*, **1995**,*16*,463

13   A. R. Siedle, B. Hanggi, R. A. Newmark, K. R. Mann, T. Wilson, *Macromol. Symp.*, **1995**, *89*, 299

14   (a) D. Coevoet; H. Cramail; A. Deffieux; C. Mladenov; J. N. Pedeutour; F. Peruch, *Polym. Int.* **1999**, *48*, 257; (b) J. N. Pedeutour, D. Coevoet, H. Cramail, A. Deffieux, *Macromol. Chem. Phys.* **1999**, *200*, 1215

15   W. A. Herrmann; J. Rohrmann; E. Herdtweck, W. Spaleck; A. Winter, *Angew. Chem. Int. Ed. Engl.* **1989**, *28*, 1511

16   (a) W. Spaleck; F. Küber; A. Winter; J. Rohrmann; B. Bachmann; M.Antberg; V. Dolle; E. F. Pauls, *Organometallics* **1994**, *13*, 954;   (b) U. Stehling; J. Diebold; R. Kirsten; W. Röll; H. H. Brintzinger; S. Jüngling; R. Mühlhaupt; F. Langhauser, *Organometallics* **1994**, *13*, 964

17   H. Sinn; W. Kaminsky;H. J. Vollmer; R. Woldt, *Angew. Chem.* **1980**, *92*, 400

18   (a) G. G. Hlatky; D. J. Upton; H. W. Turner, *PCT Int. Appl.* WO 91/09882 **1991**;   (b) H. W. Turner, *Eur. Pat. Appl.* EP 0 277 004 A1, **1988**;   (c) X.Yang; C. L. Stern; T. J. Marks, *Organometallics* **1991**, *10*, 840

19   E. Samuel; M. D. Rausch, *J. Am. Chem. Soc.* **1973**, *95*, 6263

20   M. Bochmann, S. J. Lancaster, *Angew. Chem., Int. Ed. Engl.*, **1994**, *33*, 1634

21   T. Haselwander, S. Beck, H. H. Brintzinger, *Ziegler Catalysts*, ed. G. Fink, R. Mühlhaupt and H. H. Brintzinger (Springer-Verlag, Berlin, **1995**), p. 181

22   S. Beck , M. H. Prosenc , H. H. Brintzinger , R. Goretzki , N. Herfert , G. Fink , *J. Mol. Cat. A: Chemical* **1996**, *111*, 67

23   Y. X. Chen, M. V. Metz, L. Li, C. L. Stern, T. J. Marks, *J. Am. Chem. Soc.*, **1998**, *120*, 6287

24   D. Babushkin, personal communication

24   R. Kleinschmidt; Y. van der Leek; M. Reffke; G. Fink, *J. Mol. Cat. A: Chemical* **1999**, *148*, 29

Activation of Siloxy-substituted Compounds and Homopolymerisation of Ethylene by Different Soluble Alumoxane Cocatalysts

Kalle Kallio and Jyrki Kauhanen

Borealis Polymers; Catalyst Research, R&D, P.O. Box 330 FIN-06101 Porvoo, Finland
E-mail: kalle.kallio@borealisgroup.com; jyrki.kauhanen@borealisgroup.com

Abstract. Since metallocene catalysts were discovered, a lot of efforts have been put to work out/clarify structure/property relationships of different metallocene compounds. However, surprisingly few studies on different coactivators have been published. The most well known coactivator is methylalumoxane (MAO) discovered by Kaminsky. The second important activator systems for metallocenes are boron coactivators such as tris(pentafluorophenyl)borane. Some higher alumoxanes have been studied and compared with MAO. Ethylalumoxane (EAO), tetraisobutylalumoxane (TIBAO) and hexaisobutylalumoxane (HIBAO) have been reported to be able to activate metallocene compounds in some extent, but activity of these activators has been far away from polymerisation activity obtained by methylalumoxane. In mid 1990's a new group of metallocenes was discovered in co-operation between Åbo Akademi University and Borealis. The key element of these compounds was a siloxy substituent. A siloxy substituent was found to give clear benefits compared to compounds having no functionality in substituents: catalyst stability was clearly increased, life length of heterogeneous catalyst was increased remarkably, extraordinary low Al/Zr-ratios were possible to use. By identifying these basic activation properties, the hexaisobutylalumoxane (HIBAO) and tetraisobutylalumoxane (TIBAO) were tried, and surprisingly high catalyst activity was generated. The activity of the corresponding siloxy compound was approximately 100 times higher with HIBAO than that of conventional ethylene[bisindenyl]zirconiumdichloride. These higher alumoxanes also give some other benefits, such as higher molecular weight, higher isotacticity, even lower Al/Zr-ratios when complexes are supported and a really soluble catalyst system for aliphatic solvents.

1. Introduction

1.1 Activation of metallocene compounds

Soluble catalysts have been studied since 1950's. Berslow et al. were preparing highly linear polyethylene (T_m = 137 °C) having a narrow molecular weight distribution (M_w/M_n = 3,6) by using soluble Cp_2TiCl_2/Et_2AlCl catalysts. Still, the activity of catalyst was low. [1]

First indications of increased activity of metallocene/EtAlCl$_2$ complex by water addition can be found from work performed by K. H. Reichert and K. R. Meyer. They have been studying ethylene homopolymerisation kinetics of soluble Cp_2ZrCl_2/ Et-AlCl$_2$ catalyst in toluene. They found surprisingly high catalyst activity when Al/H$_2$O ration was 20. [2]

The first proposal for new type of cocatalysts needed to activate metallocene compounds can be found by Kaminsky and Sinn et al. They discovered that by adding some water onto Cp_2ZrCl_2/Al(CH$_3$)$_3$ soluble catalyst the activity was reaching highest activity when Al/H$_2$O-ratio was among 2:1 to 5:1. The catalyst activity was 45-90g/ 2.2x 10^{-3} mol Zr (= 20 - 40 kgPE/mol Zr). [3]

MAO has several roles during the polymerisation process. Besides being effective impurity scavenger it will methylate metallocene dichlorides, it is involved in formation of cationic catalyst species, it will stabilise formed cations, it will prevent bimolecular processes of metallocenes and can improve stereochemical control. [4]. Chien has studied the role of TMA/MAO as a chain transfer agent. By adding some TMA into polymerisation system the molecular weight of produced polymer was reduced. It was also claimed that MAO itself could also act as chain transfer agent. [5]. This is clear disadvantage of coactivator because in many cases the metallocene compounds will have poor molecular weight capability.

The activity of catalyst can be affected by the structure of metallocene and several good review articles are available for structure/property comparisons. [6],[7],[8]

Structure of MAO is not known and there are several proposals for structure of other alumoxanes. In general, alumoxane will require the aluminium to have co-ordination number of at least four. This will lead to formation of cluster like three-dimensional structures.[9] Another typical feature for alumoxanes seem to be equilibriums between different structures. [10] This makes the definition of exact structures difficult as well as to define catalytically the most active components

Barron et al. has been comparing some well-defined t-Bu-Alumoxanes to MAO as a catalyst activator. In general, activities are rather low, but the interesting finding was that even the lowest Al/Zr-ratios were able to produce polyethylene with [t-Bu-Al-O]$_6$ while MAO was not able to polymerise. The activity of Cp_2ZrMe_2 was 4.25-4.50 kgPE / (mol x Zr x h) for [t-Bu-Al-O]$_6$ activated system and 26.8 kgPE / (mol x Zr x h) for MAO activation. [11] However these activities are too far away from industrially applicable systems.

1. 2 Features of heteroatom containing metallocenes

Proper complexation of metallocene together with alumoxane is crucial step in creation of active and enough stabile catalysts. Reko Leino discovered surprisingly active metallocene/MAO catalysts. The key feature of compounds was siloxy substituent. The activity of these compounds was 5300 kgPP/(mol Zr h) for propene polymerisation and 6900 kgPE/ (mol Zr h) for ethylene respectively. [12]

Siloxy compounds were showing best activation by exceptionally low Al/Zr-ratios. The best activities for PP polymerisation were reached with Al/Zr-ratios 250-500 and catalyst activity was surprisingly decreased when Al/Zr- ratio was over 1000. Same trend was seen also in polyethylene polymerisation. Even very reasonable activity of 2700 kgPE/ (mol x Zr x h) was reached by Al/Zr- ratio 100. [13]

This good activity and exceptional activation should not be self-evident because some electron withdrawing substituents have resulted in decreased catalyst activity and drastic decrease in molecular weight of polyethylene. The polyethylene polymerisation with bis[4,7-CH_3O-indenyl]Zr(Bz)$_2$/MAO was giving productivity 2.3×10^3 kgPE/ (g Zr x bar x h) is only 1/10 of activity of bis[indenyl]Zr(Bz)$_2$ or bis[4,7-CH_3-indenyl]Zr(Bz)$_2$. Also, molecular weight was low (Mw = 29000) with CH_3O- substituent compared to M_w 320 000 of reference compounds having H or CH_3 in 4,7 position. One explanation was that especially CH_3O- substituent can cordinate onto MAO's [Al(CH_3)-O] units. [14]

1. 3. Higher alumoxanes as cocatalysts

MAO has clearly some drawbacks: Structure is not well defined, high amount of Al is needed to activate metallocene, MAO can work as chain transfer agent and reduce Mw of the polymer, it can cordinate onto functional groups of metallocene. TMA used in preparation of MAO is more difficult to prepare and because of that price of MAO is rather high. There are also reports indicating gel formation of MAO that may cause drawbacks during metallocene catalyst heterogenisation. [15]

Using higher alumoxanes like hexaisobutylalumoxane (HIBAO) or tetraisobutylalumoxane (TIBAO) can avoid many drawbacks of MAO. Because tri-isobutyl-aluminium is used as raw material, the price of higher alumoxanes will be much lower. Also, chain transfer from i-butyl group of HIBAO will be such smaller compared to chain transfer of CH_3- of MAO. The main problem with higher alumoxanes has been the low catalytic activity with several metallocene compounds.

Luigi Resconi has studied TIBAO cocatalyst with many metallocene compounds. They have found rather good productivity for some metallocenes. The best compound being activated was *rac*-EBDMI-ZrCl$_2$ that gave 1205 kgPE/(g Zr h) with Al/Zr = 5000. It was concluded that TIBAO was able to activate indenyl-substituted metallocenes but for unsubstituted compounds it was bad choice. The explanation of this was proposed to be too low Lewis acidity of TIBAO to abstract X$^-$ from Cp$_2$ZrX$_2$[16]

S. Rinivasa Reddy has studied the Al/Zr-molar ratio and polymerisation temperature in ethylene polymerisation catalysed by Cp2ZrCl2/MAO and Cp2ZrCl2/tetraisobutyldialuminiumoxane (TIBDAO). The activity of Cp2ZrCl2 /TIBDAO catalysts has been 0,5 - 5,6 kgPE/(g Zr h atm that is very low compared to activity of Cp2ZrCl2 /MAO being 273-763 kgPE/g Zr h atm). It was also found that molecular weight of TIBDAO activated catalyst was higher compared to MAO activated. This is due the fact that TIBADAO has lower chain transfer capability. [17]

Kurokawa has made comparison of HIBAO and TIBAO in ethylene and propylene polymerisation. It was found that the best activity was reached with HIBAO and no activity was seen with TIBA. By using $Me_2Si(H_4Ind)_2ZrCl_2$ and Al/Zr = 1500 the activity of 50 kgPE/(g compound) was reached. Also, alkylated compound $Me2Si (H_4Ind)_2ZrMe_2$ was compared but the activity was similar to dichloride compound. [18]

Probably higher alumoxanes will have different mechanism during the activation process. P. Möhring and co-workers have studied the activity of different metallocenes with Ethylalumoxane (EAO) by comparing steric and electronic parameters of metallocene compounds. It was indicated that in case of EAO it was estimated that 80% of the activity change is due to electronic parameters. The result was somehow contras what has been seen with MAO activated catalyst systems where steric parameters of has been found to be the most feasible reason for activity differences. [19]

2. Experimental part

2.1 Catalyst handling

All catalyst manipulations were performed inside MEGAPLEX GR 60A inert atmosphere glow box containing max. 3 ppm O_2 in nitrogen that is measured continuously by TELEDYNE model 311 oxygen meter.

2.2 Raw materials

Metallocene compounds were produced in University of Åbo Akademi by methods presented in ref. [12],[13],[20],[21]. MAO 30w% in toluene was commercial product of Albemarle Corporation, USA. HEXAISOBUTYLALUMINOXANE (HIBAO) (4.2 w%Al) in hexane, TETRAISOBUTYLDIALUMINOXANE (TIBAO) (5.5 w%Al) in cyclohexane, and TA2677 were commercial products of Witco, Germany. Oxygen and moisture free toluene was used as a solvent. Polymerisation grade ethylene and nitrogen were coming from AGA and they were purified further by removing O_2, H_2O, CO, CO_2, acetylene and sulphur containing components by using special purification set up. Pentane is polymerisation grade available form Borealis AS, Sweden. Pentane was purified further by removing O_2,

H_2O, CO, CO_2, acetylene and sulphur containing components by special purification columns. The moisture content of all raw materials has been measured continuously by PANAMETRICS moisture analyser.

2.3 Catalyst preparation and polymerisation.

Complex solution of metallocene was prepared by adding metallocene compound onto moisture and oxygen free toluene. The final metallocene stock solution has concentration of 2,5 µmol / ml. To form metallocene/alumoxane complex 1,0 ml of metallocene compound stock solution was added onto 10 ml of extra toluene into certain amount of cocatalyst was added to make wanted Al/Zr- ratio according to Table 1.

Polymerisation was carried out in a 3-litter Büchi autoclave in n-pentane at 70°C. The ethylene partial pressure was 5 bar and total pressure was 8,4 bar. Into reactor 10 ml of previously prepared complex solution was fed. The total amount of metallocene compound was 2,5 µmol and Al/Zr-ratio was varied according to Table 1. After 30 min, the polymerisation reaction was stopped by closing ethylene feed and releasing ethylene over pressure from reactor and letting the pentane boil away. The polymerisation reaction was controlled by JULABO ATS3 heath/cooling bath and during the polymerisation data from ethylene flow, temperature, cumulative ethylene consumption and stirring were collected. Ethylene was fed continuously and the ethylene partial pressure was kept constant by Honeywell Versapak 94 pressure control unit. More detailed polymerisation and catalyst handling procedures are available in ref. [22].

3. Results

Ethylene homopolymerisations were conducted in pentane using five different metallocene compounds in combination with methylalumoxane (MAO), ethylalumoxane (EAO), tetraisobutylalumoxane (TIBAO) and hexaisobutylalumoxane (TIBAO). The yield and molecular weight of produced polymer were measured and activity calculated as a kgPE/ (g Zr x h). In many cases the polymerisation was stopped after 30 min. for avoiding monomer diffusion problems onto too viscous polymerisation media. The max. yield of polymer is roughly 100g in the used reactor system when unsupported catalysts are tested. Compounds: **[1]ZrCl$_2$** = *rac*-ethylene-bis (2-tert-butyldimethylsiloxyindenyl) zirconium dichloride, **[1]ZrMe$_2$** = *rac*-ethylene- bis(2-tert-butyldimethylsiloxyindenyl) zirconiumdimethyl, **[2]ZrCl$_2$** = *rac*-ethylene-bis(3-tert-butyldimethylsiloxyindenyl)zirconiumdichloride, **[3]ZrCl$_2$** = *rac*-ethylene- bis(3-tert-butyldimethylsilylindenyl)zirconiumdichloride **TA2677** = ethylene-bis(indenyl)zirconiumdichloride, **MAO** = 30 w% Methylalumoxane in toluene, **HIBAO** = hexaisobutylalumoxane, **TIBAO** = tetraisobutylalumoxane, **EAO** = ethylalumoxane

Table 1. Polymerisations of siloxy-metallocene compounds with different alumoxanes and comparison to non-siloxy compounds

Metallocene	Cocatalyst	Al/Zr	time min	Amount of comp.μmol	Yield g	Activity kgPE/g Zr h
TA2677	MAO	500	30	0.65	25	842
		200			27	910
		100			9	302
	HIBAO	1055	60	2.75	8	64
		527	60		1	8
		264	60		3	24
[1]-ZrCl₂	MAO	500	30	0.65	73	2540
		200		2.54	120	1036
		100		-"-	12	406
	HIBAO	1000	20	2.54	60	777
		500	30	-"-	62	535
		250	30	2.4	20	180
	TIBAO	1000	40	1.2	22	296
		500	60	2.54	25	110
		250	60	-"-	8	36
	EAO	1000	60	2.54	40	175
		500	60		35	153
[1]ZrMe₂	HIBAO	500	30	2.35	118	1101
		250	30	2.35	88	704
	TIBAO	898	30	2.35	60	560
	-"-	500	30		26	260
[2]ZrCl₂	HIBAO	500	60	2.50	126	552
		250	60	2.50	90	395
[3]ZrCl₂	HIBAO	500	60	2.5	<1	4

4. Discussion

4.1 Comparison of alumoxanes

The activation properties with different alumoxanes were studied by using compound **[1]ZrCl₂** (Table 1). It was found that activation power of HIBAO was surprisingly good especially with low Al/Zr-ratios and differences between alumoxanes seem to be smaller (Fig 1). This good activation for substituted metallocenes achieved by higher alumoxanes is somehow contrary to results available in literature.[11,17]

The increase in Al/Zr ratio will effect remarkably on activation of metallocene compound also when HIBAO or TIBAO is used with siloxy substituted compounds (Fig 1) but still the used Al/Zr- ratio is extremely low and catalyst activity high compared to experiments performed by Resconi et al using Al/Zr-ratio 5000 [16] With EAO increase in Al/Zr-ratio does not affect catalyst activation indicating that it is not capable to activate the compounds enough well (Table 1 and Fig 1).

The methylation of metallocene compound was studied by comparing compounds [1]ZrCl$_2$ and [1]ZrMe$_2$ activated by HIBAO and TIBAO. Methylation will improve activity remarkably especially with lower Al/Zr-ratios. The increase in activity with HIBAO is from 180 up to 704 kgPE/g*Zr*h when Al/Zr-ratio is 250 (fig 1). With TIBAO the activity increases is from 107 up to 260 kgPE/g*Zr*h when Al/Zr-ratio is 500 (table 1). This methylation of metallocene indicates that when metallocene is already methylated the function of cocatalyst is only to cationise compound and stabiliser Zr$^+$. This is somewhat contrary to results by Kurokawa who found no difference between -Cl$_2$ and Me$_2$.[17] When no premethylated compound is used the HIBAO as well as the TIBAO had to alkylate metallocene with i-butyl groups. Then it had to cationic metallocene compound - these reactions are more difficult to perform with very bulky ligands of HIBAO compared to normal MAO having methyl groups for that purpose. This kind of activation and alkylation reaction has been studied by S. Srinivasa Reddy [16]

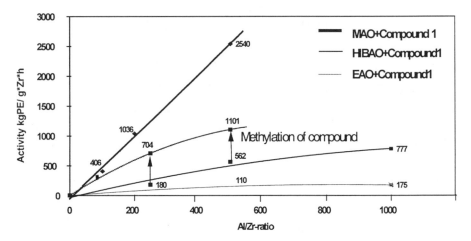

Fig. 1. Activation of [1]ZrCl$_2$ and [1]ZrMe$_2$ with different activators

4.2 Affect of metallocene structure

There are several factors on metallocene itself that are strongly affecting onto activity of compounds. The most important was effect of methylation that increased catalyst activity remarkably with HIBAO and TIBAO.

The next important factor affecting onto catalyst activity and molecular weight was position of siloxy group. With Al/Zr = 250 the activity of [1]ZrCl$_2$ was 180 kgPE/g*cat*h (siloxy substituent in 2-position) but when [2]ZrCl$_2$ (siloxy-substituent in 3-position) the activity was increased up to 395 (table 1).

The compounds TA2677 and [3]ZrCl$_2$ that have not siloxy substitution are performing very badly even with increasing HIBAO content indicating that they are not capable to form proper active sites compared to siloxy substituent in same position (see table 1 and fig 2). When there is no siloxy substituent, the activity of TA2677 is only 8 kgPE/g*Zr*h and with **[3]ZrCl₂** the activity is only 4 kgPE/g*Zr*h when HIBAO cocatalyst is used.

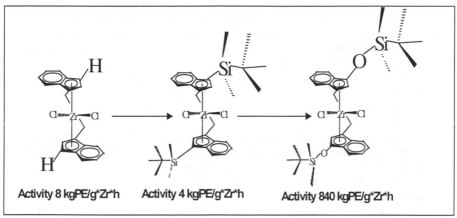

Fig. 2. Affect of metallocene structure onto catalyst activity. HIBAO as a co-catalyst, Al/Zr = 500

5. Conclusions

These siloxy-compounds have been found to be extremely interesting compounds. They have special capability to form active catalyst with higher alumoxanes like hexaisobutylalumoxane (HIBAO) and tetraisobutuldialumoxane (TIBAO) which normally are not capable to activate metallocenes enough well.

The most effective coactivator for siloxy type compound was MAO having good alkylating capability and cationisation performance. The second best is HIBAO with [1]ZrCl₂ having roughly 1/5 of the activation performance of MAO. The situation is changed remarkably when metallocene compound is methylated. The activity with HIBAO is boosted remarkably and it is very close the activity of MAO. TIBAO has also rather high activity especially with methylated siloxy compounds.

6. References

1   D. S. Breslow, N. R. Newburg; Soluble Catalysts for Ethylene Polymerisation, vol.81, p.81-86
2   K. H. Reichert, K. R. Meyer: *Die Makromolekulare Chemie* 169 (**1973**) p.166
3   H.Sinn, W. Kaminsky, *Angew.Chem.Int.Ed.Engl.*Vol 15 (**1976**), No 10, p.630-631
4   J.B.P. Soares, A.E. Hamielec; *Polymer Reaction Engineering* 3(2), **1995**, p. 1987
5   J. C.W. Chien, B.-P. Wang; *Journal of Polymer Science: Part A: Polymer Chemistry*, Vol 26, (**1988**) p.3089-3102.
6   W. Kaminsky; MetCon '95, May 17-19, 1995, Houston TX USA,

7   W. Kaminsky; *Catalysis Today* (**1994**) p.257-271.
8   W. Kaminsky; Feature Article, Huthig&Wepf. Verlag, Zug, (1996)
9   M. R. Mason, J. M. Smith, S. G. Bott, A. R. Barron; *J.Am.Chem.Soc.* (**1993**) 115, p.4971-4984
10   D. A. Atwood, J. A. Jeiger, S. Liu .; *Organometallics* 18, (**1999**) p.976-981
11   C.J. Harlan, S. G. Bott, A. R. Barron; *J.Am.Chem.Soc.* ,117, (**1995**) p. 6471
12   R. Leino, H. Luttikhedde, C.-E. Wilen, R. Sillanpää, J. H. Näsman; *Organometallics* (**1996**) p.2450-2453;
13   R. Leino, H. Luttikhedde, P. Lehmus, C.-E. Wilen, R. Sjöholm, A. Lehtonen, J. V. Seppälä , J. H. Näsman; *Macromolecules* (**1997**) p.3477-3483;
14   N. Piccolrovazzi, P. Pino, G. Consiglio, A. Sironi, M. Moret; *Organometallics*, **1990**,9 p. 3098-3105
15   Sangokoya S.A. Albemarle; European patene application EP0755936A, 29.01.1997
16   L. Resconi, U. Giannini, T. Dall'occo; Metallocene-Based Polyolefins, Ed. J. Scheirs& W. Kaminsky, Wiley & Sons Ltd. (2000) p.69-101
17   S. S. Reddy; *Polymer Bulleting* 36, (**1996**) p. 317-323.
18   H. Kurokawa, T.Sugano; *Macromol.Symp.* (**1995**) p.143-149
19   P. C. Möhring, N. J. Coville; *Journal of Molecular Catalysis*, 77, (**1992**) p.47-48
20   R. Leino, H. Lutikhedde, C-E. Wilen, J. Näsman; PCT Pat. App. WO97/28170,
21   R. Leino, H. Lutikhedde, C-E. Wilen, J. Näsman, K. Kallio, H. Knuuttila, J.Kauhanen; PCT Pat. App. WO98/46616,
22   K. Kallio, H. Knuuttila, J.Kauhanen; PCT Pat. App. WO98/32776, 30.06.1998

Methylaluminoxane as a Cocatalyst for Olefin Polymerization. Structure, Reactivity and Cocatalytic Effect

Erling Rytter[1,3], Martin Ystenes[2], Jan L. Eilertsen[2], Matthias Ott[1], Jon Andreas Støvneng[1,3] and Jianke Liu[1]

[1]Dept. of Chemical Engineering, and [2]Dept. of Chemistry, Norwegian University of Science and Technology, N-7491 Trondheim, Norway
[3]Statoil Research Centre, Postuttak, N-7005 Trondheim, Norway
E-mail: err@statoil.com; eilerts@chembio.ntnu.no; ystenes@chembio.ntnu.no

Abstract. The structure and reactivity of methylaluminoxane (MAO), used as a cocatalyst for olefin polymerization, has been investigated by *in situ* IR spectroscopy, polymerization experiments and density functional calculations. We have suggested a few $Me_{18}Al_{12}O_9$ cage structures, including a highly regular one with C_{3h} symmetry, which may serve as models for methylaluminoxane solutions. Three reactive methyl bridges, presumably the key elements in metallocene activation, are situated at the cage surfaces. Further, exchange reactions show that the methyl groups are readily exchanged with chlorine, while non-bridging methyl groups are inert. The chlorinated MAO thus formed (MAO-Cl) is unable to activate bis(pentamethylcyclopendadienyl)zirconium dichloride ($Cp*_2ZrCl_2$), even with a surplus of added trimethylaluminium (TMA). MAO and TMA are present as separate FTIR-spectroscopic entities, with TMA acting independently as chain transfer agent for this catalyst.

Introduction

Since the discovery more than 20 years ago, the structure and effect of MAO as a cocatalyst for metallocene based catalysis have been the topic of numerous investigations. The main challenge is to answer the following questions:

What is the molecular structure of MAO?
How does it work in activation of metallocene catalysts?

Or more specifically:

Which structural features of MAO are critical for obtaining an active polymerization system?

The answers hopefully can be used as a guide in optimization and invention of new metallocene activators.

Unfortunately, MAO does not readily form a crystalline solid, and one therefore is left with more indirect methods than X-ray crystallography in deducing the structure. Among the methods that frequently have been applied are NMR, freezing point depression, complexation and extraction with ethers and chemical analysis. The most important results so far seem to be:

- The molecular weight of MAO is roughly 800-1200 [1,2].
- There is a dynamic exchange of methyl groups in solution [3].
- There are 2 or 3 different methyl groups.
- There probably is more than one type of MAO molecule.
- Removing TMA from MAO gives an Me/Al ratio of ca. 1.5 [1,2].

In addition, it must be remembered that there is a vast literature on the structure of similar compounds, from the Barron type *tert*-butylaluminoxane cages (Bu/Al=1) [4], to anionic chloro- and methylaluminoxanes structurally characterized by Atwood et al. [5]. From these and similar investigations we can conclude that, in acidic systems:

- Al is preferentially 4-coordinated, but might be 3-coordinated particularly at higher temperatures or if stress or steric interactions are introduced.
- O will preferentially be 3-coordinated with 2-coordination as an option.
- To fulfill the coordination requirement of Al, Me may participate in 3-center 2-electron bonds of the Al-Me-Al type, cf. Me_6Al_2.

Further, the usual 8-electron valence shell rule (except for 3-coordinated Al) will be satisfied as well as normal oxidation number requirements. Now it easily can be deduced that, with the most stringent of the restrictions above, the only molecular formulas that fall within the MW range are

$$Me^t_{15}Me^b_3Al_{12}O_9 \quad \text{and} \quad Me^t_{20}Me^b_4Al_{16}O_{12} \tag{1}$$

where Me^t denotes terminal and Me^b bridging methyl groups. As will be rationalised later, the former of these possibilities can be fulfilled with the cage models given in Fig. 1 [6]. Note that the cages only consist of relaxed 6-rings of the types

$$\text{-O-Al-O-Al-O-Al-} \quad \text{and} \quad \text{-O-Al-O-Al-Me-Al-} \tag{2}$$

thus giving a low-energy structure according to DFT calculations.

A number of alternatives have been proposed in the literature, notably $(Me_6Al_4O_3)_4$ consisting of a cage that contains four coordinatively unsaturated aluminium atoms situated around two openings in the cage [1]. It can also be mentioned that bridging methyl groups generally are not included in papers on MAO structures, but have been suggested in connection with a proposed sheet-like structure [7].

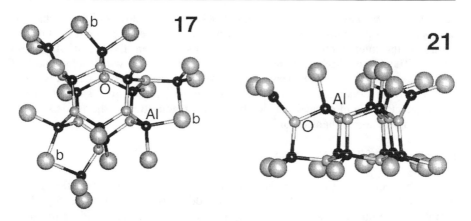

Fig. 1. Plausible $Me^t_{15}Me^b_3Al_{12}O_9$ models for MAO depleted of free TMA. **Left:** DFT model no. **17** (C_{3h}). b: Methyl bridge. **Right:** DFT model no. **21** (C_{3v}).

Experimental strategy

All chemicals, apparatus and experimental procedures employed have been well documented elsewhere [6]. Here we only would like to point out some special features of our FTIR technique [8]. First of all it is worth noticing that there is little or no relevant experimental data in the literature, probably due to the challenge of constructing adequate cells for handling the corrosive and at the same time sensitive samples. Our cell allows a liquid sample to be mixed externally and then continuously pumped through a twin FTIR cell equipped with Si or Ge windows under inert conditions. Thus it is possible to vary concentrations during the experiment and also to study reactions *in situ*.

Compared to the frequently employed NMR technique, infrared spectroscopy has an observation time several orders of magnitude shorter. This means that we obtain an instantaneous and average picture of the chemical bonding in the system, and thereby can deduce information on structure. On the other hand, data on relaxation phenomena, exchange reactions etc. cannot be gained.

In addition to the spectroscopic data and information from literature, we have added heavy use of advanced DFT quantum-chemical computation as well as polymerization experiments in order to confirm our hypotheses.

Results and discussion

From our investigation of MAO itself and its behavior as a cocatalyst, we have come up with several plausible structural models that are compatible with the general knowledge specified above as well as our new experimental and theoretical data. All of these new models are, to our best knowledge, different from what has been suggested previously. Furthermore, a major controversy exists regarding the nature of the interactions and possible reactions between TMA and MAO. An added difficulty is that the structure and properties vary with the method of preparation.[1] Besides, minute leakage of water vapor or oxygen will drastically change any observations. In the following, we will try to give support to our structural models and expand the flexibility and reactivities of the models into a more complete comprehension of MAO.

FTIR spectra of TMA/MAO mixtures.

As can be seen from Fig.2, it is striking that the difference spectra after having dried a commercial sample of MAO (c) or adding fresh TMA to such a dried sample (b) are the same as the spectrum of neat TMA (a). This result is the same for stored MAO which contains distinct methoxy groups or even for MAO at elevated temperatures to 84 °C.

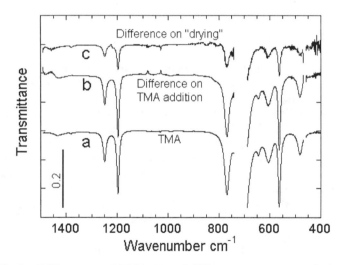

Fig.2. FTIR spectra of TMA (a) and difference spectra upon drying MAO (c) or adding TMA to the dried sample (b). Toluene samples at room temperature.

To avoid any misunderstanding: By dried MAO, TMA depleted MAO or simply by MAO we refer to a methylaluminoxane for use as a cocatalyst for olefin polymerization that has been repeatedly evacuated after being dissolved in toluene. The sample may contain up to 4 % Al as TMA, detected in FTIR as the Me_6Al_2 dimer.

If we examine the typical methyl deformation region more closely, Fig. 3, the conclusion is clear-cut. Adding TMA to MAO simply gives an increase in the characteristic TMA bands, here observed as the terminal methyl deformation at 1197 cm^{-1} compared to the similar frequencies at 1219 cm^{-1} and 1200 cm^{-1} characteristic for MAO. A close analysis shows that MAO contains two or more different types of terminal methyl groups. We can also notice that MAO has a distinct deformation mode at 1257 cm^{-1}, close to the corresponding band at 1250 cm^{-1} for TMA, that can be attributed to bridging methyls. Verification of this conclusion comes both from DFT calculations of frequency shifts, the disappearance of the band upon addition of controlled amounts of ethers like THF, and the exchange reactions described below. Adding the clear isobestic points that comply with TMA and MAO being separate components during mixing, we then conclude:

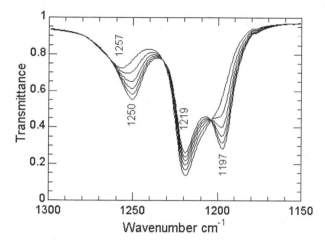

Fig. 3. The methyl deformation range as TMA is added to MAO in toluene.

✓ MAO and TMA behave as separate structural entities with next to no chemical reaction between them.
✓ There are at least 2 structurally distinct types of terminal methyl groups in MAO.
✓ MAO contains an appreciable amount of bridging methyl.

It is even possible to quantify the growing intensity of TMA as illustrated in Fig. 4. Admittedly, the accuracy of the measurements is within 1-2% of the amount of added aluminium and drying very "hard" may strip off some additional TMA, *e.g.* more or less loosely bonded to the surface of the MAO molecule. In particular, the use of donors like ethers will form complexes with additional TMA that actually may originate in the MAO molecule itself.

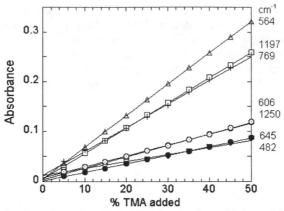

Fig.4. Quantification of growing TMA FTIR bands upon addition to MAO.

FTIR spectra of DMAC addition to MAO.

In dimethylaluminium chloride (DMAC) the chlorine atoms are located in the bridging positions of the $Me_4Al_2Cl_2$ dimer. In other words, it is energetically favorable for chlorine to enter into bridging positions whenever possible at the expense of methyl [9]. If we mix DMAC with MAO, the chlorines have the option to select bridging positions in MAO instead. In fact, this reaction appears to be more or less quantitative, as can be seen in Fig. 5:

$$2 \text{ -Al-Me-Al- } + Me_4Al_2Cl_2 \rightarrow 2 \text{ -Al-Cl-Al- } + Me_6Al_2 \qquad (3)$$
$$(\text{MAO } + \text{ DMAC } \rightarrow \text{ MAO-Cl } + \text{ TMA})$$

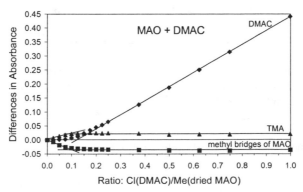

Fig.5. Chlorine exchange reactions adding DMAC to MAO.

After chlorine (as DMAC) is added in an amount of ca. 15% of the total number of methyl groups of MAO, there is no more liberation of TMA and the methyl bridges of MAO practically have disappeared from the spectrum (Fig. 6).

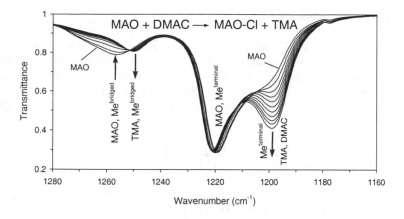

Fig. 6. The methyl deformation region of the FTIR spectrum during chlorine exchange between DMAC and MAO.

Again we notice isobestic points characteristic for a simple reaction having linear relationships between the component concentrations.

It is now possible to quantify the proportion of methyl bridges in MAO as documented in Table 1, assuming the extinction coefficient ratio of bridging relative to terminal methyls to be the same as for the TMA dimer. Thus, the previous conclusions can be extended as follows:

✓   15-20% of the MAO methyl groups are bridging.
✓   The bridging Me of MAO are more reactive towards Cl than the bridging Me in TMA.

Table 1. Quantification of methyl bridges in MAO.

System	Method	Me^b/Me^{total}
MAO	FTIR	0.20
MAO-Cl	FTIR	0.15
MAO-Cl	Chemical analysis	0.18
Average	Experimental	0.177
MAO model (Fig.1, left)	DFT	0.17

The relative Cl/Me stability is

$$-Al-Cl-Al- \text{ (MAO-Cl)} > -Al-Cl-Al- \text{ (DMAC)} > -Al-Cl^{terminal} \qquad (4)$$

pointing to one property of MAO that may be crucial in activation of metallocene catalysts.

Polymerization experiments with MAO-Cl.

Having access to a MAO-Cl where we efficiently have eliminated the methyl bridges in MAO, the obvious control experiment is to check whether this modified cocatalyst is able to initiate a polymerization reaction. From Fig. 7 (left) it is clear that this is not possible. If we now presume methylation and ionization as the two general steps in activation of the catalyst ($Cp*_2ZrCl_2$ or Cp_2ZrCl_2 in our case), this may be formulated as:[‡]

$$Cp_2ZrCl_2 + \text{-Al-Me-Al-} \rightarrow Cp_2ZrClMe + \text{-Al-Cl-Al-} \tag{5}$$
$$(\text{Cat} + \text{MAO} \rightarrow \text{Cat-Me} + \text{MAO-Cl})$$

$$Cp_2ZrClMe + \text{-Al-Me-Al-} \rightarrow \text{"}Cp_2ZrMe^+\text{"} + \text{"-Al-Me   Al-Cl}^-\text{"} \tag{6}$$
$$(\text{Cat-Me} + \text{MAO} \rightarrow \text{Cat*} + \text{MAO-Cl}^-)$$

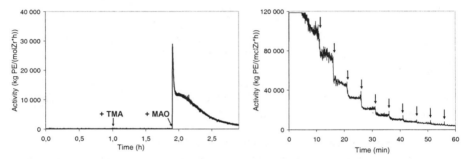

Fig.7. Left: Ethene polymerization with $Cp*_2ZrCl_2$ starting with MAO-Cl as cocatalyst. (60 °C, 2 bar, toluene, Al(MAO-Cl)/Zr = 3000, Al(MAO-Cl)/Al(TMA) = 10, Al(MAO-Cl)/Al(MAO) = 1). **Right:** Ethene polymerization with Cp_2ZrCl_2 starting with MAO as co-catalyst and adding DMAC, each arrow representing Al(DMAC)/Al(MAO)=0.035. (60 °C, 2 bar, toluene, Al(MAO)/Zr=3000).

At this point no specific structure of the activated or partly activated catalyst is implied. Reaction (5) is similar to the exchange reaction (3) with DMAC, and the former becomes impossible with MAO-Cl as the bridging positions already are occupied with chlorine. Adding TMA does not remedy the situation whereas fresh MAO naturally is able to activate the catalyst. During the activation reaction with MAO, it is confirmed by FTIR spectroscopy that the bridging methyl peak decreases, as it should. Further, TMA is able to (singly) methylate the catalyst [10]. Therefore:

✓   (Single) methylation of the chlorinated catalyst is easily performed.
✓   Ionization/activation of the catalyst depends on the (large) MAO species.
✓   The bridging Me in MAO is necessary for activation of the catalyst.

[‡] It has in separate experiments been shown that only one of the chlorines of the catalyst easily is methylated [9]

Fig. 7 (right) represents another control experiment where DMAC is added in steps of 14% chlorine compared to the starting concentration of available methyl bridges in MAO. As we gradually are saturating MAO with chlorine, forming MAO-Cl, the ionization reaction (6) is shifted to the left and the activity decreases. Repeating the experiment in Fig. 7 (left), but with the premethylated catalyst Cp_2ZrMe_2 , yields a noticable difference in that there is a small but consistent activity from the start of the run and until MAO is added, see Fig. 8. Apparently, the reverse reaction

$$Cp_2ZrMe_2 + \text{-Al-Cl-Al-} \rightarrow Cp_2ZrClMe + \text{-Al-Me-Al-} \qquad (7)$$
$$(Cat\text{-}Me_2 + MAO\text{-}Cl \rightarrow Cat\text{-}Me + MAO)$$

is able to create a small, but sufficient amount of MAO that can activate (ionize) some of the catalyst through the created methyl bridges.

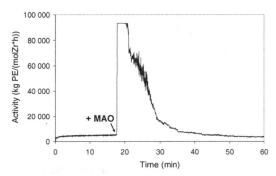

Fig. 8. Polymerization of Cp_2ZrMe_2 starting with MAO-Cl as cocatalyst. (60 °C, 2 bar, toluene, Al(MAO-Cl)/Zr=2000, Al(MAO-Cl)/Al(MAO)=1).

Polymerization experiments with added TMA.

Reverting to the two-component model for ordinary MAO cocatalyst solution, a possible consequence could be the following: There is a large variation from one metallocene catalyst to another regarding sensitivity towards termination by chain transfer to the cocatalyst. As MAO certainly is large and bulky, such chain transfer might be more efficient for TMA. In fact, this hypothesis is verified by using different TMA/MAO mixtures in ethene polymerization with $Cp^*_2ZrCl_2$. The MW in Fig. 9 (left) is inversely related to the TMA concentration at a practically constant activity level. End group analysis confirms a quantitative transition from unsaturated to saturated chain ends as TMA is added (Fig. 9, right). In other words:

✓ Chain transfer to the cocatalyst is essentially proportional to the TMA content.

An additional finding is that the relative unsaturation content of *trans*-vinylene end groups in the polymer increases. The exact nature of how TMA influences β - hydrogen transfer to a monomer, yielding vinyl, relative to the chain-end isomerization reaction proposed for *trans*-vinylene formation during ethene polymerization [11], is not clear.

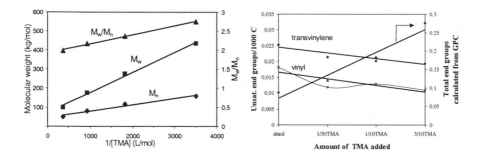

Fig. 9. Left: MW for polymerization of ethene with Cp*$_2$ZrCl$_2$ adding TMA to MAO. (60 °C, 2 bar, toluene, Al(MAO)/Zr=3000, Al(TMA)=30mol% max, GPC:2 column). **Right:** End group analysis.

DFT calculations of MAO structures.

Above it is demonstrated that MAO contains methyl bridges in good accordance with the formulas in Eqn. (1). Further, it has been well documented that cage type structures are energetically preferred [1,4,6,12]. Cages with stoichiometry (MeAlO)$_n$ all contain a mixture of 6-rings and exactly six 4-rings, and it is assumed that these –Al-O-Al-O- type 4-rings introduce strain in the system [4]. As the cage size increases, it is easily demonstrated that the energy pr. MeAlO unit drops nearly linearly with the dilution of the number of the smaller rings. Thus, one approach in constructing test structures with the appropriate Me/Al=1.5 ratio is to start with the well-known Me/Al=1 cages and add TMA by opening 4-rings. (There are several structural options for introducing the additional TMA to the 4-ring. This will not be discussed in detail here). If these 4-rings originally are adjacent to each other, more stable 6-rings are created. It turns out that for the sequence

$$Me_9Al_9O_9 + 3/2\ Me_6Al_2 \rightarrow Me_{18}Al_{12}O_9 \tag{8}$$

three pairs of adjacent 4-rings are converted into 6-rings, thus creating a structure without 4-rings at all, with correct Me/Al ratio, suitable molecular weight, fulfilment of all bonding rules and with a reasonable number of bridging methyl groups (Me^b/Me^{total}=1/6). Fortunately, the structure (Fig. 1, left) also has an energy which is among the lowest that we have calculated.

The procedure above is formalistic as the Me/Al structures have not been isolated other than for *tert*-Bu as the alkyl group. An alternative approach is to start with a smaller cage, containing bridging methyl groups at the periphery and with

Me/Al>1.5. Then a small amount of water is added, attacking the bridging bonds and liberating methane, followed by reacting the formed OH groups with the TMA present, liberating more methane:

$$Me_6Al_6O_6 + 3/2\ Me_6Al_2 \rightarrow Me_{15}Al_9O_6 \qquad (9)$$

$$Me_{15}Al_9O_6 + 3\ H_2O \rightarrow Me_{12}Al_9O_6(OH)_3 + 3\ CH_4 \qquad (10)$$

$$Me_{12}Al_9O_6(OH)_3 + 3/2\ Me_6Al_2 \rightarrow Me_{18}Al_{12}O_9 + 3\ CH_4 \qquad (11)$$

Astonishingly, one ends up with a different structure, again fulfilling all our basic requirements, and with a similar calculated energy (Fig. 1, right). However, there are no obvious methyl bridges at the surface as these then would form 4-rings. This apparently unfavourable situation with three 3-coordinated aluminium atoms, probably is compensated by a dense, almost closed ionic oxygen packing, of the cage. We are approaching the formation of alumina.

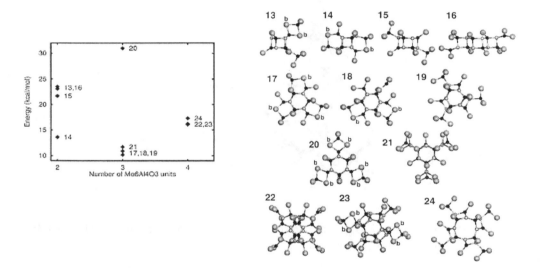

Fig. 10. Relative energies (pr. $Me_6Al_4O_3$ unit) for alternative MAO cage structures with Me/Al=1.5. (Reference energy: $Me_{18}Al_{18}O_{18}$ cage and 3 TMA dimers).

Energies for some alternative structures are depicted in Fig. 10. We have already discussed the difference between the low energy models **17** and **21**. Models **18, 19** and **20** are variations based on a similar central cage as **17**, with the high energy one containing 2-coordinated O and twice the number of methyl bridges. **18** is a more asymmetric mixed model, whereas in **19** the methyl bridges are opened restoring the original 4-rings, and creating three 4-coordinated oxygen atoms. It appears as if the molecule is preparing for a methyl ligand exchange with surrounding molecules.

For the larger cages **22-24**, including the well-documented Sinn structure [1], a mixture of 4-, 8-rings and/or 12-rings in the cage probably destabilises the molecules. Of the smaller models **13-16**, it is no. **14** that is particularly interesting. As in **17**, it contains only 3-coordinated oxygen, 4-coordinated aluminium and now, two Me-bridges. These smaller structures inevitably contain at least two 4-rings. If we include entropy effects in the calculations, some added preference for these $Me_{12}Al_8O_6$ species is expected.

At last it should be noted that by performing repeatedly the procedure in Eqs. 9-11, starting from smaller units (even from TMA), we end up with interesting candidates that have Me/Al ratios slightly different from 1.5, *e.g.* 1.4. As it is difficult to compare the energies with the other models on an equal basis, we have as yet not calculated the energies of these, but an example is given in Fig. 11.[†]

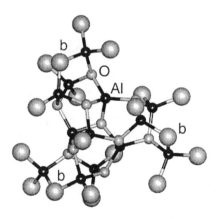

Fig. 11. $Me_{19}Al_{13}O_{10}$, an alternative MAO structure with Me/Al=1.46; Me^b/Me^{total}=0.16 (C_3 symmetry).

[†] Note that there are indications that the bridge in the model in Fig. 11 may open up similar to the situation in models **16**, **18**, **19** and **21** (Figs. 1, 10). FTIR simulations show that the influence of the nearby 3-coordinated Al atom nevertheless causes the Al-Me bond to be stretched and the methyl deformation frequency to be similar to a "true" bridge as in **17**.

Concluding remarks

Apart from analysing possible MAO structures, we have indicated how the detected methyl bridges are reactive and are likely to participate in the activation of the catalyst, Cl exchange with DMAC and in building up larger structures by reaction with minute amounts of water. Other feasible reactions at the MAO surface include:

- Methyl rearrangements and exchange.
- Cage oligomerisation and gel formation *via* methene bridges liberating methane or through "intermolecular" methyl bridges.
- Reaction with ethers and donors, probably liberating TMA from the MAO cage.
- Formation of methoxy groups with oxygen.

To conclude, our model of the cocatalyst system comprises a mixture of probable new MAO cage like structures fulfilling strict bonding rules as well as being compatible with FTIR data, polymerisation experiments, exchange reactions and DFT calculations. The surface of the species certainly is prone to a number of rearrangements, exchange with the surroundings and certain reactions, but still: The majority of TMA molecules will in commercial MAO samples (Me/Al=1.6-1.7) be present as isolated dimers.

Further work will concentrate on the activation of metallocene complexes, ref. the first step as modelled in Fig. 12. Here the initial opening of the MAO methyl bridge bond followed by saturation of the aluminium coordination by formation of the alternative Al-Me-Zr bridge is illustrated.

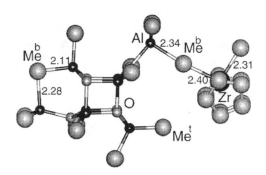

Fig. 12. The initial step in activation of Cp_2ZrMe_2 by MAO (model **14**). The calculated reaction energy for $Cp_2ZrMe_2 + Me_{12}Al_8O_6 \rightarrow Cp_2ZrMeMe^bMe_{11}Al_8O_6$ is -9.5 kcal/mol.

Acknowledgement

Financial support from the Norwegian Research Council (NFR) under the Polymer Science Program and the Programme for Supercomputing, and Borealis is gratefully acknowledged.

References

1   (a) H. Sinn, Macromol. Symp. 97 (1995) 27. (b) H. Sinn, I. Schimmel, M. Ott, N. von Thienen, A. Harder, W. Hagendorf, B. Heitmann, E. Haupt; in Metalorganic Catalysts for Synthesis and Polymerization; W. Kaminsky (Ed.), Springer, Berlin (1999) 105.

2   D.W. Imhoff, L.S. Simeral, D.R. Blevins W.R. Beard, *ACS symposium Series*, 749 (**2000**) 177.

3   (a) I. Tritto, M.C. Sacchi, P. Locatelli, S.X. Li, *Macromol. Chem. Phys.* 198 (**1997**) 3845. (b) J.J. Eisch, S.I. Pombrik, S. Gurtzgen, R. Rieger, W. Uzick; in K. Soga, M. Terano (Eds.) Catalyst Design for Tailormade Polyolefins, Elsevier, Amsterdam (**1994**) 221.

4   M.R. Mason, J.M. Smith, S.G. Bott, A.R. Barron, *J. Am. Chem. Soc.* 115 (**1993**) 4971.

5   J.L. Atwood, D.C. Hrncir, R.D. Priester, R.D. Rogers, *Organometallics* 2 (**1983**) 985.

6   M. Ystenes, J.L. Eilertsen, J. Liu, M. Ott, E. Rytter, J.A. Støvneng, *J. Pol. Sci.* 38 (**2000**) 3106.

7   S. Pasynkiewicz, *Polyhedron* 9 (**1990**) 429.

8   J.L. Eilertsen, E. Rytter and M. Ystenes, *Vibr. Spectrosc.* (**2000**) in print.

9   E. Rytter and S. Kvisle, *Inorg. Chem.* 25 (**1986**) 3796.

10 J.L. Eilertsen, E. Rytter and M. Ystenes, This proceeding.

11 K. Thorshaug, J.A. Støvneng, E. Rytter, M. Ystenes, *Macromol.* 31 (**1998**) 7149.

12 I.I. Zakharov, V.A. Zakharov, A.G. Potapov, G.M. Zhidomirov, *Macromol. Theory Simul.* 8 (**1999**) 272.

Olefin Polymerization under High-pressure: Formation of Super-high Molecular Weight Polyolefins

Noriyuki Suzuki[a]*, Yuji Masubuchi[c], Chikako Takayama[a], Yoshitaka Yamaguchi[a†], Taira Kase[b], Takeshi Ken Miyamoto[b], Akira Horiuchi[c], Takaya Mise[a]

[a]RIKEN (The Institute of Physical and Chemical Research), Wako, Saitama 351-0198, Japan, [b]Department of Chemistry, School of Science, Kitasato University, Sagamihara, Kanagawa 228-8555, Japan, [c]Department of Chemistry, Faculty of Science, Rikkyo University, Toshima, Tokyo 171-8501, Japan
nsuzuki@postman.riken.go.jp

Abstract. Polymerization of 1-hexene under high-pressures (100-1,000 MPa) was investigated using permethylated *ansa*-metallocenes/ methylaluminoxane (MAO) as catalysts. Five types of *ansa*-metallocenes, $R_2E(\eta^5-C_5Me_4)_2MCl_2$ (R = Me, Et, vinyl; E = Si, Ge, M = Zr, Hf), and their heterobimetallic derivatives $(\eta^5-C_9H_7)Rh(\eta^2-CH_2=CH)_2Si(\eta^5-C_5Me_4)_2MCl_2$ (M = Zr, Hf) were synthesized and used as catalyst precursors. These complexes exhibited remarkably enhanced catalytic activity under high pressures despite their very congested structures and gave poly(1-hexene) of unprecedented high molecular weight ($M_w = 1.02 \times 10^7$, $M_w/M_n = 3.79$, by GPC). Studies on termination processes revealed that β-hydrogen transfer to olefin is much accelerated under 500 MPa in the zirconocene complex, while the chain-tranfer reactions are little affected by pressure in the hafnocene.

1. Introduction

Since the discovery of the metallocene catalysts for olefin polymerization, their application to industry has been extensively investigated. Heterogenization of the catalysts is one practical strategy,[1] while another probable approach is utilization of high-pressure radical polymerization processes, which were originally designed to produce low-density polyethylene (LDPE), for the metallocene catalysts.[2] So metallocene-catalyzed olefin polymerization under high-pressure has recently attracted research interest. There have been some reports in this area.[3] Most of them described investigation of ethylene polymerization and ethylene/α-olefin copolymerization under less than 200 MPa at high temperature such as 100-150

---

† Current address: Department of Materials Chemistry, Faculty of Engineering, Yokohama National University, 79-5 Tokiwadai, Hodogaya-ku, Yokohama, Kanagawa 240-8501, Japan.

°C, since they used practical gas-phase equipments to simulate the real process. In a liquid phase polymerization under high pressure, on the other hand, concentration of monomer is independent of reaction pressure. Therefore it is possible to estimate pure effects of pressure, such as pressure-dependence of catalytic activity or molecular weight of polymers.

Recently we have reported high pressure polymerization of 1-hexene in a liquid phase catalyzed by metallocenes/methylaluminoxane(MAO) systems and that the metallocene catalysts showed remarkably enhanced activity under high pressure.[4] Interestingly, the catalytic activity of $(C_5H_{5-n}Me_n)_2ZrCl_2$ at 250 MPa increased in the order of the number of methyl groups ($n = 0 \approx 1 < 3 < 4 < 5$). We also examined stereospecific *ansa*-metallocenes, and reported that the pressure-effect on these complexes were not so surprising.[4c]

1: E = Si, R = Me, M = Zr
2: E = Si, R = Me, M = Hf
3: E = Si, R = Et, M = Hf
4: E = Si, R = vinyl, M = Hf
5: E = Ge, R = Me, M = Hf

+MMAO, pressure / r.t. / 2h

$$(1)$$

100 ~ 1,000 MPa $M_w = 1.02 \times 10^7$ (by GPC)
(1,000 ~ 10,000 atm)   (6.0×10^6 by MALLS)

These results prompted us to investigate high-pressure olefin polymerization using permethylated *ansa*-metallocenes **1-5**. Herein we wish to report remarkable effect of pressure on enhancement of the catalytic activity of the sterically congested *ansa*-metallocene complexes and the formation of polyhexene of very high molecular weight ($\geq 1.0 \times 10^7$ by GPC). We also discuss the effect of pressure on their termination processes in the polymerization reactions, and its difference between the zirconocene and the hafnocene.

2. Results and Discussion

2.1 Syntheses and structures of permethylated *ansa*-metallocene complexes

In this study, we employed dialkylsilylene- (**1-3**), divinylsilylene- (**4**) and dimethylgermylene-bridged (**5**) bis(tetramethylcyclopentadienyl)zirconocene and hafnocene complexes shown in equation 1.[5] Complexes **1**, **2**, **4** and **5** were prepared by the modified method of Jutzi and his co-workers,[6] and the diethylsilylene-bridged complex **3** was prepared by palladium-catalyzed hydrogenation of **4**. All the complexes were structurally characterized. Molecular structures of **1** and **2** are shown in Fig. 1. These data indicate that these two complexes are isostructural (Table 1). It was revealed, however, that these complexes showed different pressure-effects in their catalytic reactions (vide infra).

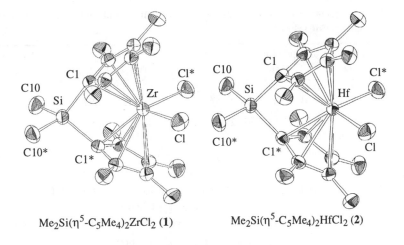

Me$_2$Si(η^5-C$_5$Me$_4$)$_2$ZrCl$_2$ (**1**)   Me$_2$Si(η^5-C$_5$Me$_4$)$_2$HfCl$_2$ (**2**)

Fig. 1. Molecular structures of **1** and **2**. Drawn with 50% probability

Table 1. Selected bond lengths (Å) and angles (deg) for **1** and **2**.

	Complex **1**				Complex **2**		
Zr-Cl	2.434(1)	Cl-Zr-Cl*	99.19(6)	Hf-Cl	2.407(2)	Cl-Hf-Cl*	98.0(1)
Zr-C1	2.473(3)	C1-Si-C1*	95.7(2)	Hf-C1	2.454(5)	C1-Si-C1*	95.1(3)
Zr-C2	2.521(3)	C10-Si-C10*	103.7(3)	Hf-C2	2.483(6)	C10-Si-C10*	102.9(5)
Zr-C3	2.623(3)	Cp(c)-Zr-Cp(c)*	128.1	Hf-C3	2.602(6)	Cp(c)-Hf-Cp(c)*	129.2
Zr-C4	2.615(4)	∠Cp-Cp*	60.9	Hf-C4	2.606(5)	∠Cp-Cp*	60.6
Zr-C5	2.505(4)			Hf-C5	2.512(5)		
Si-C1	1.884(4)			Si-C1	1.878(6)		
Si-C10	1.859(4)			Si-C10	1.850(7)		
Zr-Cp(c)	2.234			Hf-Cp(c)	2.223		
Zr⊥Cp	2.233			Hf⊥Cp	2.214		

Cp:C1-C5, Cp*: C1*-C5*, Cp(c): centroid of Cp rings.

2.2 Polymerization of higher α-olefins under high pressure

The selected data for polymerization of 1-hexene under high pressure using **1-5** /MAO as catalysts were shown in Table 2. The followings should be emphasized. (i) Catalytic activities of the *ansa*-metallocenes were remarkably enhanced due to high pressure despite their highly congested structures. It is in a sharp contrast with the results of mono-, di- and trimethyl-substituted *ansa*-metallocenes, Me$_2$Si(η^5-C$_5$H$_{4-n}$-Me$_n$)$_2$ZrCl$_2$ (n = 1-3), in which high pressures made their catalytic activity only twice of those observed at atmospheric pressure.[4c] (ii) Polyhexene of very high molecular weight was obtained under high pressures.

Table 2. Polymerization of 1-hexene under high pressures catalyzed by *ansa*-metallocenes [$R_2E(\eta^5\text{-}C_5Me_5)_2MCl_2$]/MAO.[a]

entry	catalyst R, E, M	µmol	pressure /MPa	yield /g	catalytic activity[b]	GPC M_w	GPC M_w/M_n	MALLS M_w[c]
1	**1**	0.2	0.1	0.54	1.35	133 000	2.11	
2	Me, Si, Zr	0.01	500	5.27	263.5	2 730 000	3.97	
3	**2**	0.2	0.1	0.42	1.05	385 000	1.64	
4	Me, Si, Hf	0.05	500	2.62	26.2	8 460 000	3.90	
5		0.05	750	2.24	22.4	10 200 000	3.79	$(6.0\pm0.7) \times 10^6$
6	**3**	0.2	0.1	0.27	0.68	225 000	1.57	
7	Et, Si, Hf	0.05	250	1.65	16.5	3 780 000	2.09	$(3.65\pm0.3) \times 10^6$
8		0.03	500	1.96	32.7	7 870 000	2.41	$(5.27\pm0.6)\times 10^6$
9		0.05	750	0.38	3.80	1 880 000	2.15	$(1.64\pm0.09) \times 10^6$
10	**4**	0.2	0.1	0.42	1.05	478 000	2.22	
11	vinyl, Si, Hf	0.05	500	1.37	13.7	8 450 000	3.86	
12	**5**	0.2	0.1	0.24	0.60	132 000	1.79	
13	Me, Ge, Hf	0.02	500	3.74	93.5	8 120 000	3.03	

a) Al/metal = 10,000, r.t., 2 h. b) $\times 10^6$ g-polymer/mol-cat.·h.
c) determined by multiangle laser light scattering analysis, at 25 °C in THF.

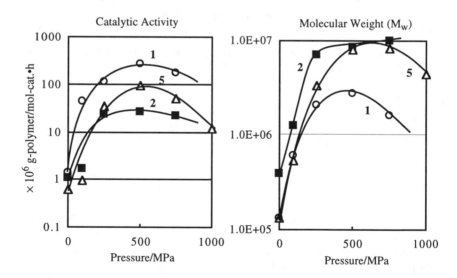

Fig. 2. Pressure-dependence of catalytic activities and molecular weight

Particularly, comlex **2** gave the polymer of $M_w = 1.02 \times 10^7$ determined by GPC, which is unprecedented high molecular weight as linear homopolymers of higher α-olefins. Multiangle laser light scattering analysis (MALLS) supported very high molecular weight of the obtained polymers. To the best of our knowledge, these results are comparable to the previously reported examples of polyolefins with super-high molecular weight produced by homogeneous Ziegler-Natta catalysts.[7] (iii) The germylene-bridged metallocene **5** exhibited much higher catalytic activity than the silylene-bridged complexes **2-4** at high pressure, although its activity at atmospheric pressure was lower than **2-4**. Pressure dependence of the catalytic activities of **1**, **2** and **5** and the molecular weight of the obtained polymer are shown in Fig. 2.

We previously reported that the catalytic activity of $(C_5H_{5-n}Me_n)_2ZrCl_2$ at 250 MPa increased in the order of the number of methyl groups.[4a] Comparison of the present results with our previous study on *ansa*-metallocenes[4c] showed a similar trend in the bridged metallocene catalyzed system. The activity of $Me_2Si(\eta^5\text{-}C_5H_{4-n}Me_n)_2ZrCl_2$ (n = 0-4) under 250 MPa was in the order *rac*-Me$_1$ $(3.0 \times 10^6$ g-polymer·mol-Zr^{-1}·h^{-1}) $\approx$ Me$_0$ (5.0) < *rac*-Me$_2$ (19.4) < *rac*-Me$_3$ (30) < *meso*-Me$_3$ (45) < Me$_4$ (111), although the order at 0.1 MPa was Me$_0$ (0.55) < *rac*-Me$_1$ (0.9) < Me$_4$ (1.35) < *meso*-Me$_3$ (6.3) $\approx$ *rac*-Me$_2$ (6.4) < *rac*-Me$_3$ (17.3). The electron-donating character of the substituents might explain the tendency under high pressure.

2.3 High pressure polymerization using heterobimetallic complexes

Recently we have reported isospecific heterobimetallic complexes and that they exhibited higher catalytic activity for olefin polymerization compared to the corresponding monometallic *ansa*-zirconocenes.[8] In this study, we prepared a heterobimetallic version of permethylated *ansa*-metallocenes (**6** and **7**, eq 2) to examine their catalytic behavior. The results of polymerization are shown in Table 3. The bimetallic complexes showed no significant advantage in their activity at normal pressure. Moreover, they were less active than the corresponding monometallic complexes under high pressure. Although these data are not so attractive, they might give some information on the effects of the bimetallic structure and the effects of high pressure.

6: M = Zr
7: M = Hf

Table 3. Polymerization of 1-hexene under high pressures catalyzed by *ansa*-metallocene/MAO.[a]

entry	catalyst	μmol	pressure /MPa	yield /g	catalytic activity[b]	GPC M_W	M_W/M_n
14	6	0.2	0.1	0.243	1.22	118,000	1.57
15		0.02	500	2.054	51.3	2,540,000	2.65
16	7	0.2	0.1	0.247	0.62	489,000	1.67
17		0.05	500	0.577	5.77	7,190,000	2.86

a) Al/metal = 10,000, r.t., 2 h; b) × 10^6 g-polymer/mol-cat.·h.

2.4 Pressure-effects on chain transfer; Zr and Hf

At 500 MPa, the zirconocene complex **1** showed 200 times higher activity relative to that at 0.1 MPa, while the molecular weight of the polymers obtained at 500 MPa had only 20-fold higher than that at 0.1 MPa. On the contrary, the hafnocene complex **2** exhibited 25-fold higher activity and produced polyhexene of 22-fold higher molecular weight at 500 MPa. These observations showed that high pressure affected differently on the zirconocene and the hafnocene complexes in the polymerization reactions. These results lead us to investigate termination processes in the olefin polymerization on **1** and **2** according to the method that Brintzinger and co-workers have reported.[9] There are two major termination processes in metallocene-catalyzed olefin polymerization: (i) β-hydrogen elimination to the metal that is a unimolecular process and its rate can be described as $k_{TM}[C^*]$, where k_{TM} is the rate constant and $[C^*]$ is the concentration of active

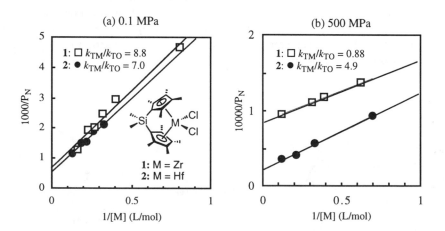

Fig. 3. Dependence of reciprocal degree of polymerization, $1/P_N$, on reciprocal 1-hexene concentration, $1/[M]$, for **1** (□) and **2** (●) under 0.1 MPa (a) and 500 MPa (b).

species; (ii) bimolecular β-hydrogen transfer to the olefin which can be depicted as $k_{TO}[C^*][M]$, where k_{TO} is the rate constant of the reaction and [M] is monomer concentration. Dependence of molecular weight on monomer concentration gives the information about these termination processes. Each complex was studied under 0.1 MPa and 500 MPa (Figure 3). The zirconocene **1** and the hafnocene **2** behave similarly under 0.1 MPa as shown in (a) in Fig. 3. The ratio k_{TM}/k_{TO} in 1-hexene polymerization by **1** and **2** was 8.8 and 7.0, respectively, indicating that, in both complexes, β-hydrogen elimination is a major termination process in diluted monomer solutions. On the contrary, their behavior under 500 MPa was significantly different (b). The ratio k_{TM}/k_{TO} drastically changed in zirconocene **1** and was 0.88. Assuming that β-hydrogen elimination is not very much retarded by high pressure, we may conclude that the β-hydrogen transfer to olefin is accelerated due to high pressure. In hafnocene **2**, on the other hand, k_{TM}/k_{TO} was 4.9 at 500 MPa, showing that β-hydrogen transfer to olefin is not accelerated so much even under high pressure. These results are consistent with the polymerization results.

It is known that the effect of high pressures on a reaction rate is governed by activation volume ($\Delta V^{\ddagger}$) of the reactions.[10] It is noteworthy that **1** and **2** have very similar molecular structure, and the difference is only the central metals (vide supra). Interestingly, the present results suggested that effects of high pressures on organometallic reactions seemed to be governed not only by ligand-controlled steric environments but also by the character of the central metal very much.

3. Experimental

3.1 General

All manipulation was carried out under an atmosphere of dry argon by using standard Schlenk techniques. 1-Hexene was distilled from sodium and stored over sodium-potassium alloy. Methylaluminoxane was purchased as a toluene solution (MMAO-3A, 5.6 wt%) from Tosoh-Akzo Corporation. $Me_2Si(\eta^5-C_5Me_4)_2ZrCl_2$ (**1**) was prepared according to the literature.[6] Preparation of the ligands and hafnocene complexes (**2-5**), their spectroscopic data, and the details on X-ray analysis were described elsewhere.[4d] The bimetallic complexes **6** and **7** were prepared similarly to the reported method.[8,11] The GPC was recorded on Shodex GPC-System 11 equipped with two columns (Shodex KF-807L × 2) whose exclusion limit is 2×10^8 using tetrahydrofuran as an eluent. The multiangle laser light scattering analysis on Wyatt Technology DAWN DSP at 25 °C. This instrument, equipped with a He-Ne laser, operates at a wavelength of 633 nm using tetrahydrofuran as a solvent. NMR on JEOL AL-300 spectrometer.

3.2 High pressure polymerization

Typically, a 25 mL Teflon® sample holder kept under argon was added ca 0.2 mL of the metallocene catalyst in toluene and a toluene solution of MAO (Al = 5.98 wt%; 10,000-fold excess). After 10 min, 1-hexene (ca. 25 mL) was added, and the sample holder was firmly closed and immediately cooled to -78 °C. High pressure was applied by a direct piston-cylinder apparatus, during warming to room temperature (Fig. 4). After 2 h the pressure was released, the reaction mixture was poured into acidic MeOH. The polymer was extracted with hexane and the volatile was evaporated. The residue was dried in vacuo at 70 °C overnight to give the polymer as a colorless viscous oil or rubber-like solid. The molecular weight was determined by GPC and multiangle laser light scattering analysis. Since the values of the polymerization rates were obtained with

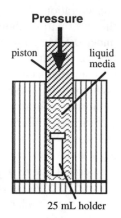

Pressure

piston   liquid media

25 mL holder

Fig. 4 . High-pressure reaction apparatus

experimental variance, the reactions were repeatedly examined to confirm their activity.

Acknowledgment

The authors thank Dr. Kenji Tabata and Dr. Hideki Abe for assistance in multiangle laser light scattering analysis. Mr. Yuji Ikegami, Mr. Akira Matsumoto and Mr. Tokuji Watanabe are acknowledged for their assistance in high-pressure experiments. CT and YY thank Special Postdoctoral Researchers Program at RIKEN (The Institute of Physical and Chemical Research). This work was supported by Grant-in-Aid for Scientific Research from the Ministry of Education, Science, Sports and Culture, Government of Japan (No. 12740370).

References and Notes

1   For recent reviews, see (a) Hlatky GG (**2000**) *Chem Rev* 100: 1347; (b) Fink G, Steinmetz B, Zechlin J, Przybyla C, Tesche B (**2000**) *Chem Rev* 100: 1377; (c) Olabisi O, Atiqullah M, Kaminsky W (**1997**) *J Macromol Sci, Rev Macromol Chem Phys* C37: 519;

2   For examples of patents, see (a) DE 4130299 A1 (BASF A.-G., **1993**); (b) EP 612768 A1 (Tosoh Corp., **1994**); (c) WO 9214766 A1 (Exxon Chemical Patents, Inc., **1992**);

3   (a) Luft G, Batarseh B, Cropp R (**1993**) *Angew Makromol Chem* 212: 157; (b) Bergemann C, Cropp R, Luft G (**1995**) *J Mol Cat A: Chem* 102: 1; (c) Bergemann C, Cropp R, Luft G (**1996**) *J Mol Cat A: Chem* 105: 87; (d) Bergemann C, Cropp R, Luft G (**1997**) *J Mol Cat A: Chem* 116: 317; (e) Bergemann C, Luft G (**1998**) *J Mol Cat A: Chem* (1998) 135: 41; (f) Götz C, Luft G, Rau A, Schmitz S (**1998**) *Chem Eng Technol* 21 12: 954; (g) Luft G, Rau A, Dyroff A, Götz C, Schmitz S, Wieczorek T,

Klimesch R, Gonioukh A (**1999**) In Metalorganic Catalysts for Synthesis and Polymerization Kaminsky W (ed) Speinger, Berlin, p.651; (h) Yano A, Sone M, Yamada S, Hasegawa S, Akimoto A (**1999**) *Macromol Chem Phys* 200: 917; (i) Yano A, Sone M, Yamada S, Hasegawa S, Akimoto A (**1999**) *Macromol Chem Phys* 200: 924; (j) Yano A, Sone M, Hasegawa S, Sato M, Akimoto A (**1999**) *Macromol Chem Phys* 200: 933; (k) Akimoto A, Yano A (**1999**) In Metalorganic Catalysts for Synthesis and Polymerization Kaminsky W (ed) Speinger, Berlin, p.180; (l) Folie B, Ruff CJ (1998) Polym Prepr 39(1): 201;

4   (a) Fries A, Mise T, Matsumoto A, Ohmori H, Wakatsuki Y (**1996**) *Chem Commun* 783; (b) Suzuki N, Mise T, Yamaguchi Y, Chihara T, Ikegami Y, Ohmori H, Matsumoto A, Wakatsuki Y (**1998**) *J Organomet Chem* 560: 47; (c) Yamaguchi Y, Suzuki N, Fries A, Mise T, Koshino H, Ikegami Y, Ohmori H, Matsumoto A (**1999**) *J Polym Sci; A: Polym Chem* 37: 283; (d) Suzuki N, Masubuchi Y, Yamaguchi Y, Kase T, Miyamoto TK, Horiuchi A, Mise T (**2000**) *Macromolecules* 33: 754; (e) Suzuki N, Yamaguchi Y, Fries A, Mise T (**2000**) *Macromolecules* 33: 4602;

5   There have been some reports that describe reactivity of permethylated *ansa*-zirconocenes. (a) Lee H, Desrosiers PJ, Guzei I, Rheingold AL, Parkin G (**1998**) *J Am Chem Soc* 120: 3255; (b) Lee H, Hascall T, Desrosiers PJ, Parkin G (**1998**) *J Am Chem Soc* 120: 5830;

6   Jutzi P, Dickbreder R (**1986**) *Chem Ber* 119: 1750; See also Fendrick CM, Schertz LD, Day VW, Marks TJ (**1988**) *Organometallics* 7: 1828;

7   (a) Miyatake T, Mizunuma K, Kakugo M (**1989**) *Macromol Chem, Rapid Commun* 10: 349; (b) Miyatake T, Mizunuma K, Kakugo M (**1993**) *Macromol Symp* 66: 203; (c) Miyatake T (1998) A proceeding of International Symposium on Metalorganic Catalysts for Synthesis and Polymerization, Hamburg, Poster 49; (d) Saßmannshausen J, Bochmann M, Rösch J, Lilge D (**1997**) *J Organomet Chem* 548: 23;

8   Yamaguchi Y, Suzuki N, Mise T, Wakatsuki Y (**1999**) *Organometallics* 18: 996

9   Stehling U, Diebold J, Kirsten R, Röll W, Brintzinger HH, Jüngling S, Mülhaupt R, Langhauser F (**1994**) *Organometallics* 13: 964

10 (a) Reiser O (1998) Koatsuryoku no Kagaku to Gijutsu 8(2): 111 (English); (b) Matsumoto K, Morrin Acheson R (eds) (**1991**) Organic Synthesis at High Pressures, Wiley-Interscience, New York;

11 Details on preparation of the permethylated heterobimetallic complexes will be reported elsewhere, Takayama C, Suzuki N, Yamaguchi Y, Wakatsuki Y in preparation

Higher-order Kinetics in Propene Polymerizations by Zirconocene Catalysts. Analysis of Alternative Reaction Mechanisms *via* a Genetic Algorithm

Frank Schaper, Hans-H. Brintzinger*
Fachbereich für Chemie, Universität Konstanz, 78457 Konstanz, Germany

Ralph Kleinschmidt, Yolanda van der Leek, Mathias Reffke, Gerhard Fink*
Max-Planck-Institut für Kohlenforschung, Kaiser-Wilhelm-Platz 1,
45470 Mühlheim, Germany

Abstract. Kinetics of propene polymerizations have been studied for four *ansa*-zirconocene catalysts, derived from $Me_2Si(2\text{-}Me\text{-benz[e]indenyl})ZrCl_2$, $Me_2C(^iPrC_5H_3)(flu)ZrCl_2$, $Me_2Si(^tBuC_5H_3)(flu)ZrCl_2$ and $Me_2C(^tBuC_5H_3)$-$(flu)ZrCl_2$ by activation with MAO. In all cases, a reaction order higher than one in propene concentration is observed at low monomer concentrations, while saturation occurs at higher monomer concentrations. By use of a genetic algorithm, alternative kinetic schemes, involving various deactivation processes and/or the participation of a second olefin in the insertion step, are analyzed with regard to ranges for their elementary rate constants which yield reasonable agreement with the experimental data.

Introduction

Despite substantial efforts to clarify the mechanisms responsible for the polymerization of α-olefins by zirconocene catalysts [1], the experimentally determined kinetics of these polymerization reactions are not completely understood: Polymerization kinetics are frequently found to deviate, e. g., from the first-order dependence on monomer concentrations [M] expected on the basis of a reversibly formed olefin-containing reaction complex.

Reaction orders in monomer concentration ranging from 1.2 to 2 have been observed in polymerizations of propene [2-9], ethene [10] and also of styrene [11]. To explain these observations, it has been proposed that the olefin-insertion step might involve more than one olefin [12,13]. This possibility has been discussed in detail by Ystenes [14,15] and, more recently, analyzed by means of density functional calculations by Prosenc et al. [16]. As an alternative, Fait et al. have pointed out that the higher-order dependence on monomer concentration might arise from an existence of two states of the catalyst with different insertion rates, where the monomer participates in the promotion of the catalyst to the more active state [17].

To gain further insights into the underlying reaction sequences, we have determined propene-polymerization rates of four MAO-activated ansa-zirconocene catalysts (Scheme 1) as a function of monomer concentrations and analyzed the resulting data with regard to their agreement with alternative reaction schemes by means of a genetic algorithm.

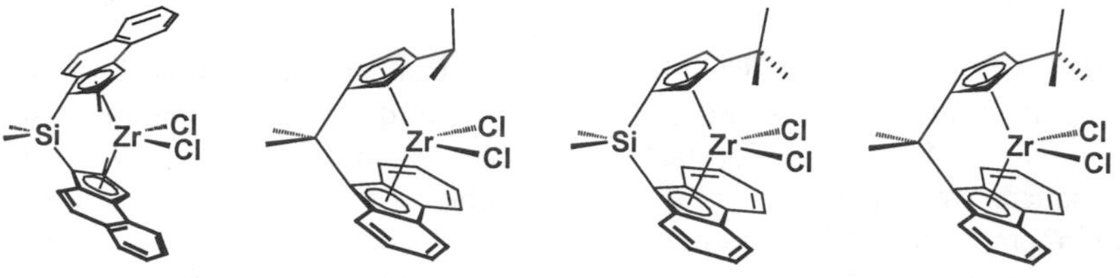

Scheme 1.

Experimental part

Polymerization kinetics

The ansa-zirconocene dichlorides **1-4** were obtained as a gift from former Hoechst AG, while MAO (10 weight-% in toluene, mean molar mass 800 g/mol) was a gift from C. K. Witco GmbH. The polymerizations (for details see ref. 4-6) were conducted in a mechanically stirred 250-mL Büchi glass autoclave, charged with 160 mL of a toluene solution of MAO. After preequilibrating at 25°C with propene at a pressure between 0.3 and 4.2 bar, the reaction was started by injecting 5 mL of a catalyst solution which contained 0.66-7.0 μmol of the zirconocene dichloride (see Table 1), preequilibrated with MAO, such that the [Al]:[Zr] ratio in the final reaction mixture was as listed in Table 1. These [Al]:[Zr] ratios had been found to give optimal reaction rates in separate experiments. While keeping the propene pressure constant with a pressure regulator, the rate of gas consumption was determined with a calibrated thermal mass-flow meter. Propene consumption rates increased for ca.10-20 min to a maximum value and declined slowly thereafter. For the following analysis, we recorded the maximal rate value at each propene pressure. Since this procedure is not affected by inhomogenities arising in these reaction systems at later times, it was judged be more reliable than an extrapolation of the rate decline to time zero, which would nevertheless yield a similar dependence on propene pressures. In repeated experiments, the maximal rates were reproduced with an estimated accuracy of ca. 10%.

Table 1. Polymerization conditions

Complex	[Zr]·L/mol	[Al]:[Zr]	Temperature
Me$_2$Si(2-Me-benz[e]indenyl)ZrCl$_2$, **1**	$4.1 \cdot 10^{-6}$	1755	25°C
Me$_2$C(^{i}PrpC$_5$H$_3$)(flu)ZrCl$_2$, **2**	$4.2 \cdot 10^{-5}$	210	25°C
Me$_2$Si(^{t}BuC$_5$H$_3$)(flu)ZrCl$_2$, **3**	$1.75 \cdot 10^{-5}$	8400	25°C
Me$_2$C(^{t}BuC$_5$H$_3$)(flu)ZrCl$_2$, **4**	$4.37 \cdot 10^{-5}$	2700	25°C

Kinetic analysis with a genetic algorithm

For the analysis of the rate data obtained in this manner, we have used a modified genetic algorithm, since the small number (6-11) of data points combined with the high numbers (4-8) of rate constants required by our models made any non-linear least-square fit impossible or at least unstable. As our fitting functions are, furthermore, highly over-parameterized, we are mainly interested in delineating acceptable ranges and systematic relations of rate constant values which reproduce the experimental data within reasonable error limits, rather than in the best set of values itself.

For this purpose, we have programmed a genetic algorithm using mostly standard evolution and mutation procedures [18]. In addition to the usually employed features, however, we have added a selection step to eliminate all solutions which do not differ in at least one parameter from another solution with a better fit value (σ) by a grouping factor.[1] In this way we obtain typically 500 solutions with distinctly different parameter combinations. Solutions with rate constants exceeding values of 10^{10} s^{-1} or 10^{10} L·(mol·s)$^{-1}$ were eliminated; all other parameters were restricted to a numerical range of 10^{-30}-10^{30} to economize CPU time. In addition to the standard deviation of observed and calculated data points, we have chosen, as the function to be minimized, appropriately weighted standard deviations of calculated and experimental first derivatives of the polymerization rates with respect to monomer concentrations, to reproduce the curvatures as well as the absolute values of the data points. This algorithm[2] can be downloaded from the internet at http://www.chemie.uni-konstanz.de/~agbri/download.html. The results obtained in this manner were analyzed in terms of the equilibrium and rate constants which connect the individual species with each other. To facilitate visualization, we have converted these results into standard free-energies of reaction or of activation by the relations $\Delta G^{\varnothing}_{XY} = -RT \cdot \ln K_{XY}$ and $\Delta G^{\ddagger}_{XY} = -RT \cdot \ln(k_{XY} \cdot h \cdot (k_B \cdot T)^{-1})$, where K_{XY} and k_{XY} are the equilibrium and the rate constants, respectively, which connect species X with species Y. They are represented in the following sections in form of tables which contain the minimal and the maximal value for each of the parameters considered, which gave fit values of $\sigma < 1.1 \cdot \sigma_{min}$.

---

[1] This factor is obtained by taking the nth root of the maximum range of the parameter considered, where n is typically 4-10.

[2] Borland Pascal 7.0, Borland International Inc. (1983,1992)

Results and Discussion

Polymerization rates

For the MAO-activated catalyst systems Me$_2$Si(2-Me-benz[e]indenyl)-ZrCl$_2$/MAO (**1**), Me$_2$C(^{i}PrC$_5$H$_3$)(flu)ZrCl$_2$/MAO (**2**), Me$_2$Si(^{t}BuC$_5$H$_3$)(flu)ZrCl$_2$/MAO (**3**) and Me$_2$C(^{t}BuC$_5$H$_3$)(flu)ZrCl$_2$/MAO (**4**) (c. f. Scheme 1), the polymerization rates per zirconium center, v_P, change in dependence of the monomer concentration as shown in Figure 1.

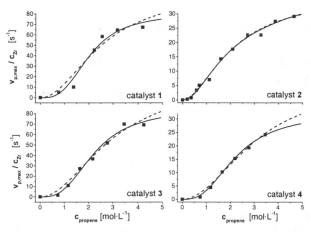

Fig. 1. Maximum polymerization rates of catalysts **1-4** in dependence of monomer concentration with best fits according to polynomial equations with maximum order in monomer of n=3 (solid lines) and n=2 (dashed lines).

For the four catalysts studied here, rates increase with monomer concentrations to an order higher than one at low monomer concentrations. At higher monomer concentrations, however, saturation occurs and the reaction order in monomer declines. This saturation effect is most pronounced with catalyst **1**, where the polymerization rate finally appears to become independent of monomer concentration. For all four catalyst systems studied, however, the first derivatives of the polymerization rates with respect to monomer concentration

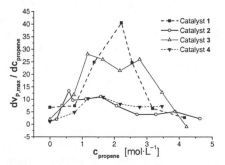

Fig. 2. First derivative of the polymerization rates v_P with respect to monomer concentration .

go to a maximum at propene concentrations of ca. 2 mol/L, followed by a decrease, which is most pronounced for catalysts **1** and **3** but noticeable also for **2** and **4** (figure 2).[3]

As the cause for this saturation effect, we can exclude precipitation of the polymer, since this is not observed in the initial reaction phase during which v_P is measured. Gas-to-liquid mass-transfer limitations are also unlikely to cause this saturation, since catalyst **1**, for which the saturation effect is most pronounced, shows the lowest absolute rate of monomer consumption, due to the low catalyst concentration used. We can thus assume that the shape of the curves represented in Figure 1 is caused by the intrinsic mechanisms of the polymerization catalysis.

Phenomenological analysis of the rate data

Before considering the underlying reaction mechanisms, we present here an analysis of the phenomenological shape of the curves in Figure 1. The initial higher-order increase and the subsequent saturation of v_P with increasing [M] is best described by a polynomial function of the general type represented in eq. 1.[4]

$$v_P = \frac{a \cdot [M]^n + b \cdot [M]^{n-1} + \ldots + c \cdot [M]}{[M]^n + d \cdot [M]^{n-1} + \ldots + f} \cdot [Zr] \tag{1}$$

Using the genetic algorithm described above, we have tried to determine which order n in equation 1 is required to arrive at an acceptable fit of the experimental data. While noticeable systematic deviations between experimental and calculated data remain even for the optimal set of parameters with an order n=2, use of equation 1 with n=3 yields an acceptable fit for most of the data.

The greater number of parameters used to fit the rate law with n=3 is not the reason for the improved agreement with the experiment: In all cases, the two most important parameters are the constants a and f. The former is limiting v_P for high olefin concentrations while the latter has to be greater than $[M]^n$ to make v_P dependent on [M] to an order greater than one at low olefin concentrations. The inflexion points of the curves in figure 1, i. e. their points of maximal slope, are essentially given by the relation $f^{1/n} = [M]$. For the other parameters, a broad range of values is possible as long as they are not greater than a or f (Table 2).

Setting all other constants to zero and refining only a and f still yields much better agreement with the experimental data for equation **1** with n=3 than for that with n=2, due to the higher order in monomer concentration. Catalyst **1**, for example, changes from higher than first-order behavior to saturation when the monomer concentration changes from 0.8 to 4.2 mol/L. Within this small concentration range, the term $[M]^3$ in the denominator can change from being

---

[3] Related saturation effects for heterogeneous titanium and vanadium systems have been reported and discussed in ref. 19.

[4] Algebraic transformations have been done using Maple V, Waterloo Maple Inc. (1997)

smaller to being greater than f by a factor of ca. 12 each way, while the term $[M]^2$ can do so only by a factor of ca. 5 each way.

Table 2. Acceptable ranges for the values of constants a-f for solutions of equation 1 with $\sigma<1.05\cdot\sigma_{min}$.

Cat.	n [a]	a	b	c	d	e	f	σ_{min}
1	2	**114-115**	10^{-5}-1			10^{-5}-0.03	**8.1-8.3**	0.225
1	3	**84-89**	10^{-5}-25	10^{-5}-10	10^{-5}-0.01	10^{-5}-0.1	**11-14**	0.169
2	2	**33-38**	10^{-5}-3			0.2-1	**2.6-3**	0.114
2	3	**49-54**	10^{-5}-1	10^{-5}-0.5	3.1-3.7	10^{-5}-0.2	**1.7-2**	0.098
3	2	**110-117**	10^{-5}-6			10^{-5}-0.3	**8-9**	0.108
3	3	**97-146**	10^{-5}-14	10^{-5}-3	10^{-5}-3	10^{-5}-5	**5-13**	0.095
4	2	**40**	10^{-5}-0.1			10^{-5}-0.01	**8**	0.092
4	3	**36-41**	10^{-5}-2	10^{-5}-0.7	1-2	10^{-5}-0.3	**8-10**	0.044

[a] maximum order in monomer

The σ-values for the rate data of catalysts **1** and **4** still decrease to some degree when the order of equation **1** is increased from n=3 to n=4, due to the strong curvatures of the data in Figure 1. Less significant further improvements are observed with regard to the data for catalysts **2** and **3** (Figure 3).

Appropriate reaction models, especially for catalyst **1**, should thus yield an overall rate law where the order in monomer is at least n=3, while rate laws with n=2 might be acceptable for other catalysts. In the following section, we will analyze a number of alternative models in this regard.

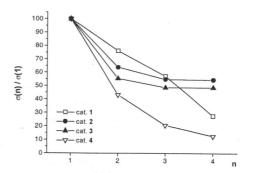

Fig. 3. Relative σ-values obtained by the genetic algorithm as a function of the order in monomer, n.

Alternative reaction models

In addition to the usually assumed reversible conversion of an olefin-free intermediate **A** to an olefin-containing reaction complex **B** as a prerequisite for olefin insertion, we want to include in a comprehensive reaction model the following elementary reaction steps, which might contribute to higher-order kinetics:

i)   Interconversion of the olefin-free precursor species **A** to a less reactive intermediate **C**, as proposed by Fait et al. [17].

ii)   Olefin insertion with participation of a second olefin, which generates, instead
       of an olefin-free species **A´** with extended polymer chain, an olefin-
       containing product species **B´**, as proposed by Prosenc et al. [16].

The first of these features (i) can be represented in general form as shown in
Scheme 2, if we assume, as usual, that the rates of individual reaction steps do not
depend on the length of the growing polymer chain (i.e. that **A´**=**A** and **B´**=**B**). This
reaction sequence can be illustrated by Scheme 3, where **A** would represent a γ-
agostic insertion product, **B** an olefin-containing reaction complex and **C** e. g. a β-
agostic deactivation product reversibly arising by isomerization from **A** or, under
loss of olefin, from **B**.

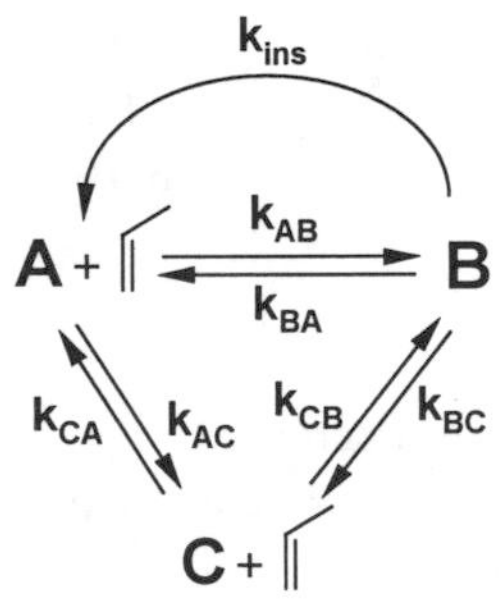

Scheme 2. 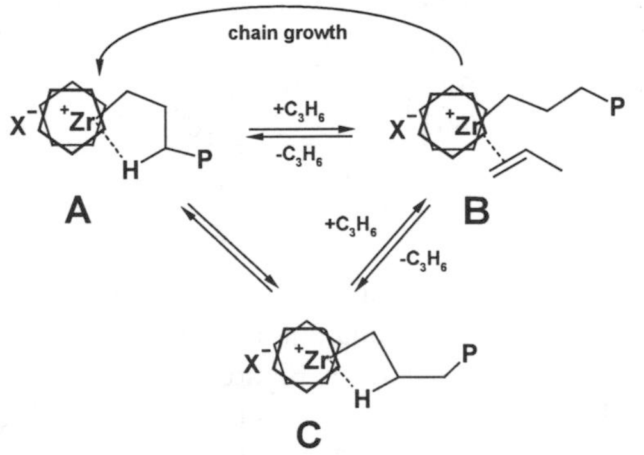   **Scheme 3.**

Scheme 3 leads, by way of the appropriate steady-state relations, to the overall
rate expression of equation **2**, with a maximum order in monomer of n=2, as noted
by Fait et al. [16]. We can relate this second-order rank to the fact that two
independent reaction paths lead, after an insertion, back to reaction complex **B**.[5]

$$
v_p = \frac{[M]^2 + a[M]}{[M]^2 + (a + b + r_i)[M] + ab + \dfrac{k_{ins}}{r_1} + \dfrac{k_{ins} K_{BA}}{r_1 K_{BC}}} k_{ins}[Zr]_{tot}
\tag{2}
$$

$$
a = \frac{k_{AC}(k_{BA} + k_{BC})}{k_{AB}k_{BC}}, \, b = K_{BA} + K_{BC}, \, r_1 = \frac{k_{CB}k_{AB}}{k_{AC}}, \, r_i = \frac{k_{ins}}{k_{AB}}, \, K_{XY} = \frac{k_{XY}}{k_{YX}}
$$

In principle, equation **2** will lead to saturation at high monomer concentrations.
To determine whether such a situation, together with higher-order behavior at low
[M] can be realized within the range 0<[M]<4.5 mol/L with plausible rate constant
values, we have investigated the latter by use of our genetic algorithm. To avoid

---

[5] This relation can be shown to hold also for more complex reaction schemes.

arbitrary restrictions, we have allowed all rate parameters in equation **2** to be varied independently. The results of this analysis (Table 3, Figure 4) lead to rather similar free-energy diagrams for all four catalysts, of which that for catalyst **1** is shown in Figure 5.

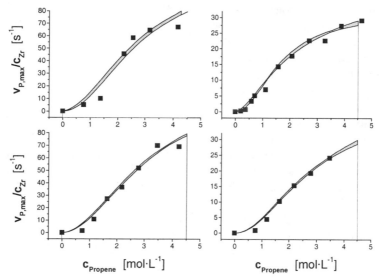

Fig. 4. Comparison of experimental and calculated polymerization rates, according to rate law **2**. The shaded areas correspond to the ranges of solutions with $\sigma < 1.1 \cdot \sigma_{min}$.

Table 3. Solutions obtained by the genetic algorithm with $\sigma < 1.1 \cdot \sigma_{min}$ for rate law 2

	catalyst **1**	catalyst **2**	catalyst **3**	catalyst **4**
σ	0.226-0.239	0.114-0.125	0.112-0.122	0.099-0.107
$\Delta G^0{}_A$ [a]	25-220	60-220	60-220	40-220
$\Delta G^0{}_C$ [a]	7-160	0-160	7-160	0-160
$\Delta G^{\ddagger}{}_{ins}$ [a]	61	64	61	64
$\Delta G^{\ddagger}{}_{AB}$ [a]	16-60	16-65	16-62	16-63
$\Delta G^{\ddagger}{}_{AC}$ [a]	30-190	20-200	30-210	30-150
$\Delta G^{\ddagger}{}_{CB}$ [a]	80-190	80-210	80-220	80-160
$\Delta G^0{}_A - \Delta G^0{}_C$	15-170	10-170	35-170	20-170

[a] in kJ·mol⁻¹, relative to species **B**.

In these diagrams, only *minimal* free energy values, relative to species **B**, are defined. Acceptable solutions are found also for higher free energy values of all intermediates and transitions states, except for the insertion transition state. This kinetic scheme does not restrict these ΔG values to an upper boundary, except for the numerical limitation to values of $10^{\pm30}$ in our algorithm.[6]

Nevertheless, we can derive the following insights from these diagrams: For all four catalysts, activation energies for the conversion $\mathbf{A}{\rightarrow}\mathbf{B}$ at $[M] \approx 1$ mol/L have to be equal to or lower than the insertion barrier of $\Delta G^{\ddagger}_{ins} \approx 60\text{-}65$ kJ/mol for saturation to be observed. Activation energies for the conversion $\mathbf{A}{\rightarrow}\mathbf{C}$, on the other hand, are found to be at least 4-12 kJ/mol higher than those for $\mathbf{A}{\rightarrow}\mathbf{B}$. This implies that deactivation of $\mathbf{A}$ to $\mathbf{C}$ is slower than its conversion to $\mathbf{B}$ by a factor of at least 20-100. Significant

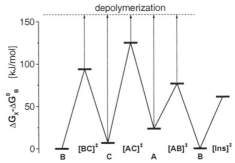

Fig. 5. Free energy diagram for intermediates and transitions states for catalyst **1**, resulting from rate law **2**.

proportions of $\mathbf{C}$ are always present, however, since its reconversion to $\mathbf{B}$ is also rather slow. Significantly, the rate constant ratio, $r_1 = k_{AB} {\cdot} k_{CB}/k_{AC}$ is rather constant for each catalyst and leads to well-defined proportions of $\mathbf{C}$ and $\mathbf{B}$ (Figure 6). With regard to its thermodynamic stability, intermediate $\mathbf{C}$ is found to be close to or higher in energy than $\mathbf{B}$. The highest $\Delta G^{\varnothing}$ value, however, is always found for species $\mathbf{A}$: In all acceptable solutions, its $\Delta G^{\varnothing}$ value is at least 10 kJ/mol higher than that of $\mathbf{C}$ and 25 kJ/mol higher than that of $\mathbf{B}$. Due to its high free energy, the steady-state proportions of $\mathbf{A}$ are much less well defined than those of $\mathbf{C}$ and $\mathbf{B}$.

The free energy relations derived from our kinetic analysis are to be compared to those obtained by DFT calculations: A γ-agostic intermediate such as $\mathbf{A}$ has been calculated to be less stable than a β-agostic species of type $\mathbf{C}$ by 10-30 kJ/mol [20,21], in reasonable agreement with the value of $\Delta G^{\varnothing}_{CB} \geq 10$ kJ/mol derived above. For propene uptake from a β-agostic species such as $\mathbf{C}$ to an olefin complex of type $\mathbf{B}$, reaction enthalpies of -25 to -50 kJ/mol have been calculated [1a],

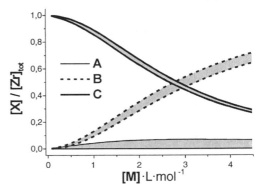

Fig. 6. Relative concentrations of species $\mathbf{A}\text{-}\mathbf{C}$ for catalyst **1** as a function of [M].

again in reasonable agreement with our results. With regard to activation

[6] While upper limits of ΔG⁰-values are of no further relevance, an upper boundary for ΔG^0_A is given by $\Delta G^0_A - \Delta G^0_B < \Delta G^{\ddagger}_{ins} - \Delta G^0_{Ins}$, with an estimate of $\Delta G^0_{Ins} \approx -100$ kJ/mol resulting in $\Delta G^0_A < 150$ kJ/mol, since the insertion reaction can be assumed to be irreversible.

enthalpies, on the other hand, the minimal values of about 20-30 kJ/mol for $\Delta G^{\ddagger}_{AC}$ for the conversion of the γ- to the β-agostic intermediate, appears too high compared to an activation enthalpy of ca. 10 kJ/mol calculated for this process by DFT methods [21]. Even more disquieting is a reaction barrier of at least 80 kJ/mol for the conversion of species **C** to **B**, which is not in line with expectations for a simple olefin coordination [22]. While the model in Scheme 2 can thus be judged to reproduce the kinetic data in general agreement with independent energy estimates, some amendments are obviously still required (vide infra).

To incorporate the involvement of a second olefin in the olefin-insertion step into our reaction scheme, we use the following model: DFT-calculations of Prosenc et al. on a gas-phase model have shown that a second olefin located at a distance of ca. 5-6 Å from the cationic Zr center leads to a small energy stabilization of ca. 5 kJ/mol, which is in the range of normal dipolar interactions [16]. It is thus unlikely that such a species represents a distinguishable intermediate, especially in hydrocarbon solutions. Even if the influence of the second olefin on the ground-state energy of the π-complex and on the activation energy of the insertion reaction is small (possibly negligible), a considerable difference arises with respect to the reaction pathways: Normally, the olefin insertion step will convert **B** to an olefin-free intermediate (**A** in Scheme 2), which can then either coordinate an olefin or undergo isomerization to other intermediates. If a second olefin is present within a radius of 5-6 Å of the Zr center, however, the insertion step will give rise directly to a new olefin π-complex **B**, due to a strong attraction exerted on this olefin by the increasingly electron-deficient reaction complex as the latter approaches the final product stage (Scheme 4) [16].

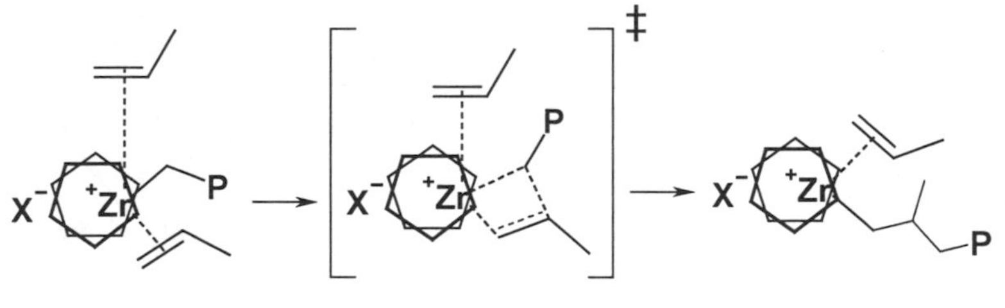

Scheme 4.

Olefin complex **B** can thus induce chain growth along two pathways, depending on whether an additional olefin is present in the vicinity of the metal center or not. In the absence of any significant stabilization, we will treat the presence of a second olefin in statistical terms. For simplicity, we will assume that the probability p^2_{ins} for the two-olefin pathway will be proportional to the probability of finding another olefin within a distance of 5-6 Å from the metal center. In 1-2 M solutions of propene, this probability just happens to be close to 50%. The probabilities for insertion occurring along the one-olefin and the two-olefin paths respectively, can thus be described by equation **3**, where K_{12} has a value of 1-2 mol/L.[7]

$$p_{ins}^1 = \frac{K_{12}}{K_{12} + [M]}, \quad p_{ins}^2 = \frac{[M]}{K_{12} + [M]} \tag{3}$$

Neglecting deactivation to species **C** for the moment and setting, for simplicity $K_{12} = 1$ mol/L, we obtain, *via* the appropriate steady-state assumptions, an overall rate law as given in equation **4**, which has, again, a maximum order in monomer of n=2, but only three adjustable parameters.

$$v_p = \frac{[M]^2 + [M] \cdot c^o}{[M]^2 + (K_{BA} + c^o)[M] + (r_i + K_{BA}) \cdot c^o} k_{ins}[Zr]_{tot}, \quad \text{with } c^o = 1\frac{\text{mol}}{\text{L}} \tag{4}$$

Adjustment of the parameters in equation **4** to the curves in Figure 1 gives a significantly worse agreement with the experimental data than for the model discussed above. Apparently, the strong curvature of the data in Figure 1 cannot be reproduced in this way. To arrive at an acceptable fit, values of K_{12} between ca. 10^{-3} and 10^{-1} mol/L have to be admitted, which correspond to unreasonably large distances of 10-60 Å at which a second olefin would be captured. In addition, an activation barrier $\Delta G^{\ddagger}_{AB}$ of at least 70 kJ/mol, required by this scheme, is not compatible with an essentially barrierless olefin addition predicted by DFT calculations. We thus have to assume that participation of a second olefin is not sufficient by itself to explain our kinetic data.

By incorporation of the two-olefin path into the deactivation scheme discussed above, further improvements in the agreement of calculated and measured values can nevertheless be obtained. Treatment of an accordingly augmented reaction scheme (Scheme 5) with the appropriate steady-state assumptions thus yields equation **5**, which is now of maximum order n=3 in monomer concentration. Use of this equation does indeed allow to fit the data in Figure 1 better than with equation **2** with n=2. This optimization (see Figure 7) yields the kinetic parameters and free energy differences listed in Table 4.

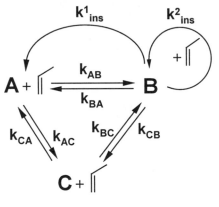

Scheme 5.

---

[7] In more strictly kinetic terms, K_{12} is given by the ratio of k_{iso}/k_{up}, in which k_{iso} is the rate constant for the isomerization to the γ-agostic complex, **B*** → **A**, where **B*** is a high-energy intermediate, formed directly by the insertion reaction but still with a vacant coordination site, while k_{up} is the second order rate constant for the olefin uptake to form **B**, **B*** + **M** → **B**.

Table 4. Solutions obtained by the genetic algorithm with $\sigma < 1.1 \cdot \sigma_{min}$ for rate law 5

	catalyst **1**	catalyst **2**	catalyst **3**	catalyst **4**
σ	0.189-0.207	0.109-0.120	0.100-0.110	0.067-0.070
ΔG^0_A [a]	50-170	20-170	50-170	20-170
ΔG^0_C [a]	9-130	−3 - 140	2-140	0-120
$\Delta G^{\ddagger}_{ins}$ [a]	62	63-64	61-62	64-65
$\Delta G^{\ddagger}_{AB}$ [a]	19-63	17-68	17-65	24-67
$\Delta G^{\ddagger}_{AC}$ [a]	70-230	30-240	25-230	50-190
$\Delta G^{\ddagger}_{CB}$ [a]	90-240	80-240	75-240	80-240
$\Delta G^0_A - \Delta G^0_C$	20-160	8-170	20-150	14-140

[a] in kJ·mol⁻¹

$$v_p = \frac{\left([M]^3 + (a + c^o)[M]^2 + ac^o[M]\right)k_{ins}[Zr]^0}{[M]^3 + (a + b + c^o)[M]^2 + \left((a + c^o)b + (a + r_i)c^o\right)[M] + \left(ab + \dfrac{k_{ins}}{r_1} + \dfrac{k_{ins}K_{BA}}{K_{BC}r_1}\right)c^o} \quad (5)$$

with a, b, r_p, r_i and c^o as defined before.

Compared to the free-energy scheme for the deactivation pathway by itself (Table 3, Figure 5), we find systematically improved agreement with the experimental data but otherwise no remarkable differences. All in all, our kinetic analysis does not exclude the participation of the two-olefin path but does not provide any clear indication for it either.

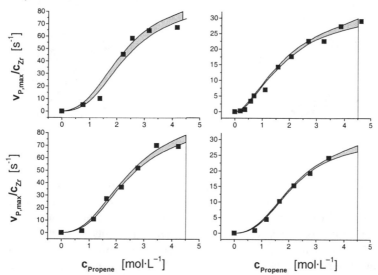

Fig. 7. Comparison of experimental and calculated polymerization rates, according to rate law **5**. The shaded areas correspond to the ranges of solutions with $\sigma < 1.1 \cdot \sigma_{min}$.

The reaction schemes in which a deactivated species participates in the catalytic cycle are, in general, in satisfactory agreement with experimental data and with ab-initio calculations, but appear to require further amendment in the following regard: Reaction barriers of ca. 70 kJ/mol for the conversion of **C** to **B** by coordination of an olefin and the barriers for the olefin insertion of $\Delta G^{\ddagger}$=60-65 kJ/mol, obtained by our algorithm, are substantially higher than the respective values predicted by ab-initio calculations [16,20,23]. This divergence points to the possibility that a major part of the catalyst is present in form of an inactive species of which only minor parts are converted to complexes **A**, **B** and **C**. It is indeed generally assumed that a resting state of such a cationic catalyst is a contact-ion pair with its counteranion, from which it is intermittently reactivated [1a].

Participation of a resting state

Attempts to introduce into our kinetic scheme the existence of a deactivated, olefin-free species, which is considerably more stable than **B**, invariably lead to substantially worsened solutions. The saturation of the polymerization rates at higher [M], observed especially for complex **1,** is not compatible with the presence of an olefin-free complex as the dominant species. As a logical consequence, we have to consider the presence of an inactive, olefin-bearing species in these reaction systems.

Density functional calculations done independently by Nifant'ev et al.[24] and by us [25] do indeed indicate that reaction of an olefin witn an alkyl zirconocene cation-anion-pair will yield, in a first step, a five-coordinated intermediate with olefin, alkyl chain and anion coordinated to the metal center (Scheme 6).

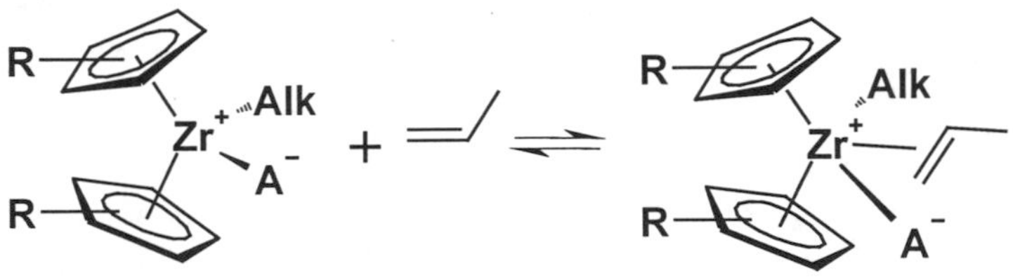

Scheme 6.

This olefin complex cannot undergo olefin insertion, however, since the insertion barrier is higher than any reasonable estimate for complete dissociation of the anion and further – conventional – insertion [25]. We have thus introduced a species **D** into our catalytic cycle, which represents an olefin-containing intermediate, which is not able to undergo insertion. Although **D** might interconvert with all other species in the catalytic cycle, we have neglected possible interconversion reactions between **A** and **D** and between **B** and **C** (Scheme 7) to avoid introduction of too many additional parameters. This reaction scheme yields the rate law shown in equation **6**.

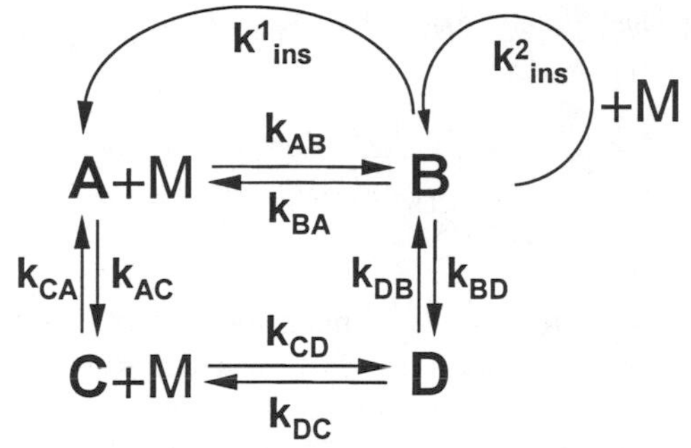

Scheme 7.

Fitting the experimental data with equation **6** gives σ-values practically identical to those obtained with equation **5** above. The solutions represented in table 5 (with graphical representations essentially identical to those shown in Fig. 7) show that the introduction of a second olefin-coordinated species, **D**, can lead to saturation by two distinct routes: At high [M] the major part of the catalyst could be present either as species **B**; this requires that $\Delta G^{\varnothing}_C \approx \Delta G^{\varnothing}_B$ and $\Delta G^{\varnothing}_D \geq \Delta G^{\varnothing}_B$ and is in essence indistinguishable from the situation discussed above. An alternative, of interest in this context, is represented by the minimal $\Delta G^{\varnothing}$ values shown in figure 8: In this case, **C** and **D** both are of lower energy than **B**, such that **C** is converted mainly to species **D** at high [M] (see table 5). While olefin coordination to **A** can still be exothermic, this is no longer required by the results of our algorithm.

Table 5. Solutions obtained by the genetic algorithm with $\sigma < 1.1 \cdot \sigma_{min}$ for rate law 6.

	catalyst **1**	catalyst **2**	catalyst **3**	catalyst **4**
σ	0.189-0.207	0.109-0.120	0.100-0.109	0.067-0.073
ΔG^0_A [a]	–10-170	–20-170	–35-170	–2-170
ΔG^0_C [a]	–50-0 [b]	–50-0 [b]	–50-0 [b]	–45-0 [b]
ΔG^0_D [a]	–50-160	–50-170	–50-170	–50-150
$\Delta G^{\ddagger}_{ins}$ [a]	16-61	16-64	16-61	17-64
$\Delta G^{\ddagger}_{AB}$ [a]	16-170	16-68	17-65	20-64
$\Delta G^{\ddagger}_{AC}$ [a]	45-230	37-240	35-230	50-220
$\Delta G^{\ddagger}_{CD}$ [a]	25-230	17-240	17-240	16-230
$\Delta G^{\ddagger}_{DB}$ [a]	17-220	20-240	16-230	17-230

[a] in kJ·mol^{-1}, [b] K_{BC} was restrained to $K_{BC} > 1$

$$v_p = \frac{\left([M]^3 + (a + c^o)[M]^2 + ac^o[M]\right) k_{ins} [Zr]^0}{\dfrac{b}{c^o}[M]^3 + \left(\dfrac{ab}{c^o} + b + c\right)[M]^2 + \left(a(b+c) + \dfrac{k_{ins}}{K_{DC}r_1} + (c + r_i)c^o\right)[M] + \left(a(c + r_i) + \dfrac{r_i r_D}{r_C} - r_i r_D\right)c^o + \dfrac{k_{ins}}{r_1}}$$

$$a = \frac{k_{AC}(k_{BA}k_{DB} + k_{DC}k_{BA} + k_{BD}k_{DC})}{k_{DC}k_{AB}k_{BD}}, b = (1 + K_{BD})c^o, c = K_{BD}K_{DC} + K_{BA}$$

$$r_C = \frac{k_{CD}}{k_{AB}}, r_D = \frac{k_{AC}}{k_{AB}}, r_1 = \frac{k_{CD}k_{AB}k_{DB}}{k_{DC}k_{AC}c^o}, r_i = \frac{k_{ins}}{k_{AB}}, c^o = 1\frac{mol}{L}$$

(6)

Introduction of an inactive, olefin-containing species **D** lowers also the minimal values for the activation barriers of the insertion step to its minimal possible value (16 kJ/mol given by the numerical restraint of $k_{xy} < 10^{10}$). This is now in agreement with the results of ab-inito calculations. Olefin uptake by species **C** to yield **D** can now also be essentially barrierless. Significantly, some of the solutions with minimal activation barriers for the interconversion **C→D** have **C** and **D** stabilized by ca. −30 kJ/mol relative to **B**, in reasonable agreement with DFT calculations [26].

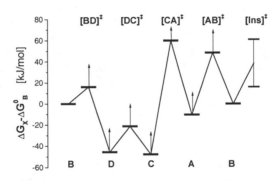

Fig. 8. Free energy diagram for intermediates and transitions states for catalyst **1**, resulting from rate law **6**.

Conclusions

Inclusion of deactivated species in the catalytic cycles appear to be essential for an acceptable reproduction of the experimental data. When only olefin-free de-activated species are considered, however, the resulting activation barriers for the olefin uptake by this species are too high compared to those expected for olefin uptake by a β-agostic complex or other comparable species probably participating in the polymerization reaction. Additional inclusion of an olefin-containing inactive species **D** allows to reproduce the observed saturation behavior even when only small parts of the catalyst centers are present as active species **A** or **B**. In this case, activation barriers for olefin coordination and insertion reactions are in agreement with results of ab-initio calculations. We thus propose that the course of olefin polymerizations in these catalyst systems is best described by the reaction paths displayed in scheme 6, where species **C** and **D** are olefin-free and olefin-coordinated contact-ion-pairs, respectively.

Participation of a second olefin in the polymerization reaction leads, by introduction of a term with $[M]^3$ into the overall rate law, to significantly improved agreement between calculated and experimental rate data. Given the limited accuracy of these data, independent verification of such a two-olefin path is still required, however.

References

1   Reviews: a) L. Resconi, L. Cavallo, A. Fait, F. Piemontesi, *Chemical Reviews* **100** (2000), 1253. b) M. Bochmann, *J. Chem. Soc., Dalton Trans.* (1996) Nr. 3, 255-270. c) H.-H. Brintzinger, D. Fischer, R. Mülhaupt, B. Rieger, R. Waymouth, *Angew. Chem.* **107** (1995), 1255. d) M. Aulbach, F. Küber, *Chem. Unserer Zeit* **28** (1994), 197. e) P. C. Möhring, N. J. Coville, *J. Organomet. Chem.* **479** (1994), 1.

2   P. Pino, B. Rotzinger, E. von Achenbach, *Makromol. Chem., Suppl.* **13** (1985), 105.

3   J. A. Ewen, M. J. Elder, R. L. Jones, S. Curtis, H. N. Cheng, in: *Catalytic Olefin Polymerization* (T. Keii, K. Soga, Ed.), Kodansha, Tokyo 1990, S. 439.

4   G. Fink, N. Herfert, P. Montag, in: *Ziegler Catalysts* (G. Fink, R. Mühlhaupt, H.-H. Brintzinger, Ed.), Springer-Verlag, Berlin 1995, S. 159.

5   a) N. Herfert, G. Fink, *Makromol. Chem., Macromol. Symp.* **66** (1993), 157. b) N. Herfert, G. Fink, *Makromol. Chem.* **193** (1992), 1359. c) N. Herfert, Dissertation, Universtität Düsseldorf (1992).

6   Y. van der Leek, Dissertation, Universtität Düsseldorf (1996).

7   S. Jüngling, R. Mülhaupt, U. Stehling, H.-H. Brintzinger, D. L. F. Fischer, *J. Polym. Sci.: Part A: Polym. Chem.* **33** (1995), 1305.

8   H. Hagihara, T. Shiono, T. Ideka, *Macromol. Rapid Comm.* **2** (1999), 200.

9   L. Resconi, A. Fait, F. Piemontesi, M. Colonnesi, H. Rychlicki, R. Zeigler, *Macromolecules* **28** (1995), 6667.

10 J. C. W. Chien, Z. Yu, M. M. Marques, J. C. Flores, M. D. Rausch, *J. Polym. Sci.: Part A: Polym. Chem.* **36** (1998), 319.

11 L. Oliva, C. Pellecchia, P. Cinquina, A. Zambelli, *Macromolecules* **22** (1989), 1642.

12 M. M. Marques, C. Costa, F. Lemos, F. R. Ribeiro, A. R. Dias, *Reaction Kinetics and Catalysis Letters* **62** (1997), 9.

13 A. Muñoz-Escalona, J. Ramos, V. Cruz, J. Martínez-Salazar. *J. Polym. Sci. A* **38** (2000), 571.

14 M. Ystenes, *J. Catal.* **129** (1991), 383.

15 M. Ystenes, *Makromol. Chem., Macromol. Symp.* **66** (1993), 71.

16 M.-H. Prosenc, F. Schaper, H.-H. Brintzinger, in: Metalorganic Catalysts for Synthesis and Polymerization (W. Kaminsky, Ed.), Springer-Verlag, Berlin 1999, S. 223.

17 A. Fait, L. Resconi, G. Guerra, P. Corradini, *Macromolecules* **32** (1999), 2104.

18 Z. Michalewicz, Genetic Algorithms + Data Structures = Evolution Programs, Springer-Verlag, Berlin 1992.

19 D. R. Burfield, *Polymer* **25** (1984), 1645 and references cited therein.

20 a) T. Yoshida, N. Koga, K. Morokuma, *Organometallics* **14** (1995), 746. b) J. C. W. Lohrenz, T. K. Woo, T. Ziegler, *J. Am. Chem. Soc.* **117** (1995), 12793. c) P. Margl, J.

C. W. Lohrenz, T. K. Woo, T. Ziegler, P. E. Blöchl, *J. Am. Chem. Soc.* **118** (1996), 4434.

21 J. A. Støvneng, E. Rytter, *J. Organomet. Chem.* **519** (1996), 277.

22 a) H. Weiss, M. Ehrig, R. Ahlrichs, *J. Am. Chem. Soc.* **116** (1994), 4919. b) T. K. Woo, L. Fan, T. Ziegler, *Organometallics* **13** (1994), 2252. c) K. Thorshaug, J. A. Støvneng, E. Rytter, M. Ystenes, *Macromolecules* **31** (1998), 7149.

23 Regardless if the rate determining step is the rotation of the alkyl chain or the actual insertion process.

24 I. E. Nifant'ev, L. Yu. Ustynyuk, D. N. Laikov, poster presented at New Millenium International Conference on Organometallic Catalysts and Olefin Polymerization, 18-22.6.2000, Oslo.

25 F. Schaper, unpublished results.

26 a) M. S. W. Chan, K. Vanka, C. C. Pye, T. Ziegler, *Organometallics* **18** (1999), 4624. b) K. Vanka, M. S. W. Chan, C. C. Pye, T. Ziegler, *Organometallics* **19** (2000), 1841.

Structures of MAO: Experimental Data and Molecular Models According to DFT Quantum Chemical Simulations

Vladimir A. Zakharov, Ivan I. Zakharov and Valentina N. Panchenko

Boreskov Institute of Catalysis, Siberian Branch of the Russian Academy of Sciences, Prospect Akademika Lavrentieva, 5, Novosibirsk 630090, Russia
E-mail: V.A.Zakharov@catalysis.nsk.su

Abstract. A DFT quantum chemical calculations have been performed in order to optimize the geometric and electronic cage structure of polymethylalumoxane (MAO) with oligomerization degree n =12 and find such structures that fit most closely the existing experimental data on the MAO composition and structure. The following peculiarities of the MAO structure were found: i) In "classic" MAO (ratio $Al:CH_3:O = 1:1:1$), which has triple-layer cage structure, the inner layer contains highly reactive bonds Al-O. ii) Reaction between "classic MAO" and trimethylaluminium (TMA) proceeds by the concerted mechanism with the insertion of $Al-CH_3$ groups into these Al-O bonds. The reaction produces "true" MAO showing ratio $Al:CH_3:O = 1:1.5:0.75$. Calculated geometry and electronic structures of "true" MAO with n= 12 are presented. iii) "True" MAO and "classic" MAO exist in equilibrium. Driving force for the formation of "true" MAO is the decrease of enthalpy, and of "classic" MAO is the increase of entropy in the equilibrium reaction between "classic" MAO and $Al(CH_3)_3$.

Introduction

Polymethylalumoxane (MAO) is the key element of the novel extremely effective catalysts containing metallocene compound and MAO as activator [1,2]. The unique feature of MAO consists in their ability to react with metallocenes forming cation-like alkyl complexes M(IV) (M=Ti, Zr, Hf) which act as the catalyst active centers [3-6].

However, the mechanism by which ionic pairs in system metallocene/MAO form is still unknown. These studies are essentially impeded by the lack of experimental data on the MAO structure. Recently reported data obtained by NMR spectroscopic methods [7] proved the MAO of composition $[Al(CH_3)O]_n$ to have three-dimensional cage structure in which Al atoms are bridged by oxygen atoms. Such structures have been identified by Barron et al. [8, 9] for tert-butylalumoxane of composition $[(t-Bu)Al(\mu_3-O)]_n$, (n=6, 9). In works [10, 11] quantum-chemical DFT method was used to calculate the geometry and electronic structure of MAO with various molecular masses (oligomerization degree n = 4-12). For MAO with n ≥ 6 the three-dimensional cage structure proved to be the most stable. Besides, it was shown that TMA could react with MAO through the cleavage of Al-O bond. This reaction is similar to the observed by Barron et al. [12, 13] for interaction of tert-butylalumoxane with alkyllithium and $(C_5H_5)_2ZrMe_2$.

Meanwhile, some reported experimental data on MAO composition and structure are poorly consistent with the calculated molecular models. In particular, as note in work [7], the structures of MAO at room temperature and temperatures above 40°C differ essentially. Reversible changes in the ^{27}Al-NMR spectra of MAO recorded at different temperatures proved this difference and suggested the presence of equilibrium MAO structures of two types. Only one structure observed at temperature above 40°C could be recognized as the "classic" cage structure of composition $[Al(CH_3)O]_n$. Another one, predominating at room temperature, is obviously a more complex structure, as indicated by numerous experimental data [14-18]. It was found that after removing of free TMA the composition of different MAO samples corresponded to $[Al(CH_3)_{1.4-1.5}O_{0.75-0.80}]_n$ rather than $[Al(CH_3)O]_n$ [14-18]. Basing on analytical data on ratio $Al:CH_3:O=1:1.5:0.75$ in MAO and molecular weight of MAO, Sinn et al. [14] suggested that the MAO composition may be represented by formula $Al_{12}O_{12}(CH_3)_{12}·[Al(CH_3)_3]_4$ ("true" MAO by Sinn's definition).

In the present work we performed the DFT quantum-chemical calculations in order to optimize three-dimensional cage structure of MAO with n=12 and develop the molecular models which are more consistent with available experimental data on the composition and structure of MAO reported in refs [7, 14-18].

Methods and details of calculation.

All the calculations were carried out with use of the Gaussian 92/DFT package [20]. The geometries of various forms of the MAO-oligomers have been calculated using density functional theory (DFT), the LANL1 effective core potential for inner shells of Al [21] and basis set single- ξ (Minimal Basis) for valence shells of Al and for the carbon, oxygen and hydrogen atoms. The Vosko-Vilk-Nusair [22] (VWN) local correlation parameters with X_α - exchange functional ($\alpha = 0.7$) were applied to evaluate the density functional (Local Density Approximation - LDA). The total charge densities of the MAO-oligomers were calculated using the Mulliken population analysis. The accuracy of the calculated geometries was tested for the Al_2Me_6 -dimeric trimethylaluminium [10].

Results and discussion

1. Optimization of cage structure of MAO with composition $[(CH_3)AlO]_n$ (n=12)

Earlier [11] we have calculated one of the possible triple-layer cage structure for MAO with n = 12. Specific feature of this structure is that Al atoms in the inner layer are penta-coordinated, while those in the outer layers are tetra-coordinated.

Barron et.al. [8,9] suggested the triple-layer cage structure for tret-butylalumoxane of composition $[(t\text{-}Bu)Al(\mu_3\text{-}O)]_9$, in which all Al atoms are tetra-coordinated. Specific feature of this structure is that Al-O bonds in the inner layer are non-equivalent. We have calculated this type structure for MAO with n = 12 (Fig. 1).

The calculated data for this structure show that the Al atoms in the inner layer (Al^2) possess a higher charge compared with the Al atoms in the outer layers (Al^1). Then bonds $Al^2\text{-}O^2$ are more polarized than bonds $Al^1\text{-}O^1$. Molecular-orbital analysis of MAO shows that the LUMO has the most contribution from the anti-bonding σ^*-orbitals of $Al^2\text{-}O^2$ bonds. Basing on the above data, it seems reasonable to suggest that the $Al^2\text{-}O^2$ bonds are the most reactive and exactly these bonds take part in the reactions of MAO with some substrates.

2. Reaction of MAO of composition $[(CH_3)AlO]_n$ with TMA as a route to obtain MAO with ratio Al:CH$_3$:O = 1:1.5:0.75

As mentioned above, experimental data of works [14-18] proved that the composition of "true" MAO at nearly room temperature corresponds to ratio Al:CH$_3$:O = 1:1.5:0.75. But some experimental data [7-9] and structure presented in Fig.1 correspond to the "classic" MAO with ratio Al:CH$_3$:O = 1:1:1. According to experimental data presented in [7] we supposed that the both forms of MAO exist in equilibrium and reaction between "classic" MAO and TMA might explain this equilibrium. The authors of work [19] used the similar assumption to explain possible transformation of cyclic structure of composition $[Al(CH_3)O]_3$ into cyclic structure of composition $Al_4(CH_3)_6O_3$. Reaction presented in Fig.2 is an adequate and simplest model for the interaction between MAO of composition $[Al(CH_3)O]_n$ and TMA.

We carried out accurate calculations of thermodynamic characteristics (ΔS and ΔH) for the reaction presented in Fig.2. For this purpose we used a DFT approach with the B3LYP exchange-correlation functional [23-25], all-electron triple-ξ quality basis (6-31G basis sets with d-functions on aluminiums/carbons/oxygens, p-functions on hydrogens) and full geometry optimization at that level. Calculated results (Tab.1) show that the driving force of the formation of methylalumoxane [-Al(CH$_3$)O-]$_3$ from hexamethyltetraalumoxane $Al_4(CH_3)_6O_3$ is the increase of entropy (ΔS = +38.4 cal/mole•K), while the driving force for the stabilization of MAO of composition $Al_4(CH_3)_6O_3$ is the decrease of enthalpy (ΔH = -18.7 kcal/mole).

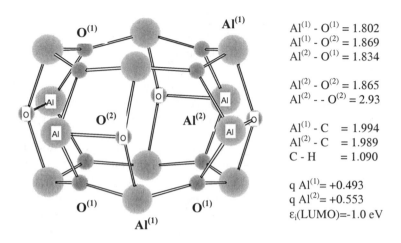

$Al^{(1)} - O^{(1)} = 1.802$
$Al^{(1)} - O^{(2)} = 1.869$
$Al^{(2)} - O^{(1)} = 1.834$

$Al^{(2)} - O^{(2)} = 1.865$
$Al^{(2)} - - O^{(2)} = 2.93$

$Al^{(1)} - C \ \ = 1.994$
$Al^{(2)} - C \ \ = 1.989$
$C - H \ \ \ \ = 1.090$

$q \ Al^{(1)} = +0.493$
$q \ Al^{(2)} = +0.553$
$\varepsilon_i(LUMO) = -1.0 \ eV$

Fig. 1. Side view of a triple-layer cage structure $(MeAlO)_{12}$ with the C_{4h} - symmetry, calculated at the DFT/LANL1MB level (E_{total} = -1390.48053 a.u.). Methyl groups of MAO are omitted for clarity, Al^2 and O^2 atoms in the inner layer are shown by element symbols.

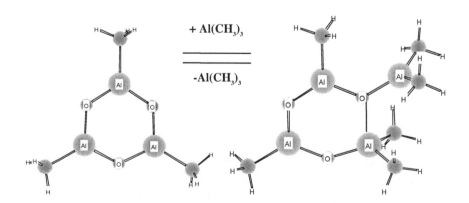

Fig. 2. A possible reversible reaction between methylalumoxane ($-Al(CH_3)O-)_3$ and hexa-methyl-tetra-alumoxane ($Al_4(CH_3)_6O_3$).

Table 1. Thermodynamic characteristics of reaction between TMA and MAO (n=3) (Fig. 2), calculated at the B3LYP/6-31G(d,p) level.

Calculated molecular systems	B3LYP/6-31G(d,p) total energies (in àu)	Calculated entropy values (in cal/(mole • K))	Calculated enthalpy values (in kcal/mole)
$Al(CH_3)_3$ +	-362.19984	73.5	
$[Al(CH_3)O]_3$ ↑↓	-1073.09073	109.8	
$Al_4O_3(CH_3)_6$	-1435.32059	145.87 ($\Delta S = 38.41$)	-18.72

Thus, we supposed that the "true" MAO showing ratio $Al:CH_3:O = 1:1.5:0.75$ forms in the equilibrium reaction between "classic" MAO ($Al:CH_3:O = 1:1:1$) and TMA. Basing on this idea, we carried out quantum-chemical DFT calculations of the reaction between TMA and "classic" MAO. As noted above, bonds Al^2-O^2 are the most reactive in "classic" MAO (Fig.1). Therefore we calculated the MAO structures which form as the TMA incorporates exactly through these bonds. Fig.3 presents calculated structure produced by the reactions between TMA and "classic" MAO with n= 12.

$Al(CH_3)_3$ reacts with the Al^2-O^2 bond via concert mechanism with the cleavage of bonds Al^2-O^2 in the MAO molecule and bond $Al-CH_3$ in the TMA molecule followed by simultaneous formation of new bonds Al^2-CH_3 and $O^2-Al^3(CH_3)_2$. Reaction between TMA and "classic" MAO with n=12 proceeds with the participation of four Al^2-O^2 bonds and binding of four molecules $Al(CH_3)_3$ and produces "true" MAO:

$$[Al(CH_3)O]_{12} + 4Al(CH_3)_3 \Leftrightarrow Al_{16}(CH_3)_{24}O_{12} \qquad (1)$$

Obviously, this reaction produce the MAO structure showing the ratio $Al:CH_3:O=1:1.5:0.75$ which is consistent with experimental data reported in refs. [14-18] for "true" MAO.

Specific feature of structure presented in Fig.3 is the presence of tri-coordinated atoms Al^3 in ratio 1:4 with regard to the total number of Al atoms in the "true" MAO molecule. These aluminium atoms, together with atoms Al^2 in the "classic" MAO molecule (Fig.3), may act as weak Lewis acidic centers, identified recently in work [26].

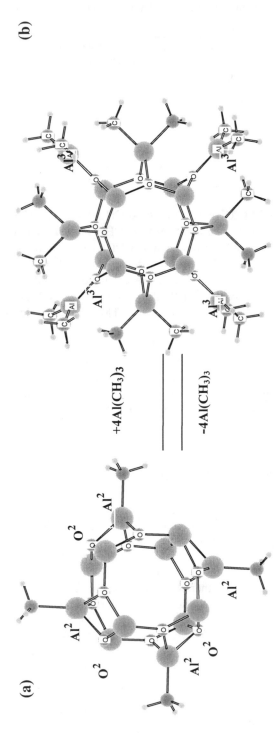

Fig. 3. A reversible reaction between triple-layer MAO [-Al(CH₃)O-]₁₂ and TMA (methyl groups of MAO in outer layers are omitted for clarity). Four Al(CH₃)₃ molecules insertion into the bonds Al^2-O^2 of inner layer of "classic"-MAO (structure **a**) with C_{4h}-symmetry. The DFT/LANL1-MB calculated structure of "true"-MAO (structure **b**) with D_{4h}-symmetry (E_{total} = -1872.69973 a.u.) has a 1:1.5:0.75 ratio for Al, CH₃ and O. The aluminum atoms Al^3 are tri-coordinated.

However, it should be taken into account that the solutions of commercial MAO, including those studied in ref. [26], always contain TMA, essential part of which presents in a free, non-connected with MAO form [7, 15]. Quite naturally, these free TMA molecules could attach to acidic centers containing tri-coordinated atoms Al^3 and thus form coordinatively-saturated structures possessing no Lewis acidity

$$
\text{MAO - O -} Al^3 {\overset{\displaystyle /CH_3}{\underset{\displaystyle \backslash CH_3}{}}} + Al(CH_3)_3 \quad \leftrightarrow \quad \text{MAO -O- } Al^3 {\overset{\displaystyle CH_3\ CH_3}{\underset{\displaystyle \backslash CH_3}{\diagup}}} Al {\overset{\displaystyle /CH_3}{\underset{\displaystyle \backslash CH_3}{}}} \quad (2)
$$

Reaction (2) may be responsible for a small number of acidic centers found in MAO at room temperature in work [26].

Methyl groups in "true" MAO could be divided into two types rated 1:2 [16]: "single" methyl groups in the outer layer (Al^1-CH_3) and "double" methyl groups in the inner layer (Al^2-$(CH_3)_2$ and Al^3-$(CH_3)_2$). As the authors of work [17] suggested, the "double" methyl groups are very reactive. The interaction of "double" methyl groups of type Al^2-$(CH_3)_2$ with neighboring tri-coordinated atoms Al^3 might explain the formation of the bridging methyl groups on the cage surface that have been found by *in situ* IR spectroscopy [27]. The following reaction between two neighboring Al-$(CH_3)_2$ groups results in the splitting off one TMA molecule (see Fig.2). As all the Al-$(CH_3)_2$ groups react, the "true" MAO transforms to "classic" MAO with the splitting off four TMA molecules (Fig.3 and equation (1)). Driving force of the reaction of TMA splitting off is the entropy increase ($+\Delta S$) in the system; the reverse reaction of the "true" MAO formation is governed by the decrease of the enthalpy ($-\Delta H$) owing to binding of the TMA molecules.

Recently we have obtained the additional experimental data on the possibility of the formation of "classic" MAO from "true" MAO via reaction (1) at heating of the solid samples of "true" MAO. The sample of solid MAO, dried under vacuum at $20°C$ ("true" MAO), has been heated under vacuum and composition and amounts of products evolving at heating have been analyzed. It was found that TMA is evolved at heating under vacuum. Amounts of TMA consist 0.02 mol $AlMe_3$/mol Al_{MAO} at temperature $50°C$ and 0.18 mol $AlMe_3$/mol Al_{MAO} at temperature $100°C$. Note that according to equation (1) the maximal amount of $AlMe_3$ evolved may consist 0.25 mol $AlMe_3$/mol Al_{MAO}. So the noticeable part of the solid "true" MAO transforms to the "classic" MAO at heating under vacuum at temperature $100°C$.

Conclusions

The results of the DFT quantum-chemical calculations presented in this work suggest the existence of MAO with three dimensional structures of two types: "classic" MAO of composition $Al:CH_3:O=1:1:1$ and "true" MAO of compositions

Al:CH_3:O=1:1.5:0.75. It is likely that the "true" MAO appears as a result of the addition reaction between trimethylaluminium and "classic" MAO, which proceeds with the cleavage of the Al-O bond in the "classic" MAO. "True" MAO and "classic" MAO exist in equilibrium.

Molecular models of both MAO structures, calculated in the present work, as well as the formulated ideas on their interrelation are most closely correspond to existing experimental data on MAO composition and structure.

"Classic" MAO and "true" MAO contain reaction centers of different types. In "classic" MAO the reaction centers are presented by highly reactive bonds Al-O in the inner layer of the three-dimensional cage structure. The reaction centers in "true" MAO are tri-coordinated aliminium atoms in groups =O-$AlMe_2$. The latter appear in "true" MAO as a result of the addition of $AlMe_3$ molecule to "classic" MAO molecule.

Acknowledgement

The present work was supported by the Russian Fund of Basic Research, grant no. 98-03-33132a. The authors are grateful to T.A Vagner for help in the manuscript preparation.

References

1   H. Sinn, W. Kaminsky, *Adv.Organomet.Chem.*18, 99 (1980)

2   W. Kaminsky, M. Miri, H. Sinn, R. Woldt, *Macromol.Chem.Rapid Commun.***4**, 417 (1983)

3   R.F. Jordan, *Adv.Organomet.Chem.* **32**, 325 (1991)

4   C. Sishta, R.H. Hathorn, T.J. Marks, *J.Am.Chem.Soc.***114**,1112 (1992)

5   X. Yang, C.L. Stern, T.J. Marks, *J.Am.Chem.Soc.***116**,10015 (1994)

6   M. Bochmann, S.J. Lancaster, *Angew.Chem., Int.Ed.Engl.*, **33**, 1634 (1994)

7   D.E. Babushkin, N.V. Semikolenova, V.N. Panchenko. A.P. Sobolev, V.A. Zakharov, E.P. Talsi, *Macromol.Chem.Phys.* **198**, 3845 (1997)

8   M.R. Mason, J.M. Smith, S.G. Bott, A.R. Barron, *J.Am.Chem.Soc.***115**, 4971 (1993)

9   C.J. Harlan, M.R. Mason, A.R. Barron, *Organometallics,* **13**, 2957 (1994)

10   I.I. Zakharov, V.A. Zakharov, G.M. Zhidomirov, Macromol.Theory Simul., **8**, 272 (1999)

11   I.I. Zakharov, V.A. Zakharov, G.M. Zhidomirov, Metalloorganic Catalysts for Synthesis and Polymerization, Ed by W.Kaminsky, Springer-Verlag, Berlin, 1999, p.128

12   C.J. Harlan, S.G.Bott, A.R.Barron, *J.Am.Chem.Soc.***117**, 6465 (1995)

13   A.R. Barron, *Macromol .Symp.* **97**, 15 (1995)

14   J. Bliemeister, W. Hagendorf, A. Harder, B. Heitmann, I. Schimmel, E. Schmedt, W. Schuchel, H. Sinn, L. Tikwe, N.von Thienen, K. Urlass, H. Winter, O. Zarncke, Ziegler Catalysts, Ed. by G. Fink, R. Mülhaupt, H.H. Brintzinger Springer-Verlag, Berlin, 1995, p.57

15   D. Imhoff, L. Simeral, S. Sangokoya, L. Peel, *Organometallics,* **17**, 1941 (1998)

16 H. Sinn, *Macromol. Symp.* **97**, 27 (1995)

17 H. Sinn, I. Schimmel, M. Ott, N.von Thienen, A. Harder, W. Hagendorf, B. Heitmann, E. Haupt, Metalloorganic Catalysts for Synthesis and Polymerization, Ed by W. Kaminsky, Springer-Verlag, Berlin, 1999, p.105

18 J. Eilertsen, E. Rytter, M. Ystenes, Metalloorganic Catalysts for Synthesis and Polymerization, Ed by W. Kaminsky, Springer-Verlag, Berlin, 1999, p.136

19 W. Hagendorf, A. Harder, H. Sinn, *Macromol.Symp.* **97**, 127 (1995)

20 M.J. Frisch, G.W. Trucks, H.B. Schlegel, P.M.W. Gill, B.G. Johnson, M.W. Wong, J.B. Foresman, M.A. Robb, M. Head-Gordon, E S. Replogle, R. Gomperts, J.L. Andres, K. Raghavachari, J.S. Binkley, C. Gonzalez,L. Martin, D.J. Fox, D.J. Defrees, J. Baker, J.J.P. Stewart, J.A. Pople, *Gaussian 92/DFT, Revision G.2, Gaussian, Inc.*, Pittsburgh PA, 1993

21 P.J. Hay, W.R. Wadt, *J. Chem. Phys.* **82**, 270 (1985)

22 S.J. Vosco, L. Wilk, M. Nusair, *Can. J. Phys.* **58**, 1200 (1980)

23 A.D. Becke, *Phys.Rev.* **A 38**, 3098 (1988)

24 C. Lee, W. Yang, R.G. Parr, *Phys.Rev.* **B 37**, 785 (1988)

25 A.D. Becke, *J. Chem. Phys.* **98**, 5648 (1993)

26 E.P. Talsi, N.V. Semikolenova, V.N. Panchenko, A.P. Sobolev, D.E. Babushkin, A.A. Shubin, V.A. Zakharov, *J.Mol.Catal.* **139**, 131 (1999)

27 M. Ystenes, J.L. Eilertsen, J. Liu, M. Ott, E. Rytter, J.A. Stovneng, J. Polymer Sci. A: Polymer Chem. **38**, 3106 (2000)

A DFT Study of Ethylene Polymerization by Zirconocene-boron Catalytic Systems

Ilya E. Nifant'ev*, Leila Yu. Ustynyuk, Dmitri N. Laikov

Chemistry Department, M.V.Lomonosov Moscow State University, 119899, Leninskie gory, Moscow, Russia
E-mail: inif@org.chem.msu.ru

Abstract. The effect of anion A^- on the energy profile of the interaction $Cp_2ZrEt^+A^- + C_2H_4 \rightarrow Cp_2ZrBu^+A^-$ ($A^- = CH_3B(C_6F_5)_3^-$, $B(C_6F_5)_4^-$) was studied. The addition of olefin to the ion pair $Cp_2ZrEt^+A^-$ is characterized by an appreciable energy barrier and even can be the rate-determining stage of the overall process. The "front-perpendicular" approach of ethylene molecule to nonagostic isomer of $Cp_2ZrEt^+A^-$ (**5c**) was found to be energetically most favorable. The results suggest that "nonagostic" reaction channels characterized by stabilization of intermediates and transition states should be growing in importance with enhancement of the nucleophilicity of the counterion.

Introduction

Density functional theory (DFT) is widely used in theoretical studies of the mechanisms of catalytic processes, in particular, polymerization of olefins by zirconocene catalysts. It is commonly accepted that active catalytic species for olefin polymerization are ion pairs $Cp_2ZrAlkyl^+A^-$, which comprise alkylzirconocene cations $Cp_2ZrAlkyl^+$ and weak nucleophilic anions A^- [1]. Usually, theoretical studies are carried out with the Cp_2ZrEt^+ cation considered as a model zirconocene catalyst. The mechanism of the interaction between ethylene molecule and Cp_2ZrEt^+ cation has been studied in detail in a series of works by T. Ziegler *et al.* [2-6]. This mechanism involves a barrierless exothermic addition of ethylene molecule to complex **1** to give an intermediate **2**, followed by rearrangement of the latter resulting in product **4** (Scheme 1).

Here, the highest energy barrier **TS-2** on the reaction pathway is overcome at the stage of isomerization of β-agostic complex **2** into an α-agostic complex **3**. The activation energy of the inner-sphere formation of the C—C bond in complex **3** *via* **TS-3** is close to zero.

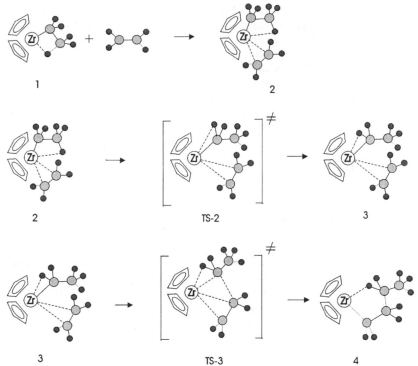

Scheme 1. Mechanism of the interaction between Cp_2ZrEt^+ cation and ethylene molecule proposed by T. Ziegler [2-6]

Recently, the first theoretical studies [7,8] devoted to consideration of the role of $Cp_2ZrAlkyl^+A^-$ ion pairs have been reported. Nevertheless, it is unclear how the anion A^- affects the course of a polymerization process.

The main purpose of this work was to study the effect of the anion on the energy profile of polymerization reaction for the mechanism presented in the Scheme 1 and to consider other possible reaction pathways in order to find the most energetically favorable reaction mechanism.

Three models of a zirconocene catalysts were used and compared in this work, namely, the ethylzirconocene cation, Cp_2ZrEt^+, and two ion pairs $Cp_2ZrEt^+A^-$ ($A^- = CH_3B(C_6F_5)_3^-$, $B(C_6F_5)_4^-$). The systems based on tris(pentafluorophenyl)boron were chosen due to the large amount of available experimental data on their structures and reactivities.

Computational Details

An original PRIRODA program developed by one of us [9] was used. Generalized gradient approximation (GGA) for the exchange-correlation functional of Perdew, Burke and Ernzerhof [10] was employed. Orbital basis sets of contracted Gaussian-type functions of size (4s)/[2s] for H, (8s4p1d)/[4s2p1d] for C, and (20s16p11d)/[14s11p7d] for Zr were used in conjunction with the density-fitting basis sets of uncontracted Gaussian-type functions of size (4s1p) for H, (7s2p2d) for C, and (22s5p5d4f4g) for Zr.

Full geometry optimization was performed for a number of stable and transition-state structures, followed by analytical calculations of the second energy derivatives. All structures were characterized by harmonic vibrational frequencies.

Results and discussion

Isomers of catalytic species

The model catalytic species considered in this work were the ethylzirconocene cation, Cp_2ZrEt^+(a), and ion pairs $Cp_2ZrEt^+A^-$, where $A^- = B(C_6F_5)_4^-$ (b) and $CH_3B(C_6F_5)_3^-$ (c). Investigation of the potential energy surfaces (PES) of these systems suggested that the catalytic species can exist in the form of isomers **1**, **5**, or **6** and that each of the isomers corresponds to a PES minimum.

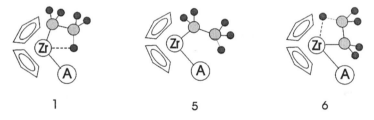

In the absence of anion, structures **1a** and **6a** are identical, since only two isomers (**1a** and **5a**) were found for ethylzirconocene cation. The values of thermodynamic characteristics of isomers **5** and **6** (relative to the corresponding parameters of complex **1** taken as zero) are listed in Table 1. Characteristic of isomers **1** and **6** are β-agostic bonds. Isomer **5** can exist both in nonagostic forms (**5b** and **5c**) and in the form with a weak α-agostic bond (**5a**).

The relative energies of isomers **1**, **5**, and **6** are drastically affected by the nature of the anion A^-. For $A^- = CH_3B(C_6F_5)_3^-$ complex **5c** has the lowest energy, whereas for $A^- = B(C_6F_5)_4^-$ and "naked" ethylzirconocene cation the β-agostic structures **1b** and **1a** have the lowest energies, respectively. For different ion pairs, the energies of the most stable forms of complexes **1** and **5** differ by no more than 2 kcal mol⁻¹. The interaction within the ion pair, which can be first of all judged from the Zr—B distance, results in stabilization of isomer **5**, where this distance is shorter than in **1**. In contrast to this, the energy of isomer **6** is appreciably higher than that of **1** (Table 1) for both $A^- = B(C_6F_5)_4^-$ and $CH_3B(C_6F_5)_3^-$ despite the shorter Zr—B distance in the former structure. This is likely due to structural distortions of the zirconocene fragment, Cp_2ZrEt^+, in **6** as compared to **1**.

Thus, one can conclude that the anion A^- and the agostic bond "force out" each other from the coordination sphere of Zr. As a result, the energies of the ion pairs are to a great extent determined by the nucleophilicity of the anion. For the stronger nucleophile, $A^- = CH_3B(C_6F_5)_3^-$, the interaction with the anion leads to substantial stabilization of the nonagostic structure **5c** as compared to agostic structure **1c**. For the weaker nucleophile, $A^- = B(C_6F_5)_4^-$, this effect is much smaller, so complex **1b** appears to be more stable than **5b**.

Table 1. Thermodynamic Data for the Isomers of the Model Catalytic Species $Cp_2ZrEt^+A^{-}$ [a]

	E kcal mol^{-1}	H_{298} kcal mol^{-1}	G_{298} kcal mol^{-1}
A^-	**"Naked" Cp_2ZrEt^+ Cation**		
5a	10.0	9.3	8.6
6a	-	-	-
A^-	**$B(C_6F_5)_4^-$**		
5b	2.0	2.1	0.9
6b	4.4	4.1	3.9
A^-	**$CH_3B(C_6F_5)_3^-$**		
5c	-1.1	-1.5	-1.7
6c	2.8	2.6	3.3

[a] The energies (E), enthalpies (H_{298}), and Gibbs free energies (G_{298}) are given relative to the corresponding values for **1**.

Thus, in the case of ion pair $Cp_2ZrEt^+A^-$ the energy of the β-agostic structure of the type **1** is comparable with that of the nonagostic structure of the type **5**. As we have found, structures **1** and **5** can undergo fast interconversion. Detailed description of these transformations will be reported elsewhere.

Interaction of Ion Pair $Cp_2ZrEt^+A^-$ with Ethylene Molecule

We considered in what extent the nature of the anion A^- is able to affect the interaction between $Cp_2ZrEt^+A^-$ and ethylene. In Scheme 2, we present main steps of such an interaction that were under our investigation.

We optimized the structures of all intermediates and transition states. The corresponding thermodynamic parameters are listed in Table 2. The energy profile of transformation $1 + C_2H_4 \rightarrow 4$ is shown in Fig. 1.

Step 1. Addition of ethylene molecule

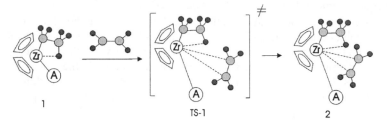

Step 2. Rearrangement of intermediate complex **2**

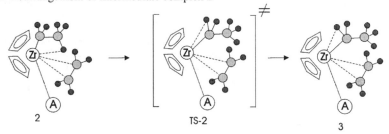

Step 3. Inner-sphere formation of aliphatic C—C bond

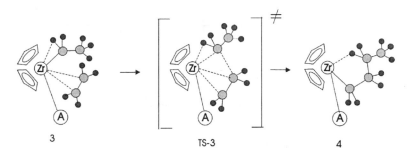

Scheme 2. Main steps of the ion pair Cp₂ZrEt⁺A⁻ interaction with ethylene molecule.

The height of the energy barrier corresponding to **TS-1** (step 1) is mainly determined by the nature of anion A⁻, *viz.*, the higher the nucleophilicity of the anion, the higher the energy barrier to step 1. We failed to determine the energy barrier to the ethylene addition to "naked" ethylzirconocene cation neither by gradient search, nor by point-by-point scanning the PES of the system under study. In the course of ethylene addition to Cp₂ZrEt⁺A⁻, the anion A⁻ is displaced to the outer coordination sphere, thus making room for the ethylene molecule in the coordination sphere of zirconium. This is accompanied by an increase in the Zr—B distance by more than 1 Å, which is the reason for the appearance of the energy barrier. Yet another distinctive feature of ethylene addition to the ion pairs Cp₂ZrEt⁺A⁻ is its lesser exothermicity as compared with the addition to "naked" Cp₂ZrEt⁺ cation. Since the addition of ethylene molecule in the systems under study is accompanied by considerable loss of entropy, the ΔG_{298} values for this reaction are positive (Table 2).

Since reorganization of the zirconocene fragment is not accompanied by appreciable changes in the Zr—B distance (Table 2), the energy profile of transformation $2 \rightarrow 4$ (steps 2 and 3, see above) is virtually unaffected by the nature of anion A⁻. The energy difference between **TS-2** and **2** is about 5.8, 6.1, and 4.6 kcal mol⁻¹ for the ethylzirconocene cation, $B(C_6F_5)_4^-$, and $CH_3B(C_6F_5)_3^-$, respectively. On the other hand, the energies of **TS-2** for the three systems, calculated relative to those of noninteracting reagents (Table 3), differ appreciably and increase as follows: **TS-2a < TS-2b < TS-2c**.

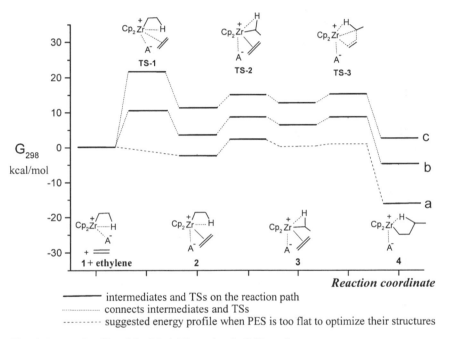

Fig. 1. Energy Profile of the Model Reaction **1**+$C_2H_4 \rightarrow$ **4**

For the "naked" Cp_2ZrEt^+ cation, the energies of α-agostic complex and **TS-3** are very close, so the geometry optimization for **3** immediately leads to the primary insertion product **4**. This was first found by T.Ziegler *et al.* [5] who estimated the energy barrier at about 0.5 kcal mol⁻¹. In this work, the structure of α-agostic complex **3a** for the system with the Cp_2ZrEt^+ cation was found by point-by-point scanning the PES of the system when moving along an approximate reaction coordinate. For the systems with ion pairs, energy barrier to transformation $3 \rightarrow 4$ is also low (the energy difference between **TS-3** and **3** does not exceed 2 kcal mol⁻¹).

Table 2. Thermodynamic Data of the Intermediates and Transition States for the Model Reaction $Cp_2ZrEt^+A^- + C_2H_4 \rightarrow Cp_2ZrBu^+A^-$.[a]

	E kcal mol^{-1}	H_{298} kcal mol^{-1}	G_{298} kcal mol^{-1}	Zr—B Å
A^-	"Naked" Cp_2ZrEt^+ Cation			
1a	0	0	0	-
TS-1a	-	-	-	-
2a	-16.2	-14.4	-2.0	-
TS-2a	-10.4	-9.8	2.5	-
3a	-13.6	-12.9	0.3	-
TS-3a	-	-	-	-
4a	-31.1	-28.3	-15.9	-
A^-	$B(C_6F_5)_4^-$			
1b	0	0	0	5.19
TS-1b	0.4	1.4	10.5	6.43
2b	-9.6	-7.2	3.8	6.65
TS-2b	-3.5	-2.4	8.7	6.42
3b	-4.6	-3.3	6.5	6.46
TS-3b	-3.8	-2.8	9.4	6.35
4b	-18.8	-16.0	-4.9	6.49
A^-	$CH_3B(C_6F_5)_3^-$			
1c	0	0	0	4.36
TS-1c	10.4	10.6	21.7	5.45
2c	-1.0	0.1	11.3	6.68
TS-2c	3.7	3.4	15.1	6.23
3c	2.2	2.6	12.9	6.29
TS-3c	3.1	3.2	15.2	6.23
4c	-11.9	-9.9	1.50	6.13

[a] The energies (E), enthalpies (H_{298}), and Gibbs free energies (G_{298}) are given relative to the corresponding values for the noninteracting reagents **1**+C_2H_4.

This stage, as well as the corresponding **TS-3**, seems to have little effect on the kinetics of the reaction. It should be noted that the energy profile of propylene polymerization changes in such a manner that the energy of **TS-3** is higher than that of **TS-2**.

Thus, we have found that ethylene addition to the ion pairs $Cp_2ZrEt^+A^-$ is not a barrierless process and that the corresponding energy barriers can be comparable in magnitude (for $A^- = B(C_6F_5)_4^-$) with or even appreciably higher (for $A^- = CH_3B(C_6F_5)_3^-$) than the energy barrier of β-agostic complex **2** isomerization into α-agostic complex **3**. The ratio of barrier heights (**TS-1, TS-2**, and **TS-3**) is determined solely by the nucleophilicity of anion A^-, *i.e.*, by the energy of the interaction between the anion and cation. The higher the nucleophilicity of the anion, the higher the **TS-1** barrier height as compared to starting reagents. It is the energy barrier **TS-1** that is responsible for the experimentally observed first-order (with respect to the monomer) kinetics of the reaction. Would this energy barrier be zero or very low, as is the case of "naked" Cp_2ZrEt^+ cation, the kinetic order of the reaction should be close to zero, taking into consideration low exotermicity of the process **1** + ethylene → **2**.

It has been found that consideration of the rearrangements within the zirconocene fragment (transformations **2** → **3** → **4**) can be performed ignoring the effect of the counterion, since they are not accompanied by appreciable changes in the Zr—B distance.

It has been shown that the effect of the counterion on the energy profile of transformation **1** + C_2H_4 → **4** manifests itself as a decrease in the overall exothermicity of the process with increasing the nucleophilicity of the anion. In other words, the energies of all intermediates and transition states of the reaction for $A^- = CH_3B(C_6F_5)_3^-$ are higher (relative to the noninteracting reagents) than for $A^- = B(C_6F_5)_3^-$.

Alternative Mechanisms for the Interaction of $Cp_2ZrEt^+A^-$ with Ethylene Molecule

As was shown above, the addition of ethylene molecule can be the rate-determining stage of the reaction $Cp_2ZrEt^+A^- + C_2H_4$ → $Cp_2ZrBu^+A^-$. Nevertheless, Scheme 2 describes only one possible pathway of ethylene addition to the ion pair $Cp_2ZrEt^+A^-$, namely the interaction of ethylene with β-agostic isomer **1**. As it was stated above, the compounds **1, 5** and **6** are able to undergo fast interconversion. Hence, it would be reasonable to consider other feasible ways of addition of ethylene molecule to complexes **1, 5**, and **6**. We performed it taking the system with $A^- = CH_3B(C_6F_5)_3^-$ as an example (Schemes 3A-C).

Generally, the addition of ethylene molecule to isomers **1, 5**, and **6** can proceed in two different manners, namely, from the front side or from the back side with respect to the position of anion A^-. In turn, there are two possibilities for the front-side addition where (i) the ethylene molecule is arranged perpendicular or nearly perpendicular to the plane which passes through the Zr atom and is equidistant from the cyclopentadienyl ligands (hereafter, front-perpendicular addition) or (ii) the ethylene molecule is in or deviates only slightly from this plane (hereafter, front-normal addition).

We optimized the structures of transition states **TS-4c** – **TS-11c** in order to compare their energies with that of **TS-1c**. The thermodynamic parameters and selected geometric parameters of these transition states are listed in Table 3. As can be seen from the data in Table 3, transition state **TS-6c** of the reaction of ethylene addition to complex **5c** has the lowest energy. Next by energy after **TS-6c** is the transition state **TS-7c** corresponding to the Cossee mechanism [11-13]. It should be noted that from two ways of addition (front-perpendicular, **TS-6c**, and front-normal, **TS-7c**), the former is the most energetically favorable owing to the shortest Zr—B distance in the transition state **TS-6c** and to stabilization of **TS-6c** hy the interaction between the anion and cation. The same also holds for other ∠ransition states (the "front-perpendicular" **TS-4c** and **TS-9c** have lower energies than the corresponding "front-normal" **TS-7c** and **TS-10c**, respectively) and intermediates of this addition reaction (the "front-perpendicular" complex **7c** has a lower energy than the "front-normal" complex **3c**).

Thus, the energy barrier to front-perpendicular addition of olefin to nonagostic complex **5c** is lower than that (**TS-1**) in the mechanism of the reaction $Cp_2ZrEt^+A^-$ + $C_2H_4 \rightarrow Cp_2ZrBu^+A^-$ considered above (Scheme 2).

Front-perpendicular addition of ethylene molecule to isomer **5c** results in the ethylene complex **7c**. The energy (E) and Gibbs free energy (G_{298}) of this complex are equal to 0.2 and 11.6 kcal mol^{-1} relative to noninteracting reagents **1c** + C_2H_4, respectively.

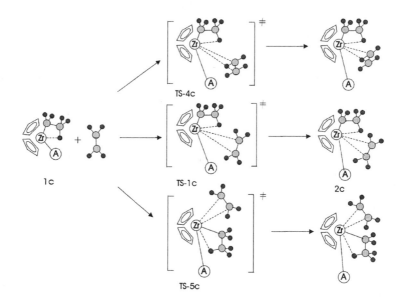

A: Isomer **1c** interaction with ethylene molecule

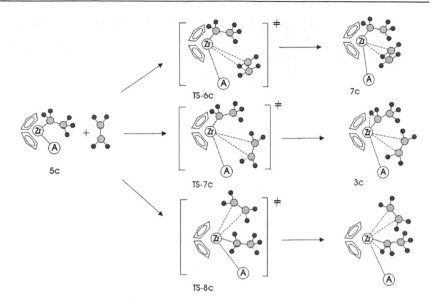

B: Isomer **5c** interaction with ethylene molecule

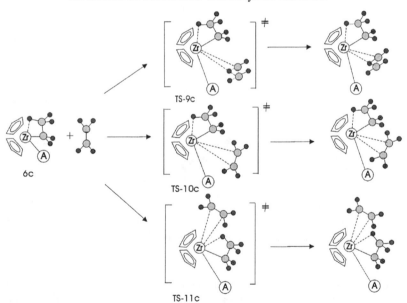

C: Isomer **6c** interaction with ethylene molecule

Scheme 3: Alternative mechanisms for the interaction of ion pair $Cp_2ZrEt^+A^-$ with ethylene molecule ($A^- = CH_3B(C_6F_5)_3^-$).

After front-perpendicular addition to complex **5c**, the ethylene molecule is rotated by about 90° in the coordination sphere of zirconium. The rotation of ethylene molecule is accompanied by simultaneous formation of an α-agostic bond, which is associated with passage through **TS-2** (see above), and the energy of the system monotonically increases.

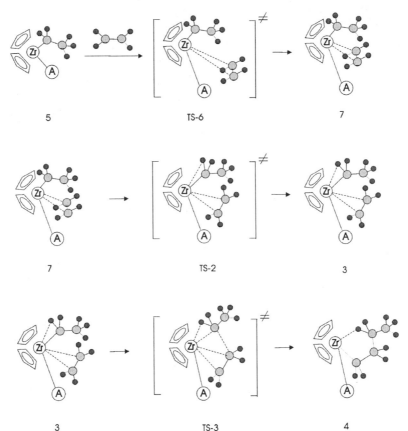

Scheme 4. Most Energetically Favorable Mechanism of the Interaction of Ion Pair Cp$_2$ZrEt$^+$A$^-$ (A$^-$ = CH$_3$B(C$_6$F$_5$)$_3^-$) with Ethylene Molecule

Comparison of the reaction mechanisms shown in Schemes 2 and 4 suggests that they differ from each other only in the first two stages of the process. It is noteworthy that in the second stage the isomerization of the corresponding ethylene complex (**2** or **7**, respectively) results in intermediate **3**, which is common to both mechanisms and undergoes then the same transformations. Of particular importance is the fact that the transformation of intermediate **7** into **TS-2** is not accompanied by the formation of new intermediate complexes or by the appearance of new transition states of the reaction. This is indicated by the fact that the geometry of **TS-2** relaxes to that of intermediate **7** upon removal of all symmetry restrictions at the stage of geometry optimization.

Table 3. Thermodynamic Characteristics and Selected Geometric Parameters of Transition States of the Reaction of Ethylene Addition to Isomers of the Model Catalytic Species $Cp_2ZrEt^+CH_3(C_6F_5)_3^{-}$ [a]

	Isomer	Direction of ethylene addition	E kcal mol^{-1}	H_{298} kcal mol^{-1}	G_{298} kcal mol^{-1}	$Zr—C$ (C_2H_4) Å	Zr-B Å
TS-4c	**1c**	Front-perpendicular[b]	10.0	10.4	20.3	3.98 4.04	5.51
TS-1c	**1c**	Front-normal[c]	10.4	10.6	21.7	4.16 4.53	5.45
TS-5c	**1c**	Back	9.5	9.6	20.6	3.51 3.65	5.29
TS-6c	**5c**	Front-perpendicular[b]	**2.0**	**3.2**	**15.8**	**3.55** **3.71**	**4.39**
TS-7c	**5c**	Front-normal[c]	8.9	7.6	21.8	4.07 4.33	4.43
TS-8c	**5c**	Back	10.6	11.0	23.2	3.15 3.31	4.61
TS-9c	**6c**	Front-perpendicular[b]	12.3	12.5	23.2	3.79 3.99	5.34
TS-10c	**6c**	Front-normal[c]	15.1	15.2	25.3	3.98 4.14	5.55
TS-11c	**6c**	Back	13.1	12.9	23.1	3.75 3.79	5.32

[a] The energies (E), enthalpies (H_{298}), and Gibbs free energies (G_{298}) are given relative to the corresponding values for the noninteracting reagents $1c+C_2H_4$
[b] Ethylene molecule has a perpendicular orientation with respect to the plane, which is equidistant from the Cp ligands
[c] Ethylene molecule is in the plane which is equidistant from the Cp ligands

The energy profiles of the transformations of isomers **1c** (from left to right) and **5c** (from right to left) into α-agostic complex **3c** are shown in Fig. 2. The mechanism analogous to that proposed by T. Ziegler for the "*naked*" Cp_2ZrEt^+ cation is characterized by the much higher energy barrier than the alternative reaction mechanism proposed in this work. The same also holds for the energy barrier **TS-7c** corresponding to the Cossee mechanism in the case of front-normal addition of ethylene molecule to complex **5c**. Taken together, this indicates that front-perpendicular ethylene addition becomes much more energetically favorable than front-normal addition in the systems with anions of moderate nucleophilicity.

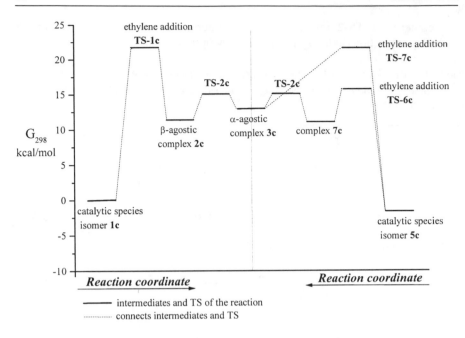

Fig. 2. Energy profiles of transformations of isomers **1c** and **5c** into α-agostic complex **3c**

Conclusion

Our study has shown that unlike the barrierless reaction between the "naked" ethylzirconocene cation and ethylene molecule, the addition of olefin to the ion pair $Cp_2ZrEt^+A^-$ is characterized by an appreciable energy barrier and even can be the limiting (rate-determining) stage of the overall process at least for the systems with rather nucleophylic A^-. In this connection, the question of the pathway of ethylene addition to the ion pair $Cp_2ZrEt^+A^-$ is of particular interest. Taking the $CH_3B(C_6F_5)_3^-$ anion as an example, we found that for counterions of moderate nucleophilicity the activation barrier to the front-side interaction of nonagostic complex **5c** with ethylene molecule is lower than the barrier to analogous interaction with participation of β-agostic complex **1c**. The front-perpendicular approach of ethylene molecule to **5c** was found to be more energetically favorable than the front-normal and back-side addition because it requires the minimal increase in the Zr—B distance. Reduction of nucleophilic properties of the anion probably would result in disappearance in some extent of this advantage of **TS-6** over **TS-1** and **TS-7**.

It seems likely that polymerization in real catalytic systems can follow several pathways simultaneously. The results obtained in this work suggest that "nonagostic" reaction channels characterized by stabilization of intermediates and transition states owing to tighter interaction within the ion pair rather than the formation of β-agostic bonds should be growing in importance with enhancement of the nucleophilicity of the counterion.

It is known that in the zirconocene—MAO catalytic system the highest activity of the catalyst is achieved at rather high Zr/Al ratios. We believe that this is due to the change in the "effective" nucleophilicity of the counterion, namely, the higher the concentration of MAO, the lower the "effective" nucleophilicity of the counterion (because of the higher ability of the anion to charge delocalization) and the lower the activation energy of the olefin addition to the catalytic species.

Main conclusions drawn in this work can be extended to polymerizations of other olefins including propylene. One can assume that front-perpendicular addition of propylene to $Cp_2ZrEt^+A^-$, similar to that of ethylene, is also possible. This would mean that stereoselectivity of the propylene addition to the catalytic species can be controlled not only by interaction of monomer with growing chain [14], but also by the interactions between the methyl group of propylene molecule and substituents in the cyclopentadienyl ligands in the **TS-6**-like transition states.

Acknowledgment

Financial support by Montell B.V. is gratefully ackowledged. The authors express their gratitude to Prof. Yu. A. Ustynyuk for fruitful discussions and help in preparation of this work.

References

1   H.-H.Brintzinger,   D.Fischer,   R.Mulhaupt,   B.Rieger   and   R.M.Waymouth, *Angew.Chem.Int.Ed.Engl.*, 1995, **34**, 1143
2   P.Margl, L.Deng, T.Ziegler, *Organometallics*, 1998, **17**, 933
3   P.Margl, J.C.W.Lohrenz, T.Ziegler and P.Blochl, *J.Amer.Chem.Soc.*, 1996, **118**, 4434
4   J.C.W.Lohrenz, T.K.Woo, T.Ziegler, *J.Amer.Chem.Soc.*, 1995, **117**, 12793
5   J.C.W.Lohrenz, T.K.Woo, L.Fan, T.Ziegler, *J.Organomet.Chem.*, 1995, **497**, 91
6   T.K.Woo, L.Fan, T.Ziegler, *Organometallics*, 1994, **13**, 2252
7   M.S.W.Chan, K.Vanka, C.C.Pye, T.Ziegler, *Organometallics*, 1999, 18, 4624
8   K.Vanka, M.S.W.Chan, C.C.Pye, T.Ziegler, *Organometallics*, 2000
9   D.N.Laikov, *Chem. Phys. Lett.*, 1997, **281**, 151
10   J.P.Perdew, K.Burke, M.Ernzerhof, *Phys.Chem.Lett.*, 1996, **77**, 3865
11   P.Cossee, *Tetrahedron Lett.*, 1960, **12**, 17
12   P.Cossee, *J.Catal.*,1964, **3**, 80
13   E.J.Arlman, P.Cossee, , *J.Catal.*,1964, **3**, 99
14   L.Cavallo, G.Guerra, M.Vacatello, P.Corradini, *Macromolecules*, 1991, **24**, 1784

Activation Reactions of Cp_2ZrCl_2 and Cp_2ZrMe_2 with Aluminium Alkyl Type Cocatalysts Studied by *in situ* FTIR Spectroscopy

Jan L. Eilertsen[1], Erling Rytter[2,3] and Martin Ystenes[1]

[1]Dept. of Chemistry, and [2]Dept. of Chemical Engineering, Norwegian University of Science and Technology, N-7491 Trondheim, Norway
[3]Statoil Research Centre, N-7005 Trondheim, Norway
E-mail: eilerts@chembio.ntnu.no; err@statoil.com; ystenes@chembio.ntnu.no

Abstract. The reactions of Cp_2ZrCl_2 and Cp_2ZrMe_2 with methylaluminoxane (MAO), trimethylaluminium (TMA) and dimethylaluminium chloride (DMAC) have been investigated by *in situ* FTIR spectroscopy. The studies have been performed in a cell that allows continuous monitoring of the reactions and stepwise additions of reactants. Most bands of the zirconocenes are unaffected by the reactions, but a strong Cp band at 803-822 cm^{-1} was found to give distinct information on structural changes in the zirconocenes. A slow formation of the monochloro-monomethyl compound $Cp_2ZrClMe$ from a mixture of Cp_2ZrCl_2 and Cp_2ZrMe_2 has been verified. Only weak complexes are formed in mixtures of zirconocenes and TMA or DMAC. The chemical potential for methylation of zirconocenes is primarily due to MAO clusters, but TMA may be important in the mechanism. Our IR data is consistent with the formation of stable compounds during activation, which we assume include methyl or chlorine bridges between zirconium and aluminium, but do not differentiate between ionic or neutral complexes. Observed disappearance of C–H stretching bands may indicate double bridges between zirconium and aluminium.

Introduction

The present study is part of a more extensive FTIR spectroscopy work where all binary mixtures in the system Cp_2ZrCl_2, Cp_2ZrMe_2, TMA, DMAC and MAO have been investigated. Earlier we have shown that MAO and TMA behave as independent stable structural entities in mixtures, that may exchange ligands, but which do not form new compounds.[1] This study revealed the presence of bridging methyl groups on the surface of stable MAO clusters, for which we recently have proposed molecular structures.[2] It was also found that MAO when reacted with DMAC loose the bridging methyl groups, and change to a chlorinated MAO that is not able to activate olefin polymerisation with $(C_5Me_5)_2ZrCl_2$.[3]

In the present work the reactions between the aluminium compounds and the metallocenes are studied, and the studies have been conducted to reveal – if possible – changes both in the metallocene and the aluminium compounds. Fig. 1 depicts all binary mixtures in the system. The figure also indicates which mixtures lead to formation of new compounds (lines), and which do not (dashes).

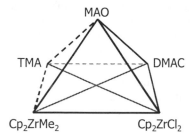

Fig. 1. Binary combinations in the system Cp$_2$ZrCl$_2$, Cp$_2$ZrMe$_2$, TMA, DMAC and MAO. Dashed lines indicate binary systems where no new compounds were observed.

The formation and the nature of the active centre for polymerisation have been much discussed in the research literature [4-11], and partly covered in a recent review.[12] In general, the formation of cationic active centres is suggested to involve a fast mono-methylation of the metallocene, if a chloride is used, followed by a slow ion pair formation by abstraction of a chloride ion.[12] Theoretical studies have shown that the ion pair formation and separation is not energetically favourable without a large anionic counterion to spread the negative charge as well as solvent or monomer coordination to the cation.[7] Several options for complexation to the cation is offered by MAO – through oxo,[4,5,22,23] methyl [11] or chlorine[6,7] bridges, or by formation of a methylene bridge.[5] In addition, the methyl-bridged binuclear cations, [(Cp$_2$ZrMe)$_2$(μ-Me)]$^+$ and [Cp$_2$Zr(μ-Me)$_2$AlMe$_2$]$^+$, have been proposed.[8,10] Mostly, these complexes have been suggested to be dormant sites, but MAO has also been suggested to be an important part of architecture of the active site.[4,5,22] Bonds between MAO and the neutral zirconocene have also been suggested. A recent NMR study indicates a fast equilibrium where Cp$_2$ZrMe$_2$ and MAO form a methyl-bridged non-ionic complex at low Al/Zr ratios.[11] Calculations on the formation of a chlorine-bridged compound Cp$_2$ZrMe(μ-Cl)AlMe$_3$ and "Cp$_2$ZrMe(μ-Cl)MAO" (model structure) have been reported.[7] The former complexes were calculated to be unstable and the latter stable.

Unfortunately, detailed information on the structural changes of MAO during activation and the interaction with zirconocenes is not easily obtained. The otherwise powerful NMR spectroscopy has not been particularly useful, partly due to extremely broad NMR peaks of MAO.[6,10,22] The low concentration and the fluxional nature of the interactions introduce further difficulties.

The study of such interactions is important in order to understand the activation process in the zirconocene/MAO system. While a large number of metallocenes have been studied by IR spectroscopy,[13] hardly any such data exist for catalyst systems used in olefin polymerisation[14].

Results

Cp$_2$ZrCl$_2$ and Cp$_2$ZrMe$_2$

Cp$_2$ZrMe$_2$ and Cp$_2$ZrCl$_2$ were dissolved in benzene, and Cp$_2$ZrMeCl was formed in high yield within two weeks at room temperature. No conversion was observed after 24 h. The proton NMR spectrum of the compound was in agreement with literature data[15]. Most of the IR spectrum of the metallocene was unaffected by the exchange, but the strong out-of-plane C–H deformation band of the Cp ring was found to be sensitive to the altered zirconium environment. The position of this band is listed for the reactants and the product in Table 1 and the main IR bands of Cp$_2$ZrMeCl are listed in Experimental section.

Table 1 Infrared band positions [cm^{-1}] for cyclopentadienyl hydrogen out-of-plane deformations (The vibration is visualised to the right).

Compound	cm^{-1}
Cp$_2$ZrCl$_2$	814
Cp$_2$ZrMeCl	809
Cp$_2$ZrMe$_2$	803

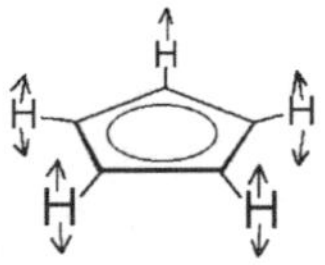

Cp$_2$ZrCl$_2$ and Aluminium Alkyls (TMA, DMAC and MAO)

Addition of TMA to Cp$_2$ZrCl$_2$ led as expected to a fast, but only partial formation of Cp$_2$ZrMeCl and DMAC. The presence of Cp$_2$ZrMeCl is seen as a shoulder at 809 cm^{-1} in Fig. 2a, and by the C–H stretching bands of zirconium-bonded methyl (2874, 2792 cm^{-1}). No evidence of di-methylation was observed, even for large excess of TMA. The methylating power of TMA was further investigated by addition of TMA to Cp$_2$ZrMeCl. Also in this case, no Cp$_2$ZrMe$_2$ was observed. Instead a new shoulder on the 809-cm^{-1} band was detected at 815 cm^{-1}, indicating formation of a bimetallic compound with a methyl or chlorine bridge. Such a shoulder was also observed in the above reaction between TMA and Cp$_2$ZrCl$_2$, but much weaker.

When DMAC was mixed with Cp$_2$ZrCl$_2$ a weak shoulder was observed at 820 cm^{-1}, as shown in Fig. 2b. Also in this case we interpret it as an indication of compound formation. Except for the increasing DMAC bands, no other notable changes were seen in the spectra.

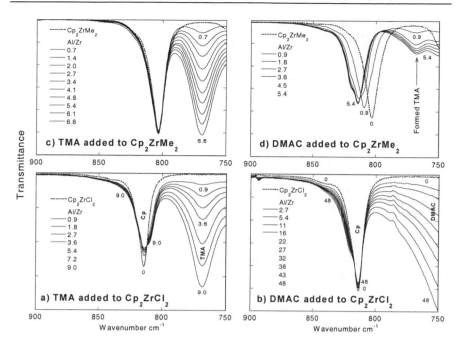

Fig. 2. Infrared spectra of binary mixtures of Cp$_2$ZrMe$_2$, Cp$_2$ZrCl$_2$, TMA and DMAC. The figure shows the region around bands attributed to out-of-plane C–H deformations of the Cp ligand. The solvent spectrum has been removed. The initial concentrations of zirconocene were a) 0.03 M, b) 0.02 M, c) 0.06 M and d) 0.05 M.

As an exchange reaction between DMAC and MAO has been demonstrated,[2] similar exchange was expected by adding solid Cp$_2$ZrCl$_2$ to a MAO solution[16]. The band at 1257 cm^{-1}, attributed to the deformation of bridging methyl groups of MAO, decreased significantly, as shown in Fig. 3. The change seen below 800 cm^{-1} and the decrease in the MAO shoulder at 840 cm^{-1} are related to the chlorination of MAO and also seen for DMAC addition. Bands attributed to methyl groups on zirconium are weak and not easily monitored due to band overlap. However, weak C–H stretching bands, probably due to the methyl in Cp$_2$ZrMeCl were identified. A new band attributed to the out-of-plane C–H deformation of Cp was clearly identified at 819 cm^{-1} after reaction with MAO. The position of this band is intriguing as it can be explained neither by the formation of Cp$_2$ZrMeCl nor by Cp$_2$ZrMe$_2$ – both have this band at lower frequencies. Therefore, an interaction between MAO and Cp$_2$ZrCl$_2$ or the formed Cp$_2$ZrMeCl seems to be evident. Unfortunately, there is overlap with the strong band at 808 cm^{-1} due to Al–O stretch of MAO; hence, a quantitative treatment is difficult. The solution turned bright yellow immediately after mixing, and slowly to red within an hour. Phase separation was also evident at this stage.

The TMA content was largely unchanged by the addition, judged by the intensity of the strong TMA-band at 564 cm^{-1} in Fig. 3.

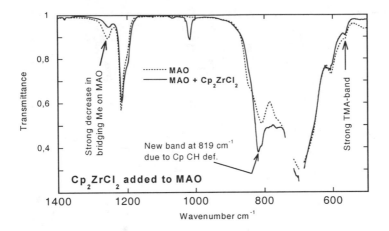

Fig. 3. Spectra of MAO and of the mixture of MAO and Cp_2ZrCl_2 in toluene solution. $c_{Al} = 1.0$ M, $c_{Zr} = 0.1$ M. The toluene spectrum has been removed.

Cp_2ZrMe_2 and Aluminium Alkyls (TMA, DMAC and MAO)

Fig. 2c shows that hardly any changes are seen in the IR spectra as TMA is added to a solution of Cp_2ZrMe_2. No notable reaction was observed, except removal of oxygen-containing impurities in the Cp_2ZrMe_2. The latter is seen from the disappearance of a band at 748 cm^{-1}. The same effect was observed for all three aluminium alkyl compounds tested.

Fig. 2d illustrates a series of changes in the IR spectra as DMAC is added to a solution of Cp_2ZrMe_2. At Al/Zr = 0.9 the formation of $Cp_2ZrMeCl$ and TMA was evident, as can be seen from the bands at 809 cm^{-1} and 768 cm^{-1}, respectively. The zirconocene part of this spectrum matched closely the spectrum of the separately prepared $Cp_2ZrMeCl$. Further addition of DMAC led to a new band at 814 cm^{-1}, which is compatible with the dichloride. Again, a clearly visible shoulder at 820 cm^{-1} indicated the presence of a new zirconocene species – probably a chorine-bridged bimetallic compound. Addition of larger amounts of DMAC, up to an Al/Zr = 25, gave a small increase in the shoulder, but no other observable changes.

Mixtures of Cp_2ZrMe_2 and MAO were studied in a range of Al/Zr from 2.5 to 60. Higher ratios could not be studied successfully due to band overlap and unfavourable intensity ratio of MAO and Cp bands. The experiments included addition of solutions, both MAO to Cp_2ZrMe_2 and *vice versa*, and addition of solid Cp_2ZrMe_2 to a MAO solution. In all experiments, a new Cp deformation band appeared at 822 cm^{-1}, as shown in Fig. 4a, and the colour of the solution immediately turned yellow. Surprisingly, bands attributed to C–H stretching of zirconium-bonded methyl groups did not appear for Al/Zr above 15. The change in the band for Al/Zr ratios below 12.5 is shown in Fig. 4b. A possible interpretation of this observation is that the methyl groups become bridged to aluminium, and thereby get their frequency shifted and become unresolved from the stronger bands of MAO (increasing bands in Fig. 4b). There was also a notable change in the profile of the set of bands attributed to the symmetrical deformation of methyl on aluminium. This composition-dependent change took place while the Al/Zr was greater than 15, whereas the profile was concentration-independent below 15. It appears to be two different composition regions – one with zirconocene excess, and the other with MAO excess. The effect is reproducible.

The positions of the out-of-plane C–H deformation bands of Cp for all the binary mixtures are summarised in

Table 2. The result of a preliminary experiment with Cp$_2$ZrMe$_2$ and B(C$_6$F$_5$)$_3$ is also included for comparison.

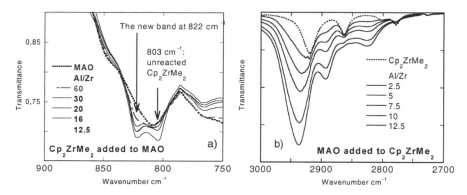

Fig. 4. Infrared spectra of a) Cp$_2$ZrMe$_2$ added to MAO, and b) MAO added to Cp$_2$ZrMe$_2$; both as toluene solution. The solvent spectrum has been removed. a) $c_{Al(start)}$= 0.4 M, scaled to maintain constant MAO intensity. b) $c_{Zr(start)}$= 0.08 M.

Discussion

The *in situ* FT-IR spectroscopic technique allows inert sample handling and the samples to be manipulated without the cell itself being moved. This allows rather tiny changes to be observed. Nevertheless, the low metallocene concentrations and high Al/Zr ratios applied in polymerisation is beyond reach of this technique.

Phase separation was observed in mixtures of MAO and zirconocenes, which introduces uncertainties regarding the distribution of the phases in the system. Phase separation typically leads to an increase in intensity of the spectrum, due to accumulation of solute between the windows. This makes quantitative treatment of the spectra difficult, but still should not preclude observation and interpretation of new bands. It was found that the disappearance of C–H stretching bands was not due such phase separation and precipitation of the zirconocene outside the IR cell.

The slow ligand exchange of Cp$_2$ZrCl$_2$ and Cp$_2$ZrMe$_2$, by which Cp$_2$ZrMeCl was formed, is in agreement with earlier reports.[17] The equilibrium constant for this exchange was reported to be 240 at 25 °C. The stability of Cp$_2$ZrMeCl is also evident from lack of conversion into Cp_2ZrMe_2 by TMA, despite the considerable energy gain by exchanging methyl bridges in TMA with chlorine bridges in DMAC. The chlorine of DMAC is always situated in the bridge position of the dimer due to this difference.[18,19]

Mono-methylation of Cp_2ZrCl_2 by TMA is in agreement with earlier reports.[5,6] It has been proposed that TMA is the alkylating agent in commercial MAO solutions.[6] For mono-methylation, this could be true, but the question is less relevant as MAO fast and quantitatively converts the formed DMAC back to TMA.[2,3] Consequently, the TMA content is largely constant on addition of solid Cp_2ZrCl_2 to the MAO solution, while MAO suffers a substantial loss of bridging methyl groups (see Fig. 3).

Weak shoulders are observed on the high-frequency side of the out-of-plane deformation band of Cp, both when TMA is added to $Cp_2ZrMeCl$ and to Cp_2ZrCl_2.[20] As such a shift in the band position is consistent with a change in the coordination environment around zirconium, and as formation of Cp_2ZrMe_2 is excluded, we suggest these shoulders to be due to formation of a bimetallic and probably chlorine-bridged complex. We suggest $Cp_2ZrMe(\mu\text{-}Cl)AlMe_3$ as a reasonable candidate.[21] Such a shoulder is also seen in the mixture of Cp_2ZrCl_2 and DMAC. Since no ligand exchange is expected, we suggest this species to be $Cp_2ZrCl(\mu\text{-}Cl)AlClMe_2$. However, positive formation energies have been calculated for these complexes,[7,25] which indicate that only small amounts may exist in equilibrium with the reactants.

No shoulder was observed when TMA was mixed with Cp_2ZrMe_2. Apparently, the double bridge in the TMA dimer is preferred to the formation of single methyl bridges between aluminium and zirconium. Note that this is not in contradiction to the finding that the methyl groups in this system are completely scrambled within minutes.[10] Hence, short-lived transition states have to exist.

When DMAC is added to Cp_2ZrMe_2 the solution soon change into a mixture of TMA, DMAC, Cp_2ZrCl_2 and $Cp_2ZrMeCl$, due to ligand exchange. Hence, at Al/Zr equal 2, where the two binary mixtures TMA/Cp_2ZrCl_2 and DMAC/Cp_2ZrMe_2 are equal in composition, the spectra in Fig. 2a,d are similar (Al/Zr = 1.8). As above, we interpret the shoulder at 820 cm^{-1} to be caused by a binuclear chlorine-bridged compound. However, the shoulder is considerably stronger in this case than in the systems TMA/Cp_2ZrCl_2 and DMAC/Cp_2ZrMe_2. This indicates that the situation might be more complex. It is possible to formulate plausible ion pairs, as depicted in Scheme 1b) and c), which might exist in this system. All chorines in this ion pair are in the preferred bridging positions, while the methyl groups are terminal. Such an ion pair would not be active in catalysis, due to the strength of the metal chorine bonds.

In the case of Cp_2ZrCl_2 and MAO, the band at 819 cm^{-1} is more pronounced than the relatively weak shoulders observed with TMA and DMAC. Several more options have to be considered with MAO, as it is capable of forming ion pairs and oxo-bridges.[5,22] Such an Al_2OZr moiety would be difficult to identify in IR spectra, as it's stretching bands are expected to overlap with the strong MAO bands. Oxo-bridges has been reported for model compounds,[9,23] but our results in the TMA-MAO system suggest that the oxygen in MAO is not very Lewis basic, as no reaction with TMA is observed.[1] A slow formation of methylene bridges has been shown,[5] but as the band appeared instantly on mixing, it is not likely due to such a product. We suggest the new band to be due to chlorine- or methyl-bridged compounds.

Table 2 Band positions of the out-of-plane C–H deformation of Cp for initial solution and resulting mixture.

Binary mixture	Band positions [cm^{-1}]			
	Initial band	Bands in mixture		
Cp_2ZrCl_2 + TMA	814 s		809 m	814 s	820 vw
$Cp_2ZrMeCl$ + TMA	809 s		809 s		815 w
Cp_2ZrCl_2 + DMAC	814 s			814 s	820 w
Cp_2ZrCl_2 + MAO	814 s			(814)^a	819 s
Cp_2ZrMe_2 + TMA	803 s	803 s			
Cp_2ZrMe_2 + DMAC	803 s		(809)^a	814 s	820 m
Cp_2ZrMe_2 + MAO	803 s	(803)^a			822 s
Cp_2ZrMe_2 + $B(C_6F_5)_3$	803 s	803 m			819, 825 s

^{a} Bands in brackets were not present at high Al/Zr ratios

Addition of MAO to Cp_2ZrMe_2 resulted in similar bright yellow colour as with Cp_2ZrCl_2 and a sharp band at 822 cm^{-1}. In this case, the system contains no chlorine, so changes due to methyl-chlorine exchange are not present in the spectra.

During addition of MAO to Cp_2ZrMe_2, the C–H stretching bands of zirconium-bonded methyl groups disappeared (Fig. 4b). Such a band disappearance is expected if both methyl groups become bridged to aluminium. The bonding in such a doubly bridged complex is visualised in Scheme 1 by the earlier proposed cation $Cp_2Zr(\mu\text{-}Me)_2AlMe_2^+$ and a chlorine analogue.[8] In the case of a complex with MAO the aluminium would be attached to the MAO cage.

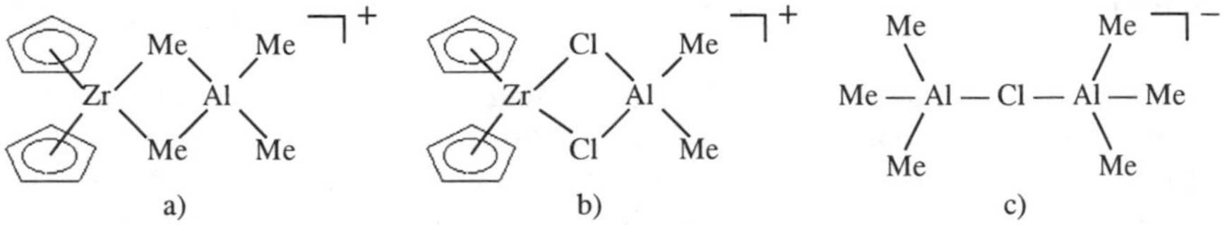

Scheme 1

When 1-hexene was added to such a MAO/Cp_2ZrMe_2, mixture decrease in the 822-cm^{-1} band was observed and a deep red colour appeared within minutes.[24] This indicates a coordination of the olefin to zirconium. It has been shown that only minute amounts of ion pairs are formed at this low Al/Zr, and most likely with a binuclear methyl-bridged cation,[10,11] ion pair separation should be greatly enhanced by olefin coordination.[7]

To assist our understanding of the observed bands, the IR spectra of several zirconocenes and binary complexes were calculated,[25] using density functional theory. The trend shown in Table 1 was well described by the calculations. The results indicate that Cp_2ZrMe^+, $Cp_2Zr(\mu\text{-}Me)_2AlMe_2^+$, and methyl- and chlorine-bridged non-ionic bimetallic species all have their Cp out-of-plane deformation frequency close to 820 cm^{-1}. The sensitivity of this band to changes at the metal centre is a generally observed phenomena for metallocenes, and the mechanism has been subject to some controversy.[26]

It has been shown that ionic pairs are formed quantitatively in Cp$_2$ZrMe$_2$/tris(pentafluorophenyl)boron mixtures at -78 °C.[27] A strong band was expected in the 820 cm^{-1} region if our predictions were right. In a preliminary experiments a solution of tris(pentafluorophenyl)boron was added to Cp$_2$ZrMe$_2$ in the *in situ* IR apparatus at 25 °C (Reaction at low temperature was reported to be prerequisite, but can not be achieved with this cell). A new band was immediately observed at 819 cm^{-1} (B/Zr =0.3). Another new band, at 825 cm^{-1}, were observed at higher boron to zirconium ratios (B/Zr =0.6-1.7). The bands were strong and well resolved. A tentative suggestion is that these bands might originate in to di- and mononuclear cations, respectively.

Conclusions

A slow formation of the monochloro-monomethyl compound Cp$_2$ZrClMe from a mixture of Cp$_2$ZrCl$_2$ and Cp$_2$ZrMe$_2$ has been verified. Only weak complexes are formed in mixtures of zirconocenes and TMA or DMAC. The chemical potential for methylation of zirconocenes is primarily due to MAO clusters, but TMA may be important in the mechanism. Our IR data is consistent with the formation of stable compounds during activation, which we assume include methyl or chlorine bridges between zirconium and aluminium, but do not differentiate between ionic or neutral complexes. Observed disappearance of C–H stretching bands may indicate double bridges between zirconium and aluminium.

Experimental

All liquids and solids were handled according to standard techniques for air sensitive compounds. Dry argon (AGA, 99.999%) was used as inert gas. Toluene and benzene were dried by reflux over sodium and benzophenone. MAO solution (Albemarle Corp., 10wt% in toluene) was evaporated to dryness under reduced pressure (0.08 mbar, 25 °C) and the MAO stored in solid state (Me/Al =1.55 ±0.05). The amount of free TMA in the MAO was 3-5% (by mole Al), estimated from the 564 cm^{-1} band of the TMA dimer. The experiments were performed with freshly prepared benzene or toluene solutions. TMA (Aldrich Co.), DMAC (Aldrich Co.), Cp$_2$ZrCl$_2$ (Boulder Scientific Comp.), Cp$_2$ZrMe$_2$ (Strem Chemicals, Inc.) were used as received.

Cp$_2$ZrMeCl was prepared from Cp$_2$ZrMe$_2$ and Cp$_2$ZrCl$_2$ in benzene at room temperature in 2 weeks.[17] ^{1}H-NMR (d^8-toluene) 0.45, 5.78; Main IR bands: 3110w, 2933w, 2874w, 2792w, 1442m, 1412w, 1017m, 809s, 457w.

The IR spectra were obtained with a special built flow cell with silicon and germanium windows. The path length was 50-75 μm. The cell was mounted in a

flow loop that includes a micro pump and a mixing tank, as shown in Fig. 5. Details
have been describes elsewhere.[28]

The spectra were recorded on a Bruker IFS 66v FTIR spectrometer, 2-10
minutes after each addition. Separate experiments have established that good
mixing of additives is achieved within a minute. All systems investigated were also
checked for slow reaction by repeated spectrum recordings at longer intervals.
Only systems including MAO and metallocenes showed slow reactions, as also
evident by colour change and phase separation.

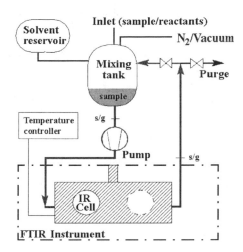

Fig. 5. Sketch of the apparatus. The dash-dot line indicates the sample chamber of the IR
instrument. s/g indicate steel/glass junctions. The dashed circle indicates the second cell,
which was not used in this work (see ref. [28]).

Acknowledgement

Financial support from the Norwegian Research Council (NFR) under the Polymer
Science Program is gratefully acknowledged.

References and Notes

1   a) J.L. Eilertsen, E. Rytter, M. Ystenes: In situ FTIR spectroscopy shows no evidence
     of reaction between MAO and TMA, In Metalorganic Catalysts for Synthesis and
     Polymerization; W. Kaminsky, Ed.; Springer: Berlin, **1999**, p 136 b) J.L. Eilertsen, E.
     Rytter, M. Ystenes, *Vibrational Spectroscopy*, in print

2   M. Ystenes, J.L. Eilertsen, Jianke Liu, M. Ott, E. Rytter, J.A. Støvneng, *J. Polym. Sci.
     A*, 38 (**2000**) 3106

3   E. Rytter, M. Ystenes, J.L. Eilertsen, M. Ott, J.A. Støvneng, J. Liu, This proceeding.

4   E. Giannetti, G.M. Nicoletti, R. Mazzocchi, *J. Polym. Sci. Polym. Chem.* 23 (**1985**) 2117

5   W. Kaminsky, R. Steiger, *Polyhedron* 7 (**1988**) 2375

6   D. Cam, U. Giannini, *Macromol. Chem.* 193 (**1992**) 1049

7   R. Fusco, L. Longo, F. Masi, F. Garbassi, *Macromolec.* 30 (**1997**) 7673

8   M. Bochmann, S.J. Lancaster, *Angew. Chem. Int. Ed. Engl.* 33 (**1994**) 1636

9   C.J. Harlan, S.G. Bott, A.R. Barron, *J. Am. Chem. Soc.* 117 (**1995**) 6465

10   I. Tritto, R. Donetti, M.C. Sacchi, P. Locatelli, G. Zannoni, *Macromolec.* 30 (**1997**) 1247

11   D.E. Babushkin, N.V. Semikolenova, V.A. Zakharov, E. Talsi, *Macromol. Chem. Phys.* 201 (**2000**) 558

12   E.Y.-X. Chen, T.J. Marks, *Chem. Rev.* 100 (**2000**) 1391

13   E. Maslowsky, Vibrational spectra of organometallic compounds, John Wiley & sons, New York, **1977**

14. L.A. Nekhaeva, S.V. Bondarenko, S.V. Rykov, A.I. Nekhaev, B.A. Krentsel, V.P. Mar'in, L.I. Vyshinskaya, I.M. Khrapova, A.V. Polonskii, N.N. Korneev, *J. Organomet. Chem.* 406 (**1991**) 139

15   J.R. Surtees, *J. Chem. Soc. Chem. Comm.* 22 (**1965**) 567

16   Addition of solid was chosen to avoid dilution of the initial MAO solution as a consequence of the low solubility of Cp_2ZrCl_2. The target composition was unity ratio of added chlorine to methyl bridges.

17   R.F. Jordan, *J. Organomet. Chem.*, 294 (**1985**) 321

18   K. Brendhaugen, A. Haaland, D.P. Novak, *Acta Chem Scand A* 28 (**1974**) 45

19   E. Rytter, S. Kvisle, *Inorg Chem* 25 (**1986**) 3796

20   The IR spectrum of crystalline Cp_2ZrCl_2 contains bands at 840-850 cm^{-1}, but they become inactive in solutions.

21   The possibility that MAO has been formed in situ due to contamination must also be considered. The experiments with Cp_2ZrMe_2, where an oxygen-containing contaminant was present indicate that this is not the reason for the shoulder.

22   A.R. Siedle, W.M. Lamanna, R.A. Newmark, J.N. Scoepfer, *J. Molec. Catal. A* 128 (**1998**) 257

23   G. Erker, M. Albrecht, S. Werner, C. Krüger, *Z. Naturforsch.* 45b (**1990**) 1205

24   Work in progress.

25   J.A. Støvneng, unpublished results.

26   E. Diana, R. Rossetti, P.L. Stanghellini, S.F.A. Kettle, *Inorg. Chem.* 36 (**1997**) 382

27   X. Yang, C.L. Stern, T.J. Marks, *J. Am. Chem. Soc.* 116 (**1994**) 10015

28   Ø. Bache, M. Ystenes, *Appl. Spectrosc.* 48 (**1993**) 985

Single Component Zirconocene Catalysts for the Stereospecific Polymerization of MMA

Holger Frauenrath, Helmut Keul and Hartwig Höcker*

Institut für Technische Chemie und Makromolekulare Chemie, RWTH Aachen,
Worringer Weg 1, 52064 Aachen, Germany
E-mail: Hoecker@dwi.rwth-aachen.de

Abstract. The stereospecific polymerization of methyl methacrylate (MMA) with single component cationic zirconocene catalysts has been investigated, yielding highly isotactic poly(methyl methacrylate) (PMMA) with $Me_2C(Cp)(Ind)Zr(Me)(thf)^+BPh_4^-$ **1** as a catalyst, and syndiotactic PMMA at low temperatures with $Me_2C(Cp)_2Zr(Me)(thf)^+BPh_4^-$ **2**. Similar cationic complexes with other ligands have been found to be inactive for MMA polymerization under similar reaction conditions. On the basis of polymerization kinetics and stereospecificity control in MMA polymerization with **1** and **2** a possible polymerization mechanism is discussed.

Introduction

The copolymerization of 1-olefins with functionalized monomers such as (meth)acrylates or vinyl esters is still an unsolved problem. At least, achieving a degree of control of catalyst activity, molecular weight, polymer microstructure or of comonomer as has been achieved for the metallocene polymerization of 1-olefins for the homopolymerization of functionalized monomers would be a big step forward.

In 1992 *Collins et al.* reported the first application of zirconocenes to the polymerization of MMA, generating the catalyst system from Cp_2ZrMe_2 and $B(C_6F_5)_3$ [1-3]. This paved the way for the investigation of the influence of catalyst structure on polymerization behaviour and polymer properties. The polymerization mechanism has been shown to involve two types of zirconocene complexes [2]. A cationic methyl complex activates a monomer as an acceptor, while the growing chain is a neutral ester enolate methyl complex, thus being activated as a donor. The carbon bond formation proceeds in a Michael-type addition reaction between activated monomer and activated growing chain. This mechanism with the carbon bond formation taking place far away from the zirconium centres of both components rendered the possibility of stereocontrolled polymerization unlikely. Consequently, the use of chiral *ansa*-zirconocene based catalyst systems has been briefly mentioned [3], but has not been reported in a systematic study for many years. Recently, we have been able to show that catalyst systems consisting of chiral cationic and ester enolate complexes also yield basically atactic, syndio-enriched PMMA as do the achiral systems described by *Collins et al.* , while the application of the chiral cationic complex [Me$_2$C(Cp)(Ind)Zr(Me)(thf)][BPh$_4$] **1** alone yields highly isotactic PMMA[4-6].

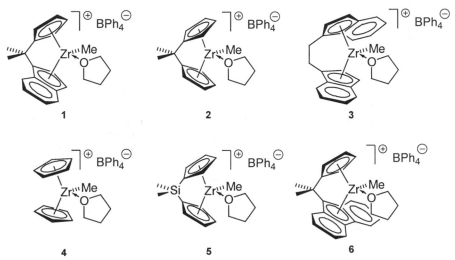

Fig. 1. Different cationic zirconocene complexes used in this study.

Soga et al. describe the use of zirconocene based catalyst systems and zinc alkyls as cocatalysts for the stereospecific polymerization of MMA [7-8]. With the catalysts Me$_2$Si(Ind)$_2$ZrMe$_2$ and Me$_2$Si(Ind)(Cp)ZrMe$_2$ activated with B(C$_6$F$_5$)$_3$ and a large excess of zinc alkyls in toluene they have been able to produce isotactic PMMA [8]. The methodology appears to be similar to that of the Collins-type systems, however, the role of the zinc cocatalysts is not fully understood and an explanation for the stereospecificity of these systems has not been put forward.

Yasuda et al. investigated the syndiospecific polymerization of MMA with samarocene catalysts [9-12]. They have been able to prove that starting from samarocene hydrides the active species is actually a neutral ester enolate samarocene complex that is able to complex a monomer molecule. Syndiospecificity has been found to be chain end controlled. *Marks et al.* applied chiral *ansa*-samarocenes and obtained isotactic PMMA [13]. Isospecificity is enantiomorphic site controlled in this case, which was explained in terms of the mechanism proposed by *Yasuda et al.*.

Recently, *Collins et al.* published the first stereospecific polymerization of MMA with a single component cationic ester enolate zirconocene complex [14]. The active species is proposed to be the cationic ester enolate zirconocene complex itself. Thus, the polymerization mechanism is closely related to the polymerization with the isoelectronic neutral samarocene enolate complexes.

Gibson et al. studied the MMA polymerization with various cationic zirconocenes [15]. The activation of the zirconocenes appears to be important. While cationic zirconocenes generally have been found to be inactive for the polymerization of MMA in the absence of an ester enolate complex [1-3], some cationic zirconocenes appear to be active when synthesized in situ from the corresponding dimethyl zirconocenes and an equivalent amount of $B(C_6F_5)_3$. However, the presence of a Collins-type catalyst system cannot be fully excluded for this activation method.

Experimental Part

General Procedures. All experiments were carried out in a nitrogen atmosphere using standard Schlenk techniques. Nitrogen was passed over an activated copper catalyst (BASF R3-11), molecular sieves and over K/Al_2O_3 for purification. Me_3SiCl (Fluka, >99%), MeLi (Fluka, 1.6 M solution in DE), BuLi (Aldrich, 2 M solution in pentane), $NaBPh_4$ (Fluka, >98%) and $ZrCl_4$ (Fluka) were used without further purification. All other chemicals were purified using the standard methods. $[HNBu_3][BPh_4]$ [16], $Zr(NEt_2)_4$ [5], $Me_2C(Cp)(Ind)ZrCl_2$ [5], $Me_2C(Cp)(Ind)ZrMe_2$ [5], $[Me_2C(Cp)(Ind)Zr(Me)(thf)][BPh_4]$ [5], $Me_2C(Cp)_2ZrCl_2$ [6], $Me_2C(Cp)_2ZrMe_2$ [6], and $[Me_2C(Cp)_2Zr(Me)(thf)][BPh_4]$ [6] were prepared according to published procedures. Toluene and THF were dried over sodium/benzophenone and CH_2Cl_2 was dried over P_2O_5. All solvents were stored in a nitrogen atmosphere. MMA was stored at 0°C over CaH_2 and freshly distilled before use. All deuterated solvents were degassed and stored over molecular sieves (4Å).

All ^1H and ^{13}C NMR spectra were recorded on a Bruker DPX 300 FT-NMR spectrometer at 300 MHz and 75 MHz, respectively. All chemical shifts are given in ppm with tetramethyl silane as a reference. Quantitative ^{13}C spectra for the analysis of PMMA microstructure were recorded with at least 16000 scans using an inverse gated decoupling pulse sequence.

All gel permeation chromatography (GPC) analyses were carried out at room temperature using a HPLC pump (Waters 510) and a refractive index detector. The eluent was THF stabilized with 250 mg/mL 2,6-di-*tert*-butyl-4-methylphenol. The flow rate was 1.0 mL/min. Four columns were applied (PS-DVB gel from MZ Analysentechnik), with pore sizes of 100 Å, 100 Å, 1000 Å and 10000 Å. Number average molecular weights and polydispersities of PMMA are given relative to PMMA standards.

MMA Polymerization. MMA (2.5 g, 25 mmol) and 7.5 mL CH_2Cl_2 containing approximately 3% (w/w) hexylbenzene were placed in a 50 mL Schlenk tube with a septum and stirred at the desired polymerization temperature for 30 min. The polymerization was started by addition of a solution of the catalyst in 2.5 mL of CH_2Cl_2. In the kinetic experiments samples of approximately 0.05 mL were taken from the reaction mixture. Relative polymer yield was determined by GPC via integration of the polymer peaks relative to hexylbenzene as an internal standard. Absolute polymer yield was determined gravimetrically by terminating the reaction just after taking off the last sample by addition of 2 mL MeOH/HCl (aq.)/hydroquinone (90/9.9/0.1% w/w), precipitation of PMMA in MeOH, filtration and drying in vacuum.

Results and Discussion

The results of MMA polymerization with the cationic complexes $Me_2C(Cp)(Ind)Zr(Me)(thf)^+BPh_4^-$ **1** and $Me_2C(Cp)_2Zr(Me)(thf)^+BPh_4^-$ **2** are shown in Table 1. Both complexes are highly active for the polymerization of MMA.

Table 1. Typical results of MMA polymerization with $Me_2C(Cp)(Ind)Zr(Me)(thf)^+BPh_4^-$ **1** and $Me_2C(Cp)_2Zr(Me)(thf)^+BPh_4^-$ **2**[a]

Cat.	$\dfrac{T_p}{°C}$	$\dfrac{t_p}{min}$	$\dfrac{yield[b]}{\%}$	$\overline{M}_n{}^c$	$\dfrac{\overline{M}_w}{\overline{M}_n}{}^c$	$I*^d$	$\dfrac{mm^e}{\%}$	$\dfrac{mmmm^f}{\%}$
1	30	20	54	40500	1.43	0.33	84.3	74.9
1	20	40	62	45700	1.42	0.34	87.0	79.6
1	10	60	94	51000	1.42	0.46	89.0	82.1
1	0	90	97	58800	1.34	0.41	90.4	84.5
1	-10	180	97	55100	1.30	0.44	92.1	87.6
1	-20	180	93	55300	1.24	0.42	93.5	n. d.
1	-30	240	77	57300	1.21	0.34	94.7	91.5

Cat.	$\dfrac{T_p}{°C}$	$\dfrac{t_p}{min}$	$\dfrac{yield[b]}{\%}$	$\overline{M}_n{}^c$	$\dfrac{\overline{M}_w}{\overline{M}_n}{}^c$	$I*^d$	$\dfrac{rr^e}{\%}$	$\dfrac{rrrr^f}{\%}$
2	30	20	65	66600	1,64	0.12	69.0	52.1
2	20	30	61	93000	1,60	0.08	72.9	57.6
2	10	40	59	104100	1,38	0.07	75.1	59.2
2	0	60	81	90600	1,32	0.11	78.8	61.9
2	-10	105	86	123200	1,30	0.09	80.1	67.2
2	-20	180	71	110500	1,35	0.08	82.4	69.1
2	-45	1080	98	78500	1,31	0.16	89.0	76.9

[a] Reaction conditions in polymerizations with **1**: [MMA] = 2 mol/L, [**1**] = 8 mmol/L; solvent CH_2Cl_2; reaction conditions in polymerizations with **2**: [MMA] = 4 mol/L, [**1**] = 16 mmol/L; solvent CH_2Cl_2.
[b] Determined gravimetrically.
[c] Determined by GPC relative to PMMA standards.
[d] Catalyst efficiency $I* = M_{n,exp} / M_{n,theor}$ with $M_{n,theor} = [MMA]_0 / [Zr]_0 \cdot$ conversion.
[e] Determined by ^1H NMR spectroscopy.
[f] Determined by ^{13}C NMR spectroscopy.

While the polymerization with **1** yields highly isotactic PMMA, **2** yields syndiotactic PMMA at low temperatures. This is the first example for the rational control of PMMA microstructure via catalyst symmetry. The results are consistent with the finding of *Gibson et al.* [15] who used in situ generated catalysts similar to **1** in their studies.

However, with our methodology we found the cations **3-6** to be inactive for the polymerization of MMA, whereas according to *Gibson et al.* in situ generated catalysts similar to **3** are active for MMA polymerization. The activation methodology may be important as outlined by *Gibson et al.*, but given the similarity of the complexes that are active for MMA polymerization this seems to be only part of the problem. The nature of the ligand may also be crucial for the generation of an active catalyst. The most obvious difference is the large ligand aperture with a Cp-Zr-Cp angle of only 116.6° in $Me_2C(Cp)_2ZrCl_2$ [17], as opposed to 129.3° in Cp_2ZrCl_2 [18] or 125.4° in $Me_2Si(Cp)_2ZrCl_2$ [19]. Possibly this geometric feature alters the reactivity compared to 3-6 by either changing the steric demand of the complexes or their electronic properties. This might again have consequences for a suitable method of activation.

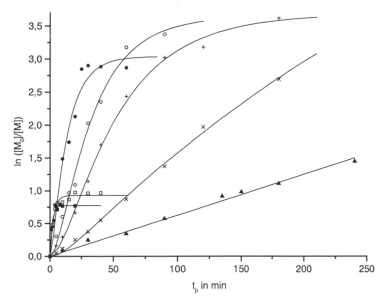

Fig. 2. Kinetics of MMA polymerization with 1 as the catalyst at different polymerization temperatures: (■) 30°C, (□) 20°C, (●) 10°C, (○) 0°C, (+) -10°C, (×) -20°C, (▲) -30°C; lines from function fitting.

Kinetics of MMA polymerization with $Me_2C(Cp)(Ind)Zr(Me)(thf)^+BPh_4^-$ **1** at different polymerization temperatures are shown in Fig. 2. Quantitative yields are obtained at polymerization temperatures of 0°C and below. At these temperatures the dependence of monomer conversion on polymerization time approaches to what is expected for first order kinetics. A short initiation period can be observed. Apparently, the cationic complex **1** itself is not the active species. The active species is also subject to a termination reaction. The curves shown in Fig. 2 have been fitted from a model function assuming first-order kinetics with respect to the monomer, and time-dependent catalyst concentrations (with $[M(t)]$ monomer concentration, $[M_0]$ initial monomer concentration, $[Zr^*(t)]$ concentration of active species, $[Zr_0]$ initial Zirconocene concentration, k_p rate constant of propagation, k_i rate constant of initiation and k_t rate constant of termination):

$$-\frac{d[M(t)]}{dt} = k_p \cdot [Zr*(t)] \cdot [M(t)] \tag{1}$$

$$-\frac{d[M(t)]}{dt} = k_p \cdot [M(t)] \cdot [Zr_0] \cdot \frac{k_i}{k_t - k_i} \cdot \left(e^{-k_i t} - e^{-k_t t}\right) \tag{2}$$

$$\ln\frac{[M_0]}{[M(t)]} = k_p [Zr_0] \cdot \frac{k_i}{k_t - k_i} \cdot \left\{ \frac{1}{k_t}\left(e^{-k_t t} - 1\right) - \frac{1}{k_i}\left(e^{-k_i t} - 1\right) \right\} \tag{3}$$

The activation energies have been determined from an Eyring plot of the respective rate constants derived from the fitting of the experimental data (Fig. 3) and can be estimated to be $E_{a,p}$ = 28 (±2.3) kJ/mol for the propagation reaction, $E_{a,i}$ = 13 (±8.9) kJ/mol for the initiation reaction and $E_{a,t}$ = 57 (±6.9) kJ/mol for the termination reaction.

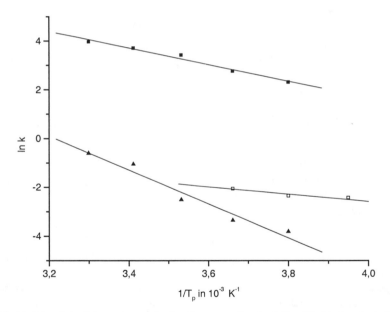

Fig. 3. Eyring plot of the rate constants of propagation k_p (■), of initiation k_i (□) and of termination k_t (▲) at different polymerization temperatures in MMA polymerizations with **1** determined by function fitting of the experimental kinetic data.

The presence of the initiation and the termination reactions are also reflected by the shape of the GPC curves (Fig. 4) and by the polydispersities (Fig. 5). At low polymerization temperatures only a slight tailing is observed. The polydispersity is below 1.2 for moderate monomer conversion, but increases when the polymerization is driven to completion. At high polymerization temperatures low molecular weight shoulders are visible in the GPC curves, and the polydispersities are larger than 1.3. The termination reaction may be due to the reaction of the zirconocene with the solvent CH_2Cl_2. Decomposition of the zirconocene complexes **1** and **2** is observed in the NMR spectra in CD_2Cl_2 after a couple of minutes.

A plot of the molecular weight vs. MMA conversion (Fig. 6) reveals a linear dependence up to high conversions, indicating the absence of chain transfer reactions. Remarkably, the experimentally determined number average molecular weights relative to PMMA standards are by more than a factor of 2 larger than the molecular weights determined from the ratio of monomer and initiator concentration. Catalyst efficiences I* determined from the ratio of experimental and theoretical number average molecular weight are below 0.5 (Table 1). This is consistent with *Gibson et al.* who found a factor of around 2 between experimentally determined and theoretical molecular weight. Possibly, the active species is generated in a bimolecular reaction from **1** and **2**, e.g. due to the formation of binuclear complexes. Such species have also been proposed by *Gibson et al.* [15], as well as by *Yasuda et al.* in the case of the samarocene catalyzed MMA polymerization [20].

However, there may be a more straightforward explanation. As mentioned above, decomposition of the zirconocene complexes **1** and **2** is observed in the NMR spectra in CD_2Cl_2 after a couple of minutes. If one assumes that this decomposition reaction proceeds with the rate constant k_{dec} parallel to the initiation reaction with the rate constant k_i, then only part of the initial amount of the zirconocene complexes $[Zr_0]$ is converted to the active species, while the other part is not active for MMA polymerization.

It is the sum $[Zr*]_{conv}$ of all zirconocene complexes converted to the active species that determines the molecular weight of PMMA:

$$[Zr*]_{conv} = \lim_{t \to \infty}\left(\int_0^t dZr* \right) = \lim_{t \to \infty}\left(k_i \cdot [Zr_0] \cdot \int_0^t e^{-(k_i + k_{dec})t}\, dt \right) \qquad (4)$$

$$[Zr*]_{conv} = \frac{k_i}{k_{dec} + k_i} \cdot [Zr_0] \qquad (5)$$

Thus, larger theoretical molecular weights and lower catalyst efficiences should be expected. If the decomposition was similar to the termination/decomposition reaction of the active species and thus $k_{dec} \approx k_t$ were true, then values of I* = 0.39 (20°C), 0.71 (0°C) and 0.92 (-20°C) would be derived. This example that the catalyst efficiencies can well be explained without the assumption of binuclear complexes.

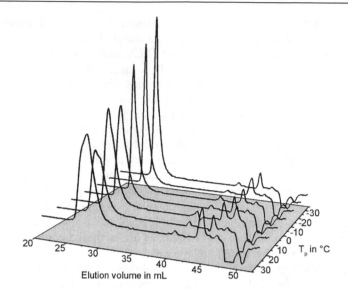

Fig. 4. GPC curves of PMMA obtained with **1** as the polymerization catalyst at different polymerization temperatures after precipitation and workup.

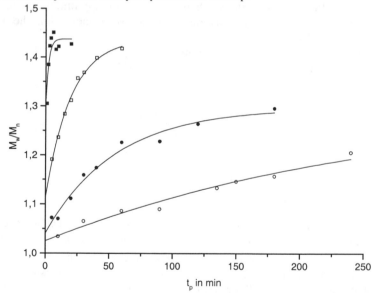

Fig. 5. Polydispersity of PMMA polymerized with **1** as the catalyst as a function of time at different polymerization temperatures: (■) 30°C, (□) 10°C, (●) -10°C, (○) -30°C.

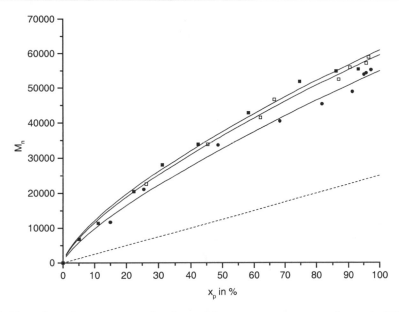

Fig. 6. Plot of number average molecular weight vs. monomer conversion x_p in MMA polymerization with **1** as the catalyst at different polymerization temperatures: (■) 0°C, (□) -10°C, (●) -20°C; dashed line represents theoretical molecular weight $M_{n,theor}$ = ([MMA]$_0$ / [Zr]$_0$) · x_p.

MMA polymerization with Me$_2$C(Cp)(Ind)Zr(Me)(thf)$^+$BPh$_4^-$ **1** is highly isospecific, while MMA polymerization with Me$_2$C(Cp)$_2$Zr(Me)(thf)$^+$BPh$_4^-$ **2** is syndiospecific at low polymerization temperatures, as can be seen from the pentad and triad abundances in PMMA (Table 1). Examples of ^{13}C NMR spectra are displayed in Fig. 7. From the relative diad abundances (calculated from the triad abundances determined by ^{1}H NMR spectroscopy) at different polymerization temperatures, the differences of the activation energies that control the stereospecificity can be calculated to be $E_{a,r}$ – $E_{a,m}$ = 12.2 (±0.6) kJ/mol in the case of **1** and $E_{a,r}$ – $E_{a,m}$ = -9.5 (±0.4) kJ/mol in the case of **2** [6]. The pentad analysis of the polymers reveals that isotacticity in the case of **1** is enantiomorphic site controlled (*mmmr : mmrr : mrrm* ≈ 2 : 2 : 1, over the whole temperature range), whereas syndiospecificity of **2** appears to be chain end controlled (*rrrr : rrmr* ≈ 2 : 1, over the whole temperature range) [6]. Remarkably, these findings are very similar to the results of *Yasuda et al.* [11] and *Marks et al.* [13] in the case of samarocene catalyzed syndiospecific/isospecific MMA polymerization.

In order to explain the experimental results we propose the mechanism displayed in Scheme 1 as a working hypothesis. We postulate a cationic ester enolate zirconocene complex as the active species, which is isoelectronic to the neutral ester enolate samarocenes active in MMA polymerization reported by *Yasuda et al.*.

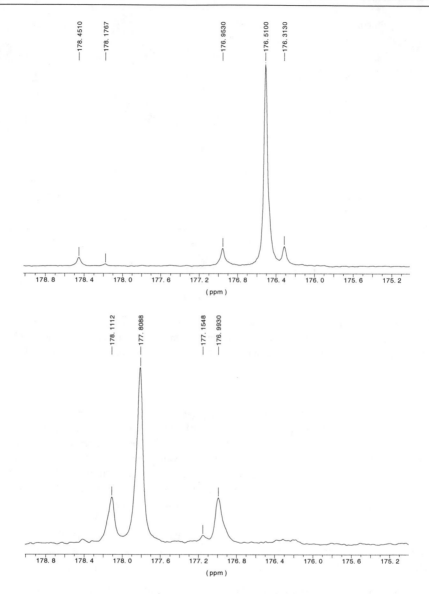

Fig. 7. Quantitative ^{13}C NMR spectra of isospecific PMMA obtained with **1** ($T_p = 0°C$, above) and of syndiospecific PMMA obtained with **2** ($T_p = -20°C$, below).

The active species is formed from the ziconocene cation by the transfer of a methyl group to a coordinated MMA molecule. Like the neutral samarocene ester enolates it activates the growing chain end as a donor and complexes an incoming monomer, thus activating it as an acceptor. The carbon bond formation proceeds intramolecularly via a cyclic intermediate. The possibility of such a monometallic mechanism has already been evaluated from theoretical considerations by *Sustmann et al.* [21]. It is further supported by the recent findings of *Collins et al.* who reported the isospecific polymerization of MMA with a cationic ester enolate zirconocene complex [14].

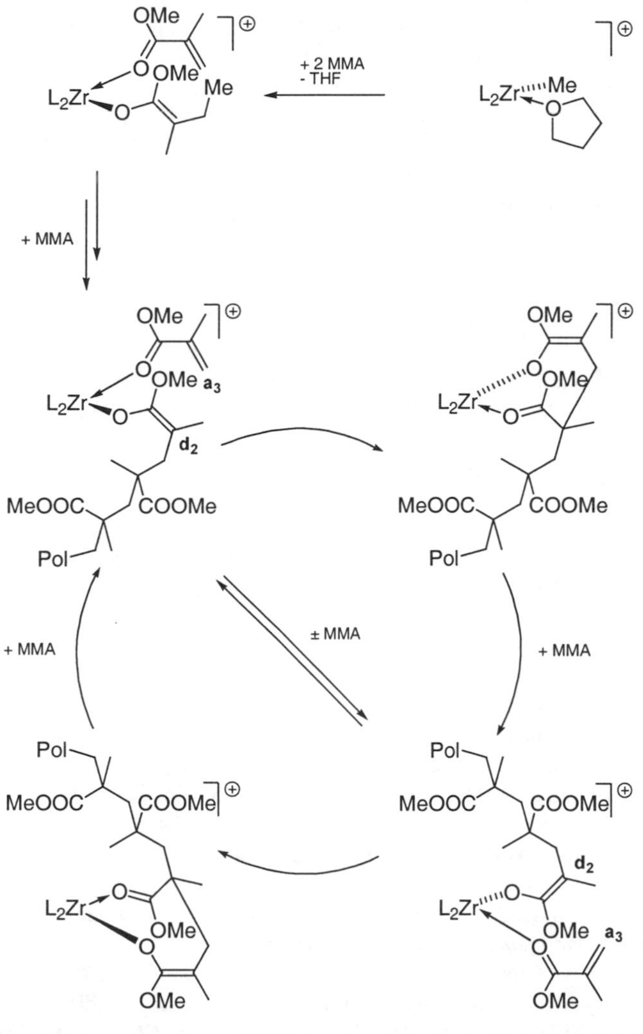

Scheme 1. Proposal for a mechanism of MMA polymerization with **1** [L$_2$ = Me$_2$C(Cp)(Ind)] and **2** [L$_2$ = Me$_2$C(Cp)$_2$] as catalysts.

Taking into account the C$_1$-symmetry of **1**, the enantiomorphic site controlled isospecificity of **1** can only be due to the polymerization step always taking place on the same side of the catalyst. Consequently, after each polymerization step a fast active site epimerization reaction must occur with a strong preference for the chain oriented to one side of the complex and not the other one. This explanation is analogous to the findings of *Marks et al.* regarding the isospecificity of the C_1-symmetric ansa-samarocenes applied in their study [13]. In the case of the C_s-symmetric catalyst **2** no preference for either coordination site is to be expected. Nevertheless, two subsequent orientations of the chain are diastereomeric on the level of the cyclic intermediate. Consequently, **2** should yield basically atactic PMMA with growing syndiotacticity at lower temperatures, when active site epimerization is slowed down.

In conclusion, we have presented the first example for a rational approach to catalyst design in order to control PMMA microstructure. Our findings parallel similar findings in the field of samarocene catalyzed MMA polymerization. The proposed mechanism still has to be proved. The nature of the active species and of the termination reaction are subject to further experiments.

References

1   Collins, S.; Ward, D. G. *J. Am. Chem. Soc.* **1992**, *114*, 5460-5462.
2   Collins, S.; Ward, D. G.; Suddaby, K. H. *Macromolecules* **1994**, *27*, 7222-7224.
3   Collins, S.; Li, Y.; Ward, D. G.; Reddy, S. S. *Macromolecules* **1997**, *30*, 1875-1883.
4   Stuhldreier, T. Dissertation, RWTH Aachen, **1999**.
5   Stuhldreier, T.; Keul, H.; Höcker, H. *Macromol. Rapid Commun.*, in press.
6   Frauenrath, H.; Keul, H.; Höcker, H. *Macromolecules*, submitted.
7   Deng, H.; Shiono, T.; Soga, K. *Macromolecules* **1995**, *28*, 3067-3073.
8   Saegusa, N.; Saito, T.; Shiono, T.; Ikeda, T.; Deng, H.; Soga K. in "Metalorganic Catalysts for Synthesis and Polymerization", Kaminsky, W. (ed.), Springer Verlag, Berlin, **1999**, 583-589.
9   Yasuda, H.; Yamamoto, H.; Yokota, K.; Miyake, S.; Nakamura, A. *J. Am. Chem. Soc.* **1992**, *114*, 4908-4910.
10 Yasuda, H.; Yamamoto, H.; Yamashita, M.; Yokota, K.; Nakamura, A.; Miyake, S.; Kai, Y.; Kanehisa, N. *Macromolecules* **1993**, *26*, 7134-7143
11 Yasuda, H.; Ihara, E. *Macromol. Chem. Phys.* **1995**, *196*, 2417-2441.
12 Ihara, E.; Morimoto, M.; Yasuda, H. *Macromolecules* **1995**, *28*, 7886-7892.
13 Giardello, M. A.; Yamamoto, Y.; Brard, L.; Marks, T. J. *J. Am. Chem. Soc.* **1995**, *117*, 3276-3277.
14 Nguyen, H.; Jarvis, A. P.; Lesley, M. J. G.; Kelly, W. M.; Reddy, S. S.; Taylor, N. J.; Collins, S. *Macromolecules*, **2000**, *33*, 1508-1510.
15 Cameron, P. A.; Gibson, V. C.; Graham, A. J. *Macromolecules* **2000**, *33*, 4329-4335.
16 Amorose, D. M.; Lee, R. A.; Petersen, J. L. *Organometallics* **1991**, *10*, 2191-2198.
17 Shaltout, R. M.; Corey, J. Y.; Path, N. R. *J. Organomet. Chem.* **1995**, *503*, 205-212.
18 Corey, J. Y.; Zhu, X.-H.; Brammer, L.; Path, N. P. *Acta Crystallogr., Sect. C* **1995**, *51*, 565-567.
19 Bajgur, C. S.; Tikkanen, W. R.; Petersen, J. L. *Inorg. Chem.* **1985**, *24*, 2539-2546.
20 Desurmont, G.; Li, Y.; Yasuda, H.; Maruo, T.; Kanehisa, N.; Kai, Y. *Organometallics*, **2000**, *19*, 1811-1813.
21 Sustmann, R.; Sicking, W.; Bandermann, F.; Ferenz, M. *Macromolecules* **1999**, *32*, 4204-4213.

Modeling Methylaluminoxane (MAO)

E. Zurek, T.K. Woo, T.K. Firman, T. Ziegler*

Department of Chemistry, University of Calgary, Calgary, Alberta T2N 1N4, Canada
E-mail: edzurek@ucalgary.ca; ziegler@ucalgary.ca

Abstract. Density Functional Theory (DFT) has been used to calculate the energies of 36 different methylaluminoxane (MAO) cage structures with the general formula $(MeAlO)_n$, where n ranges from 4 to 16. These energies in conjunction with frequency calculations based on molecular mechanics have been used to estimate the finite temperature enthalpies, entropies and free energies for MAO structures within the temperature range of 198.15K-598.15K. Formulae were fitted and used to predict the free energy values for structures where n ranges from 17 to 30. From free energy values the percentage of each n found at a given temperature was calculated, giving an average n value of 18.41, 17.23, 16.89 and 15.72 at 198.15, 298.15, 398.15 and 598.15K, respectively. Topological arguments have been used to show that the MAO cage structure contains a limited amount of square faces as compared to octagonal and hexagonal ones. It is also suggested that the limited number of square faces with their strained Al-O bonds explain the high molar Al:(Catalyst) ratio required for activation.

Introduction

In 1980, Sinn and Kaminsky discovered that the addition of water to systems such as $Cp_2ZrMe_2/AlMe_3$ caused this rather inactive reaction system to become highly active in ethene polymerization [1]. It was suspected that partial hydrolysis of $AlMe_3$ (TMA) brought about the formation of methylaluminoxane (MAO). It was further postulated that the role of the MAO/TMA mixture in this system was to act as co-catalyst. Equation **1** illustrates the commonly accepted role of MAO as the catalyst activator. Equation **2** indicates the possible role of the product cation as a catalyst in olefin polymerization.

$$(n_5\text{-}C_5H_5)_2ZrMe_2 + MAO \rightarrow [(n_5\text{-}C_5H_5)_2ZrMe]^+ + [(MAO)Me]^- ; M=Ti, Zr \qquad (1)$$

$$[(n_5\text{-}C_5H_5)_2ZrMe]^+ + n[CH_2{=}CH_2] \rightarrow [(n_5\text{-}C_5H_5)_2Zr\text{-}[CH_2\text{-}CH_2]_n\text{-}CH_3]^+ \qquad (2)$$

The high activity imparted by MAO has caused it for many years to be one of the most industrially important activators in single site or metallocene catalyzed olefin polymerization. Yet, despite this fact the structure (structures) of MAO remain largely unknown. The characterization of MAO by NMR spectroscopy has been hindered by disproportionation reactions at high temperatures and association in solution yielding a mixture of different oligomers with multiple equilibria. Moreover, the characterization cannot be carried out using X-ray diffraction due to the fact that it is not possible to isolate crystalline samples [2].

The determination of the structure of MAO can be linked to the determination of the structures of alumoxanes in general. Alumoxanes are intermediates in the hydrolysis of organoaluminum compounds to aluminum hydroxides. They were originally proposed as consisting of a linear or cyclic chain structure which were composed of alternating three-coordinate aluminum and two-coordinate oxygen atoms [3]. The first crystallographic evidence for the presence of four-coordinate aluminum atoms was given by Atwood and co-workers in their structural determination of the $[Al_7O_6Me_{16}]^-$ anion [4]. This result encouraged many groups to propose structures consisting of fused four or six membered rings, or both for that of MAO [3]. While these structures were more reasonable, they still contained a peripheral aluminum atom which remained three-coordinate. Methyl bridges and/or the presence of trimethylaluminum groups were suggested [3], but these resulted in structures whose chemical formula substantially deviated from the generally accepted formula of 'pure' MAO, $(MeAlO)_n$ where n is an integer.

Replacement of the methyl substituents in MAO with bulkier t-butyl groups made the first structural determination of alkylalumoxanes possible. Barron and co-workers synthesized a series of compounds, $[(^tBu)Al(\mu_3\text{-}O)]_n$, where n=6, 8, 9 and 12 [2,5]. The synthesis of these compounds led to the suggestion that MAO has a three-dimensional cage structure. Within these cage structures four-coordinate aluminum centers bridged by three-coordinate oxygen atoms were thought to predominate [2].

Moreover, Barron's work supported the accepted proposal that MAO consists of an equilibrium mixture of oligomeric species [2]. This proposed equilibrium for MAO can be seen in the following equation, where x, y and z are integers:

$$(MeAlO)_x \leftrightarrow (MeAlO)_y \leftrightarrow (MeAlO)_z \qquad (3)$$

It was known that species of exceptional Lewis acidity are found in MAO solutions, but four co-ordinate aluminum centers are not thought of as being exceptionally Lewis acidic. Barron and co-workers found that indeed they are and developed the concept of Latent Lewis Acidity (LLA) which is a consequence of the ring strain present in the cluster [6, 7].

As has been mentioned previously, the characterization of the structure of MAO via NMR spectroscopy has not been successful. Yet, NMR and other spectroscopic methods have been used to give further clues as to the structure(s) and role of MAO. In most cases these methods have been used to give an estimate of the size range for a typical MAO oligomer. For example, the linewidths of ^{27}Al NMR have predicted that for $[AlOMe]_n$, n ranges between 9 and 14 at high temperatures and between 20 and 30 at ambient conditions [8]. EPR studies have been done via the addition of a spin probe to a MAO solution. Once again, linewidths coupled with line intensities were used to find the radius of a MAO structure. This method found that n ranges between 14 to 20 [9].

It is well known that there exists residual TMA (trimethylaluminum) in all MAO solutions. It is accepted that TMA participates in an equilibrium with different MAO oligomers [1]. However, we shall first establish a model for a pure (TMA free) MAO solution, although such a system has not been established experimentally. Finally, we shall comment on MAO containing TMA as well as possible mechanisms for the activation of Cp_2ZrMe_2 by MAO .

The main objective of this study is to establish the percent abundance of different MAO structures. Ultimately, it is the Gibbs Free Energy which determines the stability of a given structure. The Gibbs Free Energy is given as:

$$G_T(n) = H_T(n) - TS_T(n) \tag{4}$$

where $H_T(n)$ is the finite temperature enthalpy at T for $(AlOMe)_n$ and $S_T(n)$ the corresponding entropy.

Within this article first we discuss different structural alternatives (sheets, cages, fused cages) showing that caged structures are energetically the most stable. Secondly, we derive formulae which are important in determining the topologies of caged structures. Energetic considerations are investigated and a method is proposed which can be used to predict them. The same is then done for enthalpic corrections and entropies. Next, we examine the Gibbs Free Energy and percent abundance of different MAO structures. Finally, we consider MAO-TMA interactions and modes of activation.

Computational Details

The density functional theory calculations were carried out using the Amsterdam Density Functional (ADF) program version 2.3.3 developed by Baerends et. Al. [10] and vectorized by Ravenek [11]. The numerical integration scheme applied was developed by te Velde et. al. [12] and the geometry optimization procedure was based on the method of Verslius and Ziegler [13]. For total energies and gradients the gradient corrected exchange functional of Becke [14] and the correlation functional of Perdew [15] was utilized in conjunction with the LDA parametrization of Vosko et al [16]. The electronic configurations of the molecular systems were described by a double-zeta STO basis set with polarization functions. A 1s frozen core was used for carbon and oxygen, while an [Ar] frozen core was used for aluminum. A set of auxiliary s, p, d, f and g STO functions centered on all nuclei was used to fit the molecular density and represent Coulomb and exchange potentials in each SCF cycle [17]. Single-point numerical differentiation of energy gradients were used for frequency calculations.

UFF2 [18, 19] was used to calculate entropic and finite temperature enthalpy corrections to the Gibbs Free Energy. It was necessary to reparametrize the force field for our specific system. Details can be found elsewhere [20].

Results and Discussion

Energetics of Sheet/Caged/Fused Caged Structures

Despite the fact that experiment has provided evidence that MAO consists of three-dimensional cage structures, it was decided that a preliminary investigation on the relative stability of sheet, caged and fused caged structures ought to be performed. These results would then allow us to test the consistency of our theoretical calculations with this belief. Figure 1 shows a selection of the sheet and fused caged structures.

First of all, it must be noted that during the geometry optimization of the fused caged structure the bonds corresponding to five coordinate Al and four coordinate O atoms broke giving a caged structure. This shows that such structures are unstable alternatives for MAO.

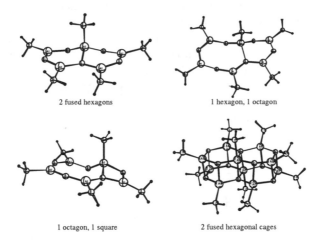

2 fused hexagons   1 hexagon, 1 octagon

1 octagon, 1 square   2 fused hexagonal cages

Fig. 1. A selection of Sheet and Fused Caged Structures

The electronic binding energy per monomer unit is defined as:

$$BE(n) = 1/n \; (\; E[(AlOMe)_n] - n \; E[AlOMe]) \qquad (5)$$

It gives the energy which is gained per monomer (AlOMe) when a certain geometry is formed from n monomers. The lower the binding energy per monomer, the more stable the given structure is. We examined the binding energy per monomer for ring and fused ring structures and compared it with that of caged alternatives finding that even for a very strained structure such as $(AlOMe)_4$, the binding energy per monomer is approximately 9 kcal/mol lower than for any of the sheet structures. Hence, this preliminary investigation indicates that caged MAO structures consisting of three coordinate oxygen and four coordinate aluminum atoms are much more energetically stable than sheet or fused caged structures.

Accordingly, in our investigation of possible MAO geometries it was decided to focus on three dimensional caged structures comprised of square, hexagonal and octagonal faces.

Mathematical Relationships

In order to construct MAO cage structures, it is first necessary to gain some insight into the construction of such polyhedrons in general. Within this section, we propose a mathematical method by which this may be done. Next, we shall give a formula which relates the number of square faces to the number of hexagonal faces found within a polyhedron. This result will prove useful in explaining the large ratio of Al:Catalyst needed in order for polymerization to occur. Finally, we will give mathematical relationships which we used in order to construct large MAO cages.

The most effective way by which one can construct polyhedrons is via the drawing of Schlegel diagrams [21]. A Schlegel diagram is a projection of a three dimensional object onto a plane surface. An example of such a diagram is shown in Figure 2. Within this study, the MAO cage structures were constructed via the use of Schlegel diagrams.

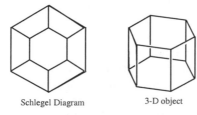

Schlegel Diagram   3-D object

Fig. 2. A Schlegel Diagram of a Hexagonal Prism

Despite the fact that it is not possible to derive all of the possible connectivities present in a polyhedron corresponding to a given number of atoms, some assertions can be made. The first deals with the relationship between the number of square (S), hexagonal (H) and octagonal (O) faces comprising a given polyhedron. We have derived the following equation [20]:

$$S = O + 6 \tag{6}$$

Equation **6** shows that the minimum amount of square faces which can exist in such a polyhedron is six and that this occurs when the number of octagonal faces is zero, that is when the polyhedron is made up solely of square and hexagonal faces.

Other relationships which have been derived [20] apply to the case when only square and hexagonal faces are present. Within such a polyhedron there are only four environments within which each atom (point) may be found. They are the following: a = the number of atoms which are part of 3 square faces; b = the number of atoms part of 2 square and 1 hexagonal face; c = the number of atoms part of 1 square and 2 hexagonal faces; d = the number of atoms part of 3 hexagonal faces. The relationships between these are given in equations **7** and **8**.

$$3a + 2b + c = 24 \tag{7}$$

$$b + 2c + 3d = 6H \tag{8}$$

Take into consideration a MAO cage which simply consists of square and hexagonal faces. As has been shown, the number of square faces present in this instance is always equal to six. Then, as the cage grows the number of hexagonal faces increases while the number of square faces stays the same. Thus, for a large cage it can be imagined that the probability that an atom is bonded to three square faces becomes very small. Similarly, so does the probability that an atom is bonded to 2 square and 1 hexagonal face. Hence, in equations **7** and **8** a and b can be put to zero. Thus, it becomes trivial to solve for the types of connectivities present in a MAO cage. Later on, this method shall be used to generate large MAO structures.

Energetic Considerations

The potential energies of thirty-six different $(AlOMe)_n$ structures, where n ranges between 4 and 16 were determined via DFT level calculations. Some representative structures composed of square and hexagonal faces are shown in Figure **3**. For $(AlOMe)_{14}$ only the most stable structural alternative is shown.

It was determined that the stability of a given MAO is heavily dependent upon the structure of the cage, and specifically upon the bonding environment present. Hence, a least squares fit was done to predict the energies of the MAO structures, using the bonding environments as an index. This fit resulted in the following energy expression for any given MAO structure:

$$E = -373.57(3S) -377.49(2S+H) -381.13(2H+S) -381.80(3H) -377.14(2S+O) - \tag{9}$$
$$380.59(2O+S) -381.03(H+O+S) -378.86(2H+O) -365.51(2O+H)$$

where (3S) denotes the number of atoms belonging to three square faces, (2S+H) the number of atoms belonging to two square and one hexagonal face and so on. None of the structures which we have considered contained an atom in a (3O) environment, due to the fact that in order for such an environment to be present, the cage would have to be quite large. Hence, the coefficient for (3O) is missing in the fit above. The root-mean square deviation of this fit for the total energy is 4.70kcal/mol.

Moreover, it must be noted that the coefficients pertaining to each specific bonding environment provide a means by which one can gauge the stability of a particular environment. The more negative the coefficient, the more stable the environment. Hence, the order of stability is, in decreasing order, 3H > 2H+S > H+O+S > 2O+S > 2H+O > 2S+H > 2S+O > 3S > 2O+H.

It is also interesting to note that the structures which are composed simply of square and hexagonal faces have the lowest energies for a given n with the exception of $(AlOMe)_{10}$ where another structure is 0.38kcal/mol more stable. This can be accounted for by considering equation **6**. Octagonal faces increase the amount of square faces in a MAO cage. These square faces exhibit a large amount of ring strain therefore destabilizing the structure. Hence, the structures with the least amount of square faces present for a given n, are most stable, energetically.

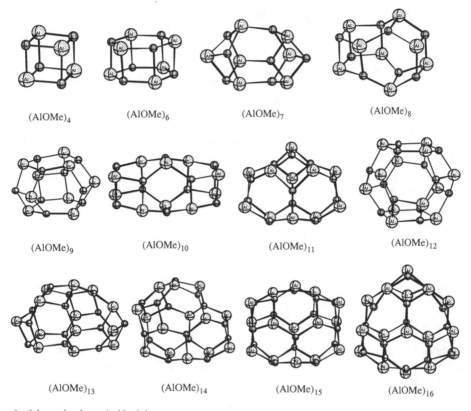

$(AlOMe)_4$ $(AlOMe)_6$ $(AlOMe)_7$ $(AlOMe)_8$

$(AlOMe)_9$ $(AlOMe)_{10}$ $(AlOMe)_{11}$ $(AlOMe)_{12}$

$(AlOMe)_{13}$ $(AlOMe)_{14}$ $(AlOMe)_{15}$ $(AlOMe)_{16}$

*methyl groups have been omitted for clarity

Fig. 3. MAO Cage Structures Composed of Square and Hexagonal Faces Only for $(AlOMe)_4$-$(AlOMe)_{16}$

Enthalpic Considerations

Finite temperature enthalpies and entropies can be calculated from standard expressions [22] provided that all the vibrational frequencies are known. UFF2

[18,19] was reparametrized in order so that the frequencies agreed with those calculated with ADF for selected structures. The total enthalpy is given as:

$$H_T(n) = E(n) + H_{EC}(n) \tag{10}$$

where $E(n)$ is the energy and $H_{EC}(n)$ is the finite temperature enthalpy correction given by:

$$H_{EC}(n) = H_{rot} + H_{trans} + H_{vib} \tag{11}$$

Here H_{rot}, H_{trans} and H_{vib} are the rotational, translational and vibrational finite temperature enthalpy corrections, respectively. Formulae were fitted so as to predict these contributions. If H_0 is the zero-point energy, they are:

$$H_0 = 25n \text{ kcal/mol} \tag{12}$$

$$H_{rot} = H_{trans} = 3/2RT \tag{13}$$

$$H_{vib} = H_0 + (0.0028T - 0.3548)n \times \ln(T) \tag{14}$$

Equation **12** gives an rms deviation of 1.16kcal/mol, equation **13** is exact while the rms deviation for **14** is 3.28, 0.78, 1.32 and 3.36kcal/mol at 198.15K, 298.15K, 398.15K and 598.15K, respectively.

At all temperatures $H_{EC}(n)/n$ is approximately equal for all MAO structures. This shows that for a given disproportionation reaction $\Delta H_{EC}(n)$ will be nearly zero and does not contribute to the relative stability of the MAO oligomers.

Entropic Considerations

Entropic values were calculated via the parametrized UFF2 code. The total entropy is given as:

$$S_T(n) = S_{trans} + S_{rot} + S_{vib} \tag{15}$$

where S_{trans}, S_{rot}, S_{vib} are the translational, rotational and vibrational contributions to the entropy. The following equations predict the different entropic contributions at 298.15K.

$$S_{trans} = 0.351n + 41.168 \tag{16}$$

$$S_{rot} = 0.573n + 30.573 \tag{17}$$

$$S_{vib} = 7.91(3S) + 8.30(2S+H) + 10.20(2H+S) + 8.49(3H) + 10.41(2S+O) + \tag{18}$$
$$9.50(2O+S) + 10.45(S+O+H) + 7.32(2H+O) + 0(2O+H)$$

Above n corresponds to $(AlOMe)_n$, (3S) the number of atoms bonded to three square faces and so on. The rms deviation for $S_T(n)$ is 1.78kcal/mol at 298.15K. Entropic corrections are temperature dependent and hence we devised equations which could be used to predict entropies at different temperatures given those at 298.15K. They are the following:

$$S_{2trans} = S_{1trans} + T_2/T_1 + (0.014)T_2 - 5.47 \tag{19}$$

$$S_{2rot} = S_{1rot} + T_2/T_1 + (0.007)T_2 - 3.28 \tag{20}$$

$$S_{2vib} = (T_2/T_1 - ((0.0006T_2^2 - 0.5353T_2 + 108.85)^{-1}) S_{1vib} \tag{21}$$

where S_{2trans} is the translational entropy at temperature T_2 and so on. Equations **19** and **20** are nearly exact while equation **21** gives an rms deviation of 0.27, 1.70, and 4.09kcal/mol at 198.15, 398.15 and 598.15K, respectively.

We found that at low temperatures $-TS_T(n)/n$ is not very significant. Yet, as the temperature increases $-TS_T(n)/n$ becomes more important in stabilizing smaller structures.

The Gibbs Free Energy

The Gibbs free energy per Monomer as a function of n at different temperatures is plotted in Figure **4** for structures composed of square and hexagonal faces only. It was found that these structures give the lowest Gibbs free energy for a given n, with one exception. That is of $(AlOMe)_{10}$ where another structure is 2.59kcal/mol more stable. Error bars are present for n =17, 18, 20, 21, 25 and 30 where equations **7** and **8** were used to find the connectivities and equations **10-21** were used to estimate the Gibbs free energy.

As can be noted, the same general trend is followed at all temperatures, with $(AlOMe)_{12}$ being the most stable structure. At lower temperatures, $(AlOMe)_{16}$ is almost as stable as $(AlOMe)_{12}$, while at high temperatures the difference increases. This can be attributed to entropic effects which were shown to be more important at higher temperatures. Local maxima at n=7, 10 and 13 are present at all temperatures. They are due to the fact that these structures contained a greater amount of atoms in strained (3S or 2S+H) environments than their neighbours. Thus, energetically they were less stable.

Equation **6** shows us that there are only six square faces present in a MAO structure composed of square and hexagonal faces only. Above we see that these structures have the lowest Free Energy for a given n. Moreover, the structures which are most stable do not have atoms in (3S) or (2S+H) environments, that is they do not contain square-square edges. Due to the fact that these bonds are more strained and less stable they ought to be the sites with greatest Latent Lewis Acidity. Thus, we have shown that there are not many acidic or active sites in MAO. This topological consequence could be used to explain the high Al:Zr ratio which is necessary for polymerization to occur.

Gibbs Free Energy per Monomer Vs. n at Different Temperatures

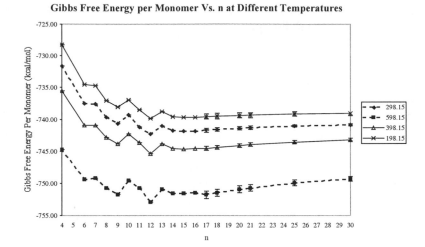

Fig. 4. Gibbs Free Energy per Monomer Vs. n at Different Temperatures

The Gibbs free energy of $(AlOMe)_n$ relative to n monomeric units is given by:

$$\Delta G_0(n) = G_T(n) - nG^0_T \tag{22}$$

If $\Delta G_0(n)$ is defined in such a way, then equation **23** may be used to calculate equilibrium constants between $(AlOMe)_n$ and n monomer units. It must be noted that for n =17, 18, 20, 21, 25 and 30 these were obtained via using the estimated Gibbs free energies for $(AlOMe)_n$. For n = 19, 22, 23, 24, 26, 27, 28 and 29 the Gibbs free energies were found via interpolation and next the equilibrium constants were calculated. As a check, structures were found via drawing a Schlegel diagram for n = 19, 22 and 24 then the Gibbs free energy were predicted using the formulae given above. All of the Gibbs free energies found in such a manner fell within the error bars given when interpolation was done. Hence, this shows that it is valid to find the equilibrium constants for n = 26, 27, 28 and 29 in such a manner.

$$K_{eq} = \exp(-\Delta G_0(n)/RT) \tag{23}$$

Next, it is possible to find the percent abundance of a given structure according to equation **24**.

$$\%(AlOMe)_n = (K_{eq}(n)/\sum_i K_{eq}(i)) \times 100 \tag{24}$$

Figure **5** shows the percent abundance of the MAO's at different temperatures. The most abundant species at all temperatures is n=12 which ranges between 16-22%. Higher temperatures stabilize smaller species (which can be seen in the large increase of n=9 at 598.15K), while lower temperatures stabilize larger species (which can be seen in the increase of n ≥ 15 at 198.15K). This can also be seen when the weighted average of n is determined, which is 18.41, 17.23, 16.89 and 15.72 at 198.15, 298.15, 398.15 and 598.15K, respectively. These values agree well with experimental data which estimates that n ranges between 14 to 20 [10] especially considering the fact that only TMA-free MAO was examined in this study. However, it must be noted that it is not clear exactly what the experimental results show. That is, we are not sure if they report n for $(AlOMe)_n$, or n+m for $(AlOMe)_n(TMA)_m$.

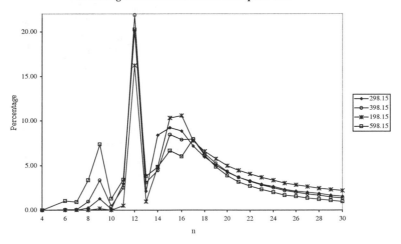

Fig. 5. Percentage of Each n at Different Temperatures

TMA-MAO Interactions

We have studied a number of possible alternatives to how TMA interacts with MAO, the most favorable of which is shown in Figure **6a** for $[(AlOMe)_6 \bullet TMA]$. An Al-O bond is broken resulting in $AlMe_2$ bonding to the O and a methyl transferring to the Al. Moreover, it must be noted that the bond which was broken (the most acidic) belonged to two square faces. The atoms comprising the bond were both in a (2S+H) environment. We have examined the acidity of different bonds in MAO's composed of square and hexagonal faces only. The most acidic bond consisted of an O in a (2S+H) or (3S) environment and an Al in a (2S+H) or (2H+S) environment. In all cases, but one, the bond which was most acidic belonged to two square faces. For $(AlOMe)_9$, it belonged to one square and one hexagonal face. Hence, in general, in order for a bond in MAO to be active, it must belong to two square faces. Since we have shown that the number of such faces is limited, this points to the conclusion that the number of acidic sites in MAO is limited and thus a high Al:catalyst ratio is necessary for activation to occur.

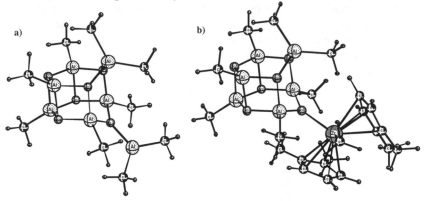

Fig 6. a) How TMA interacts with $(AlOMe)_6$. **b)** How $Cp_2Zr(CH_3)_2$ interacts with $(AlOMe)_6$

Moreover, since the interaction of TMA with MAO is an exothermic one, the presence of TMA in solution stabilizes the MAO structures which have a greater amount of strained bonds. Thus, it shifts the equilibrium towards smaller MAO's. Currently we are conducting a thorough study on this equilibrium shift and hope to estimate the Me:Al ratio in a 'real' MAO solution [23].

Possible Methods of Activation

The addition of $Cp_2M(CH_3)_2$ to pure MAO leads to the breakage of a strained Al-O bond with the formation of $[Cp_2M(CH_3)]^+ \bullet [CH_3MAO]^-$ where the metal, typically zirconium, is bound to oxygen and one methyl has migrated to the aluminum. This is analogous to the reaction of TMA with MAO and is shown in Figure **6b**. Due to the very strong O-M bond present in $[Cp_2M(CH_3)]^+ \bullet [CH_3MAO]^-$ the ΔH_{IPS} (ion-pair-separation) for this species is quite high. In fact, it is too high for this to be the active species in polymerization. Hence, we propose instead the active species as being $[Cp_2M(CH_3)]^+ \bullet [TMACH_3MAO]^-$, shown in Figure 7. Note the bonding of $[Cp_2M(CH_3)]^+$ through a methyl group to the MAO. This bond is not nearly as strong as the O-M bond, making this structure a feasible activator. Two possible routes to the formation of this species and thus to activation of $Cp_2M(CH_3)_2$ by MAO (M = Zr, Ti, etc.) are shown in equations **25** and **26**. The latter route has been suggested by another group [24].

$$Cp_2M(CH_3)_2 + MAO \rightarrow [Cp_2M(CH_3)]^+ \bullet [CH_3MAO]^- \qquad (25)$$

$$[Cp_2M(CH_3)]^+ \bullet [CH_3MAO]^- + TMA \rightarrow [Cp_2M(CH_3)]^+ \bullet [TMACH_3MAO]^-$$

$$MAO \bullet TMA + Cp_2M(CH_3)_2 \rightarrow [Cp_2M(CH_3)]^+ \bullet [TMACH_3MAO]^- \qquad (26)$$

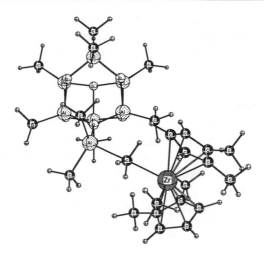

Fig 7. [Cp$_2$M(CH$_3$)]$^+$•[TMACH$_3$MAO]$^-$; the active species in polymerization

Conclusions

Within this study we have first of all proposed a method which can be used in theoretical structure determination, drawing Schlegel diagrams and from them constructing the appropriate three-dimensional object. We have also shown from topological arguments that in the most stable MAO structures there are only six square faces present and hence few square-square edges. These faces exhibit high ring strain and hence such edges/bonds would be the most acidic and thus active sites. We propose the low abundance of these faces and thus even a lower abundance of square-square edges to explain the high Al:catalyst ratio required for polymerization to occur.

Moreover, we have performed a complete study on pure MAO structures which consist of three-coordinate O and four-coordinate Al atoms. In so doing, we have fitted formulas which may be used to predict the energies, enthalpies and entropies of any given MAO structure within the temperature range of 198.15K-598.15K effectively. Finally, we have calculated the percent distribution for each n within this temperature range. The weighted average gives n as 18.41, 17.23, 16.89 and 15.72 at 198.15, 298.15, 398.15 and 598.15K, respectively. These values agree well with experimental data. However, it is not clear whether the experimental data gives values for n in (AlOMe)$_n$, or for n+m in (AlOMe)$_n$(TMA)$_m$. Using some of the results obtained here, we shall in a forthcoming study discuss the structure of TMA containing MAO.

Currently our calculations on 'real MAO' show that TMA binds to the square-square edges of less stable MAO oligomers thereby stabilizing them. For large MAO's there are more stable structural alternatives which do not contain these strained bonds. Thus, TMA interacts predominantly with small MAO structures shifting the equilibrium in their direction. We are currently investigating the Me:Al ratio present in a 'real MAO' solution at different temperatures.

Acknowledgements

This study was supported by the Natural Science and Engineering Research Council of Canada (NSERC) and by Novacor Research and Technology (NRTC) of Calgary, Alberta, Canada. We would like to thank Dr. Clark Landis of the University of Wisconsin for supplying us with UFF2 and Dr. Ted Bisztriczky of the University of Calgary Mathematics Department for his suggestion to use Schlegel Diagrams for the construction of our MAO cages.

References

1   Sinn, H.; Kaminsky, W.; Vollmer, H.J.; Woldt, R.; *Agnew. Chem.* **1980**, *92*, 396
2   Mason, M.R.; Smith, J.M.; Bott, S.G.; Barron, A.R.; *J. Am. Chem. Soc.* **1993**, *115*, 4971.
3   Pasynkiewicz, S.; *Polyhedron*, **1990**, *9*, 429.
4   Atwood, J.L.; Hrncir, D.C.; Priester, R.D.; Rogers, R.D.; *Organomet.* **1983**, *2*, 985.
5   Harlan, C.F.; Mason, M.R.; Barron, A.R.; *Organomet.* **1994**, *13*, 2957.
6   Harlan, C.J.; Bott, S.G.; Barron, A.R.; *J. Am. Chem. Soc.* **1995**, *117*, 6465.
7   Koide. Y.; Bott, S.G.; Barron, A.R.; *Organomet.* **1996**, *15*, 5514.
8   Babushkin, D.E.; Semikolenova, N.V.; Panchenko, V.N.; Sobolev, A.P.; Zakharov, V.A.; Talsi, E.P.; *Macromol. Chem. Phys.* **1997**, *198*, 3845.
9   Talsi, E.P.; Semikolenova, N,V.; Panchenko, V.N.; Sobolev, A.P.; Babushkin, D.E.; Shubin, A.A.; Zakharov, V.Z.; *J. Molecular Catalysis A: Chemical, 1999*, *139*, 131.
10   (a) Baerends, E.J.; Ellis, D.E.; Ros, P. *Chem. Phys.* **1973**, *2*, 41. (b) Baerends, E.J.; Ros, P. *Chem. Phys.* **1973**, *2*, 52.
11   Ravenek, W. Algorithms and Applications on Vector and Parallel Computers; te Riele, H.J.J.; Dekker, T.J.; vand de Horst, H.A.; Eds. Elservier: Amsterdam, The Netherlands, 1987.
12   (a) te Velde, G.; Baerends, E,J. *Comput. Chem.* **1992**, *99*, 84. (b) Boerringter, P.m,; te Velde, G.; Baerends, E.J. *Int. J. Quantum Chem..* **1998**, *33*, 87.
13   Verslius, L.; Ziegler, T. *J. Chem. Phys.* **1988**, *88*, 322.
14   Becke, A.D. *Phys. Rev. A.* **1988**, *38*, 3098.
15   Perdew, J.P. *Phys. Rev. B.* **1986**, *33*, 8822.
16   Vosko, S.H.; Wilk, L.; Nusair, M. *Can. J. Phys.* **1980**, *58*, 1200.
17   Krijn, J.; Baerends, E.J.; Fit Functions in the HFS-Method; Free University of Amsterdam, 1984.
18   Casewit, A.K.; Colwell, K.S.; Rappe, A.K.; *J. Am. Chem. Soc.* **1992**, *114*, 10046.
19   Casewit, C.J.; Colsell, K.S.; Rappe, A.K.; *J. Am. Chem. Soc.* **1992**, *114*, 10035.
20   Zurek, E.D.; Woo, T.K.; Firman, T.K.; Ziegler, T.; Submitted to Inorganic Chemistry.
21   Coxeter, H.S.M.; *Regular Polytopes*, 2nd ed., Macmillian Company, 1963, New York.

22) Hehre, W.J.; Radom, L.; Schleyer, P.V.R.; Pople, J.A.; *Ab Initio Molecular Orbital Theory*, John Wiley & Sons, 1986, pg. 251, 259., New York.

23) To be submitted.

24) Rytter, E.; Ystenes, M., Eilertsen, J.L.; Ott, M., Andreas, J.A.; Liu, J.; *Proceedings of the Conference Organometallic Catalysts and Olefin Polymerization in Oslo*, 2000.

2. Non Group IV Catalysts

Phosphinoalkyl-Substituted Cyclopentadienyl Chromium Catalysts for the Oligomerization of Ethylene

A. Döhring, V. R. Jensen*, P. W. Jolly*, W. Thiel, J. C. Weber

Max-Planck-Institut für Kohlenforschung, 45470 Mülheim a.d. Ruhr, Germany
E-mail: jolly@mpi-muelheim.mpg.de; jensen@mpi-muelheim.mpg.de

Abstract. Heterogeneous chromium catalysts are used to prepare much of the polyethylene manufactured industrially. Related homogeneous catalysts have received less attention and have never found commercial use. Reported here is a new family of phosphinoalkyl-substituted cyclopentadienyl chromium compounds which, after activation with methylalumoxane, give highly active homogeneous catalysts for the oligomerization and polymerization of ethylene whereby the degree of oligomerization is controlled by the size of the substituents on the P-atom. By varying the substitutent the catalysis can be directed to give mainly dimers and trimers, oligomers or polymer. Density functional calculations indicate that this effect is the consequence of an unusually selective steric destabilization of the transition state for termination which proceeds by β-hydrogen transfer to the incoming ethylene in a reaction involving spin inversion.

1 Introduction

Polyethylene (PE), in its various forms (low-density, high-density, linear-low-density), is the highest volume, man-made polymer with a global 'consumption' of around 4kg per person per year. Most of this is prepared by polymerizing ethylene in the presence of heterogeneous catalysts based upon titanium (Ziegler-Natta catalysts) and chromium (Phillips and Union Carbide catalysts) [1, 2].

Shortly after the introduction of the heterogeneous chromium catalysts, G. Wilke *et al.* [3] discovered that solutions of tris(η^3-allyl)chromium are also active and although this system did not go into production, it was presumably the first example of a single-component, homogeneous catalyst for the polymerization of an olefin. Later, additional examples were reported, e.g. Cp(η^3-allyl)$_2$Cr, (Me$_5$C$_5$)Cr(CH$_2$SiMe$_3$)$_2$, and their chemistry has been suggested to give some insight into the mechanism of the heterogeneous systems [2].

The spectacular renewal of interest in the polymerization of olefins has been associated with the discovery of the methylalumoxane (MAO)-activated *ansa*-zirconocene [4] and amido-substituted cyclopentadienyl titanium [5] catalysts. Analogous chromium-based systems have only been reported recently and include compounds stabilized by N,O- and N,N-chelating ligands [6, 7]. The most active systems, however, contain an amino-substituted cyclopentadienyl ligand: for example, the **I**-MAO system (Scheme 1) converts ethylene into linear high-density PE with an activity of up to 30,000 kgPE/mol Cr·h at 21°C and 2bar [8].

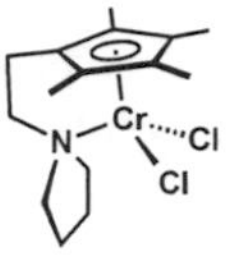

I

Scheme 1.

2 Results and Discussion*

We have extended the above described investigations to the related phosphinoalkyl-substituted analoga (**II**) and have observed a remarkable dependence of the product composition of the reaction with ethylene upon the steric properties of the substituent (R) on the P-atom. The chromium complexes (**II**) were prepared by treating spiro[2.4]hepta-4,6-diene with the appropriate alkalimetal-phosphide followed by reaction with Cr(THF)$_3$Cl$_3$ (Scheme 1) [9, 10].

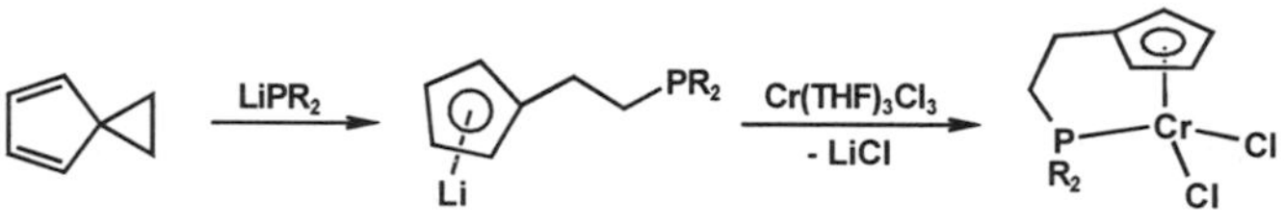

II

Scheme 2.

The molecular structure, with a metal-bonded P-atom, has been confirmed by X-ray crystallography for the compounds where R is phenyl (Ph) and cyclohexyl (Cy, Figure 1) as well as for (Cy$_2$PC$_2$H$_4$C$_5$H$_4$)CrMe$_2$ (prepared by reacting **II** (R = Cy) with MeLi)[10]†. Treatment of the (R$_2$PC$_2$H$_4$C$_5$H$_4$)CrCl$_2$ complexes (**II**) with MAO leads to the generation of highly active catalysts for the oligomerization and polymerization of ethylene whereby the degree of oligomerization and catalytic

---

* Preliminary results have been submitted as a communication.
† Crystal structure data for **II** (R = Cy): C$_{19}$H$_{30}$Cl$_2$CrP, M = 412.3, monoclinic, P2$_1$/n (no. 14), a = 8.396(2), b = 20.064(3), c = 12.398(3) Å; β = 105.44(2)°, v = 2013.0(7) E^3, Z = 4, μ = 0.911 mm^{-1}, crystal data 0.11x0.53x0.60 mm, T 20°C, R$_1$ = 0.0434, wR^2 = 0.1197.

activity are dependent upon the nature of the substituent (R) at the P-atom (Table 1).

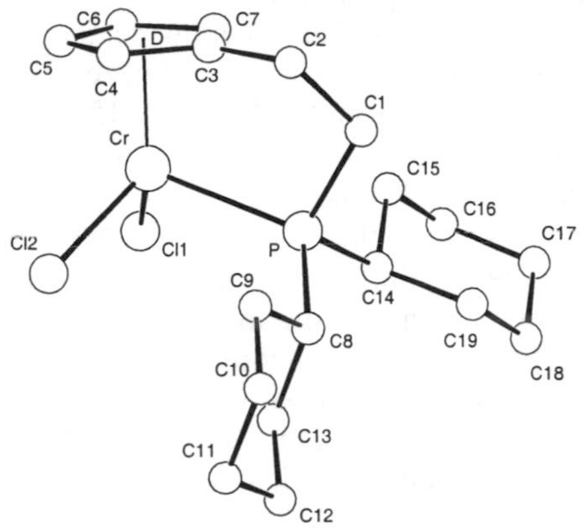

Fig. 1. The molecular structure of **II** (R = Cy): P–Cr 2.459(1), D–Cr 1.886, Cr–Cl1 2.276(1), Cr–Cl2 2.279(1) Å; Cl1-Cr-Cl2 95.9(1), P-Cr-Cl1 98.3(1), P-Cr-Cl2 102.2(1), P-Cr-D 109.8°.

Table 1. The $(R_2PC_2H_4C_5H_4)CrCl_2$ (**II**)-MAO catalyzed oligomerization and polymerization of ethylene.[a]

R	Product composition (%) [b]	Activity (kg prod./mol Cr·h)	θ (PR$_3$) (degrees)
Me	C$_4$ (83.6), C$_6$ (13.0), C$_{8+}$ (3.5)	4620	118
Et	C$_4$ (21.3), C$_6$ (26.5), C$_8$-C$_{14}$ (49.5), C$_{16+}$ (2.7)	4450	132
Bu^i	C$_4$ (10.6), C$_6$ (12.5), C$_8$-C$_{10}$ (25.1), C$_{12+}$ (51.8)	4230	143
Pr^i	C$_4$-C$_{10}$ (3.1), PE (96.9, M$_w$ 6.3.10^4)	3590	160
Cy	C$_6$-C$_{10}$ (3.1), PE (96.9, M$_w$ 1.8.10^5)	1450	170
Bu^t	C$_6$-C$_8$ (15.7), PE (84.4, M$_w$ 1.3 x 10^5)	590	182
Ph	C$_4$-C$_8$ (9.5), C$_{10}$-C$_{14}$ (7.0), C$_{16+}$ (81.0), PE (2.2)	4950	145
o-tolyl	C$_6$-C$_{10}$ (4.1), PE (95.9)	2330	194

R, substituent on phosphorus; θ (PR$_3$), Tolman cone angle [11].
[a] Standard condition: solvent, toluene; P(C$_2$H$_4$) 2 bar; Cr:MAO = 1:100; t = 4 min; T 21°C, ΔT ≤ 4°C.
[b] C$_4$ = 1-butene; C$_6$ = 1-hexene etc.

That these effects are largely steric in origin is apparent from a comparison with the cone angle (θ) of the related triorganophosphine (PR$_3$) compounds as defined by Tolman [11] and is underlined by the results for the phenyl- and o-tolyl-substituted derivatives shown in Table 1.

An effect as dramatic as that shown here has not been reported previously for a chromium catalyst. It has, however, been mentioned that the polymerization of ethylene using MAO-activated, diimine complexes of nickel [12] and pyridinediimine complexes of iron and cobalt [13–15] having aryl-substituents (e.g. **III** and **IV** in Scheme 3) can be tailored to give higher oligomers by decreasing the size of the ortho-substituents on the aryl group.

In addition, it has been reported that the degree of oligomerization of ethylene catalyzed by $(\eta^3\text{-}C_3H_5)Ni(PR_3)X\text{–}Et_3Al_2Cl_3$ is controlled by the nature of the tertiary phosphine bonded to the nickel atom: the main product with PMe_3 is the dimer while in the presence of $PBu^t_2Pr^i$ higher oligomers are also formed [16]. Finally we should mention that the classical SHOP ethylene oligomerization catalyst of the type $(Ph_2PCH{:}CPhO)Ni(PPh_3)R$ [17] can be converted into a polymerization catalyst by introducing a salicylaldimine ligand having bulky groups in the ortho-position of the ligand [18].

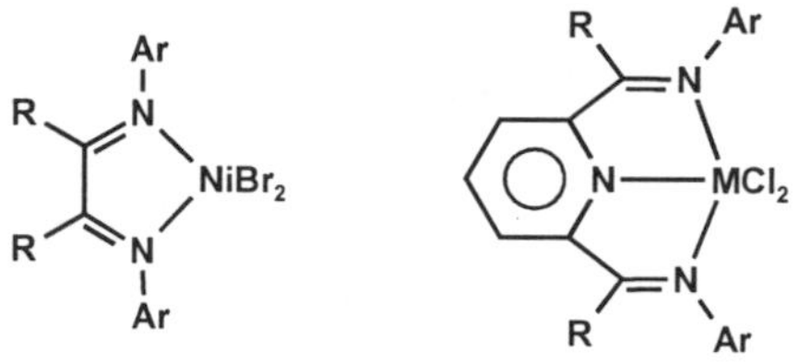

III   IV (M = Fe,Co)

Scheme 3.

Of particular industrial interest in our case is presumably the formation of a 1-butene/1-hexene mixture where R is Me, of a mixture of long chain oligomers where R is Ph and of polyethylene where R is Cy. The dimer/trimer-mixture could potentially be copolymerized directly with ethylene using the amino-substituted cyclopentadienyl chromium system [8] or the $(Pr^i_2PC_2H_4C_5H_4)CrCl_2\text{–}MAO$ catalyst described above to LLDPE while the wax-like higher olefin mixture could form the basis of a SHOP-type [17] long chain alcohol production.

The reaction conditions shown in Table 1 were chosen to facilitate comparison and are by no means optimal. Considerably higher activities can be obtained by increasing the MAO concentration and by introducing methyl-substituents into the cyclopentadienyl ring. For example, the activity of **II** (R = Et) with a Cr:MAO ratio of 1:500 is almost three times (13400 kg product/mol Cr·h) that shown in Table 1 while that of $(Ph_2PC_3H_6C_5Me_4)CrCl_2\text{–}MAO$ is approximately four times that of $(Ph_2PC_3H_6C_5H_4)CrCl_2$ under the same standard conditions (5680 vs. 1490 kg product/mol Cr·h.). A comparison of the $(Me_2DC_2H_4C_5H_4)CrCl_2\text{–}MAO$ systems where D is N or P suggest that the P-containing species are more active than the analogous N-containing systems (activity, D = N 1680 kg/mol Cr·h) [8, 10]

The remarkable dependence of the product composition in the Cr-catalyzed oligomerization of ethylene upon varying the substituent on phosphorus prompted us to investigate the propagation and termination reactions using first principle methods. In recent years, density functional theory (DFT) has become the 'workhorse' for computational investigation of the mechanisms of transition metal catalyzed olefin polymerization reactions (see, for example, [19–22]), and for the current work we selected a gradient-corrected DFT functional (BPW91[‡]) in combination with DZP basis sets for geometry optimizations[§] and TZD2P bases for single-point energy evaluations[**].

Two catalysts (**II,** R=Me and Bu^t) spanning the product distribution from predominantly short oligomers to long-chain PE were selected for the computational investigation. The active center was assumed to be an alkyl chromium cation of the type $[(R_2PC_2H_4C_5H_4)CrPr]^+$ generated by reaction of the chromium dichloride precursor with methylalumoxane (MAO) followed by one ethylene insertion. For group 4 metallocene-based catalysts, the corresponding metallocene alkyl cations have been identified as the active species [23]. Identification of the active species of Cr(III)-based catalysts has not yet been reported but there are a number of indications [22] that the active species have a related structure, viz. Cp(donor)CrR$^+$ (where Cp is a (substituted) cyclopentadienyl ligand).

The preferred routes of propagation and termination as indicated by the calculations are depicted in Figure 2 (see footnote[††] for labels) and the associated relative energies of the intermediates and transition states are listed in Table 2.

[‡] Local part from Vosko *et al.*[28], exchange corrections from Becke [29], correlation corrections from Perdew and Wang [30]. Extensive studies [31, 32] show that the BPW91 functional is capable of providing accurate energy profiles for the monomer insertion step during metal-catalyzed olefin polymerization. All calculations were done at the unrestricted level (UBPW91).

[§] A contracted [10s,8p,3d] basis was applied for Cr [31]. P, N, C and H were described by standard Dunning-Hay [33] valence double-ζ basis sets, with a scale factor of 1.2 (1.15) applied for the inner (outer) exponents of H (exception: Bu^t-methyl groups represented by STO-3G sets [34]). Polarization functions were added to C atoms (α_d=0.75) being part of the ethylene or the polymer chain as well as to P and N (α_d=0.60 and 0.80, respectively). The Gaussian 98 [35] defaults were applied for integration grid and convergence criteria. For each stationary point, the curvature of the surface was verified through analytical calculation of the second derivative matrix. Zero-point and thermal corrections were computed from the harmonic frequencies using standard procedures.

[**] The basis sets defined above[§] were improved for the single-point calculations as follows: C and H atoms forming part of the ethylene or the polymer chain were described by augmented Dunning triple-ζ sets denoted TZD2P [32] to account for known basis set sensitivities [32], the Bu^t-methyl groups were represented by the standard double-ζ basis [33], and polarization functions[§] were included for the C atoms of the Cp rings.

[††] A minimum is labeled by a number, followed by a Greek letter indicating the presence of a metal–H agostic interaction. Transition states are identified by a bracket containing the reactant and product separated by an arrow, followed by a suffix "‡". The molecular term (^{4}A or ^{2}A) is also given.

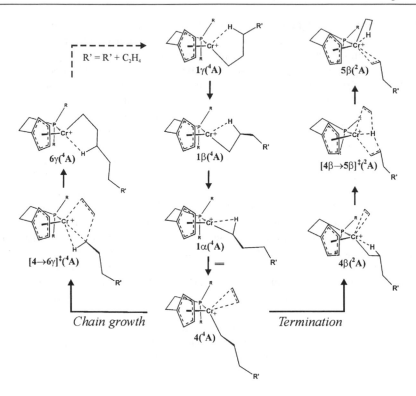

Fig. 2. The preferred routes for propagation and termination from DFT calculations.[‡‡] **1γ(⁴A)** and **6γ(⁴A)** differ by an inversion at the metal center due to chain-migratory insertion: another insertion (or an intermediate chain migration (back-skip)) is needed to complete the propagation cycle. This is indicated by the dashed arrow (top left). Elimination of the olefinic chain in **5β(²A)** completes the termination sequence.

The latter are of direct relevance to the chain-length of the products: the activation free energies ($\Delta G^{\ddagger}_{298}$) of insertion ($[4\rightarrow6\gamma]^{\ddagger}(^4A)$) and termination through β-hydrogen transfer (BHT, $[4\rightarrow5\beta]^{\ddagger}(^2A)$) are calculated to be almost equal (±1 kcal/mol) for R=Me, whereas termination is more than 10 kcal/mol less favorable for R=Bu^i.[§§]

[‡‡] The reported energies were obtained for the conformations shown in Figs. 2 and 3. Another set of conformers, with very similar energies, can be generated by rotation around the C–C bond of the ethano bridge to give negative Cp-C-C-P torsion angles.

[§§] A competing chain termination process is initiated by transfer of a β-hydrogen atom from the chain to the metal (β-hydrogen elimination, BHE) to form a metal(olefin)–hydride complex. The subsequent elimination of the olefin is associated with an entropic barrier and is the rate-determining step of the BHE termination. For R=Me, even the separated elimination products are of higher free energy than the barriers to both ethylene insertion and BHT and the latter reaction is clearly the preferred route to chain termination. For R=Bu^i, the BHE barrier ($\Delta G^{\ddagger}_{298} \approx 25$ kcal/mol) is actually lower than that of BHT, but is still significantly higher than that of chain propagation.

In other words, the rates of insertion and termination are predicted to be of the same order for R=Me with the resulting product being dominated by short-chain oligomers, whereas for the Bu^t-analogue, the rate of termination should be several orders of magnitude lower than that of insertion, and therefore predominantly long-chain PE is formed.

Table 2. Calculated enthalpies and free energies for the species shown in Figure 2, relative to ethylene + $[(R_2PC_2H_4C_5H_4)CrPr]^+$ containing a β-agostic H atom ($1\beta(^4A)$).[a]

R	Me		Bu^t	
Species[††]	ΔH_{298} (kcal/mol)	ΔG_{298} (kcal/mol)	ΔH_{298} (kcal/mol)	ΔG_{298} (kcal/mol)
$1\gamma(^4A)$ [b]	7.4	7.5	8.8	8.4
$1\beta(^4A)$ [b]	0.0	0.0	0.0	0.0
$1\alpha(^4A)$ [b]	8.5	7.2	6.4	5.7
$[1\alpha{\rightarrow}4]^{\ddagger}(^4A)$ [c]	8.4	13.0	9.3	17.8
$4(^4A)$	-2.5	8.7	4.0	15.5
$[4{\rightarrow}6\gamma]^{\ddagger}(^4A)$	4.0	16.5	8.9	21.6
$4\beta(^2A)$	-3.7	11.1	11.4	26.5
$[4\beta{\rightarrow}5\beta]^{\ddagger}(^2A)$	1.4	16.8	19.2	34.6

[a] From single-point total energies[**] at optimized geometries[§], with zero-point energies and temperature corrections included[§].

[b] Interconversion between different agostic conformers of **1** is facile ($\Delta G^{\ddagger}_{298}$<11 kcal/mol for R=Me).

[c] Ethylene coordination to β or γ-agostic complexes is less favorable. A very shallow minimum assignable to a charge-induced dipole metal–ethylene complex is located prior to $[1\alpha{\rightarrow}4]^{\ddagger}(^4A)$. The thermochemical data for this complex have been omitted since they are not important for the discussion.

These numerical results can be rationalized by considering the electronic structure of chromium. In the alkene-free chromium–alkyl cations (**1**), the Cr-atom has 15 valence electrons; the agostic hydrogen atom is counted here as a dative ligand. However, as a consequence of the presence of three unpaired electrons (high-spin, ^4A), the total number of doubly and singly occupied valence orbitals (9) is the same as that for an 18-electron complex which would be expected to be stable. In many respects these high-spin 15-electron complexes can therefore be regarded as electronically saturated [22]. The high-spin species $1\beta(^4A)$ has the lowest calculated free energy and thus represents the 'resting-state'. Formation of the high-spin metal–alkene complex ($4(^4A)$) requires the breaking of the agostic metal–H bond interaction which is accomplished in a low-barrier, near-thermoneutral process. Chain propagation also proceeds on the high-spin surface, the corresponding TS of insertion ($[4{\rightarrow}6\gamma]^{\ddagger}(^4A)$) being significantly lower in energy than its ^2A counterpart. However, no TS of termination through BHT could be located on the high-spin surface since it is energetically unfavorable for the metal to coordinate the additional migrating hydrogen atom in the electronically saturated high-spin state. Spin-pairing in **4** generates a low-lying, empty d orbital at the metal which can mediate the hydrogen transfer on the low-spin surface, starting from a 17-electron metal–ethylene complex with a strong β-agostic interaction (**4β(²A)**) and leading to the BHT product (**5β(²A)**). The spin-inversion can only be efficient if the two spin states are energetically close which is therefore an essential prerequisite for a facile termination reaction. These mechanistic findings support the view that two-state reactivity may indeed be a 'key feature in organometallic chemistry' [24].

The electronically facile BHT-termination reaction can be suppressed by the introduction of larger substituents on phosphorus. The six-membered metallacycle constituting the central, reacting part of the transition state (TS) of the BHT-step requires more space than the corresponding four-membered metallacycle of the TS of insertion; these metallacycles are both arranged essentially perpendicular to the equatorial Cp-Cr-P plane (Figs. 2 and 3). An effective strategy to obtain higher oligomers would be to selectively block the axial positions (with respect to the equatorial Cp-Cr-P plane) in the coordination sphere of the Cr-atom, thereby destabilizing **[4β→5β]‡(²A)** with respect to **[4→6γ]‡(⁴A)** as has also been discussed for the (bisimine)Ni(II) catalysts [25, 26]. In our case, the distorted tetrahedral configuration of the phosphorus atom ensures that the P–R bonds are directed out of the equatorial Cp-Cr-P plane causing the substituents to block the axial positions and forcing the two olefin fragments closer together to give a more compact TS.

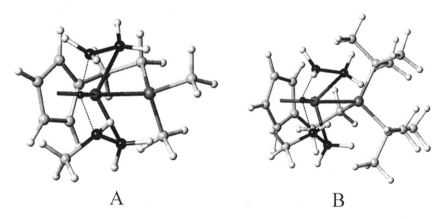

A   B

Fig. 3. The two DFT-optimized transition states (A: R=Me, B: R=Buⁱ) of hydrogen transfer to ethylene, **[4β→5β]‡(²A)**. The six-membered metallacycle constituting the central, reacting part of the TS is indicated by black color for alkene C and transferring H and dashed C–H bonds for the latter atom.‡‡.

Larger steric hindrance in the axial coordination positions can also be obtained with the related amino-substituted derivatives (e.g. **I**) for which only polymer formation is observed [8]: the N-atom is smaller than the P-atom and consequently the chromium–donor and donor–substituent separations are smaller and the associated Tolman cone-angle larger [11]. The BHT transition state in the case of the NMe₂ donor fragment is thus calculated to be more compact than that with PMe₂ and as a result the barrier for BHT ($\Delta G^{\ddagger}_{298}$=28.0 kcal/mol) is higher than that for propagation ($\Delta G^{\ddagger}_{298}$=17.1 kcal/mol). Similarly, increasing the size of the donor atom will reduce the steric hindrance which explains the observed increase in oligomer formation in the presence of the As-containing system ($Cy_2AsC_2H_4C_5H_4$)$CrCl_2$–MAO compared to the analogous P-containing system [10].

Finally it should be noted that blocking axial coordination positions in complexes such as **III** and **IV** is more difficult since the nitrogen–aryl bonds are located in the equatorial N-M-N plane and steric hindrance has to be achieved by introducing bulky substituents in the more distant (with respect to the metal) ortho-aryl positions. More importantly, the conformational flexibility arising from rotation around the nitrogen–aryl bonds presumably causes these substituents to be less selective in blocking the axial positions [26].

The results presented here indicate that the observed and calculated trend in catalytic activity has a steric origin: bulkier substituents destabilize the structures **4–6** (and thus also the TS of insertion) to a larger extent than the monomer-free resting state **1β(^{4}A)**. Therefore the activity is expected to decrease upon increasing the size of the substituents on the P-atom which is in contrast to the trend reported for the (bisimine)Ni-catalysts (**III**) [25], where the metal–alkene complex is suggested to constitute the resting state [12, 25–27]. The negative effect of steric crowding upon activity is also consistent with the observation that the P-containing catalysts are apparently more active than the analogous N-containing systems (calculated propagation barrier for R=Me, $\Delta G^{\ddagger}_{298}$=16.5 (P) and 17.1 (N) kcal/mol).

Acknowledgment

Financial support was provided by Borealis Polymers Oy and computer resources by the Rechenzentrum der Gesellschaft für wissenschaftliche Datenverarbeitung in Göttingen and the Max Planck Institute for Biological Cybernetics in Tübingen.

References

1   Gavens PD, Bottrill M, Kelland JW, McMeeking J (1982) Compr Organometal Chem 3: 475
2   McDaniel MP (1985) Adv Catalysis 33: 47
3   Wilke G, Bogdanovic B, Hardt P, Heimbach P, Keim W, Kröner M, Oberkirch W, Tanaka K, Steinrücke E, Walter D, Zimmermann H (1966) Angew Chem 78: 157
4   Resconi L, Cavallo L, Fait A, Piemontesi F (2000) Chem Rev 100: 1253
5   McKnight AL, Waymouth RM (1998) Chem Rev 98: 2587
6   Gibson VC, Newton C, Redshaw C, Solan GA, White AJP, Williams DJ (1999) J Chem Soc-Dalton Trans: 827
7   Kim WK, Fevola MJ, Liable-Sands LM, Rheingold AL, Theopold KH (1998) Organometallics 17: 4541
8   Döhring A, Göhre J, Jolly PW, Kryger B, Rust J, Verhovnik GPJ (2000) Organometallics 19: 388
9   Butenschön H (2000) Chem Rev 100: 1527
10 Weber JC (1999), Ph.D. thesis, Ruhr-Universität Bochum

11   Tolman CA (1977) Chem Rev 77: 313
12   Killian CM, Johnson LK, Brookhart M (1997) Organometallics 16: 2005
13   Bennett AMA (1999) Chemtech 29: 24
14   Britovsek GJP, Bruce M, Gibson VC, Kimberley BS, Maddox PJ, Mastroianni S,
     McTavish SJ, Redshaw C, Solan GA, Strömberg S, White AJP, Williams DJ (1999) J
     Am Chem Soc 121: 8728
15   Small BL, Brookhart M (1998) J Am Chem Soc 120: 7143
16   Bogdanovic B (1979) Adv Organometal Chem 17: 105
17   Vogt D (1996) Appl Homogen Cat Organometal Cpds 1: 245
18   Younkin TR, Connor EF, Henderson JI, Friedrich SK, Grubbs RH, Bansleben DA
     (2000) Science 287: 460
19   Margl P, Deng LQ, Ziegler T (1999) Organometallics 18: 5701
20   Musaev DG, Morokuma K (1999) Top Catal 7: 107
21   Lohrenz JCW, Bühl M, Weber M, Thiel W (1999) J Organomet Chem 529: 11
22   Jensen VR, Angermund K, Jolly PW, Børve KJ (2000) Organometallics 19: 403
23   Jordan RF, Dasher WE, Echols SF (1986) J Am Chem Soc 108: 1718
24   Schröder D, Shaik S, Schwarz H (2000) Acc Chem Res 33: 139
25   Johnson LK, Killian CM, Brookhart M (1995) J Am Chem Soc 117: 6414
26   Deng LQ, Woo TK, Cavallo L, Margl PM, Ziegler T (1997) J Am Chem Soc 119: 6177
27   Froese RDJ, Musaev DG, Morokuma K (1998) J Am Chem Soc 120: 1581
28   Vosko SH, Wilk L, Nusair M (1980) Can J Phys 58: 1200
29   Becke AD (1988) Phys Rev A 38: 3098
30   Perdew JP, Wang Y (1992) Phys Rev B 45: 13244
31   Jensen VR, Børve KJ (1997) Organometallics 16: 2514
32   Jensen VR, Børve KJ (1998) J Comput Chem 19: 947
33   Dunning TH, Hay PJ (1977) In: Schaefer HF (ed) Methods of Electronic Structure
     Theory. Plenum Press, New York p 1
34   Hehre WJ, Stewart RF, Pople JA (1969) J Chem Phys 51: 2657
35   Frisch MJ, Trucks GW, Schlegel HB, Scuseria GE, Robb MA, Cheeseman JR,
     Zakrzewski VG, Montgomery JA, Jr., Stratmann RE, Burant JC, Dapprich S, Millam
     JM, Daniels AD, Kudin KN, Strain MC, Farkas O, Tomasi J, Barone V, Cossi M,
     Cammi R, Mennucci B, Pomelli C, Adamo C, Clifford S, Ochterski J, Petersson GA,
     Ayala PY, Cui Q, Morokuma K, Malick DK, Rabuck AD, Raghavachari K, Foresman
     JB, Cioslowski J, Ortiz JV, Baboul AG, Stefanov BB, Liu G, Liashenko A, Piskorz P,
     Komaromi I, Gomperts R, Martin RL, Fox DJ, Keith T, Al-Laham MA, Peng CY,
     Nanayakkara A, Gonzalez C, Challacombe M, Gill PMW, Johnson B, Chen W, Wong
     MW, Andres JL, Gonzalez C, Head-Gordon M, Replogle ES, Pople JA (1998)
     Gaussian 98, Revision A.7, Gaussian, Inc., Pittsburgh PA

1-Hexene Polymerization Initiated by α-Diimine [N,N] Nickel Dibromide / MAO Catalytic Systems : Influence of the α-Diimine Ligand Structure

Séverine Gomont, Sylvain Boissière, Alain Deffieux, Henri Cramail*

Laboratoire de Chimie des Polymères Organiques UMR n° 5629 – ENSCPB – CNRS – Université Bordeaux 1, Avenue Pey-Berland, BP 108, 33402 Talence Cedex, France
E-mail: cramail@enscpb.u-bordeaux.fr

Abstract. 1-Hexene polymerization has been investigated, in cyclohexane at 20°C, in the presence of (R-Phenyl)$_2$ α-diimine[N,N] nickel dibromide catalysts bearing different alkyl groups (R= Me, iPr, tBu) on the phenyl ligand. Methylaluminoxane (MAO) was used as an activator. Kinetic and UV/visible spectroscopic studies show the strong influence of the phenyl alkyl substituents of the diimine ligands onto the polymerization process as well as on the polyhexenes characteristics.

Introduction

Ethylene and α-olefins polymerization initiated by late-transition-metal complexes (Ni-, Pd-, Co- and Fe-based) is the subject of intense research activity. The higher tolerance towards polar monomers of such catalytic systems compared to early-transition-metal derivatives and their ability to yield polyolefins with tailored chain architecture make them very attractive [1].

Among the well-defined catalytic systems recently developed, square planar nickel(II) derivatives bearing substituted α-diimine-[N,N] ligands (scheme 1), first described by Brookhart and coll. [2], have retained our attention for the present study.

We recently investigated 1-hexene polymerization with **1**/methylaluminoxane (MAO) system, focusing on the elementary reactions involved in the formation of polymerization active species [3]. Polymerization kinetics and UV/visible spectroscopy of the catalytic system, allowed us to show the strong and very unusual dependence of this system on the monomer concentration. Results were interpreted by the contribution of two propagating active species of different reactivity and present in various proportion according to the monomer concentration.

In this paper, a series of α-diimine[N,N] nickel dibromide derivatives bearing different alkyl groups (Me, iPr, tBu) on the aromatic ring are compared as polymerization catalysts for 1-hexene polymerization.

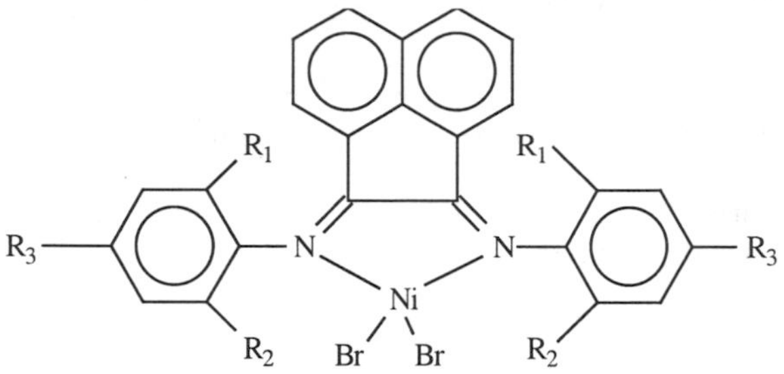

$1 : R_1 = tBu ; R_2 = R_3 = H$
$2 : R_1 = R_2 = iPr, R_3 = H$
$3 : R_1 = R_2 = R_3 = Me$

Scheme 1.

Results and discussion

Influence of the ligands on polymerization kinetics

1-Hexene polymerization in the presence of **1**, **2** or **3** activated by MAO was investigated in cyclohexane at 20°C. Kinetics of 1-hexene polymerization were followed by dilatometry.

The effect of the Al/Ni ratio, with respect to nickel, on the initial catalytic activity was first investigated for the three systems. Results are presented figure 1.

Upon the addition of increasing amounts of MAO, the initial activity first increases sharply before it reaches a plateau of maximal activity for an Al/Ni ratio (r) between 100 and 200, depending on the structure of the substituents of the diimine ligand. The presence of a plateau of activity suggests that the amount of MAO used is sufficient to complete the formation of active species. The nature of the alkyl group on the diimine ligand has a strong influence on the catalytic activity. Indeed, the 3/MAO system is four-five times more active than the two other catalytic systems. The smaller size of methyl groups compared to iPr or tBu substituents may lead to less hindered and thus more reactive active species. A higher efficiency of the MAO activation process with complex **3** can also explain the different activities observed. This point will be further discussed in the UV/visible part.

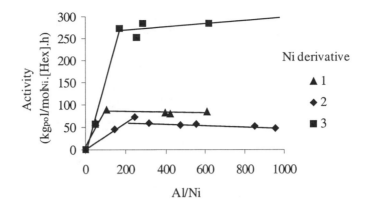

Fig. 1. Influence of [MAO]/[Ni] ratio on 1-hexene polymerization initiated by (R-Ph)$_2$ α-diimine NiBr$_2$ / MAO systems -(R=tBu (◆), R=iPr (▲), R=Me (■)) in cyclohexane at 20°C. [Ni] = 0.9×10^{-3} M; [1-hexene] = 0.83 M

The kinetic order of the polymerization with respect to MAO was determined. MAO concentration was varied from 0.045 to 0.87 M while the nickel procatalysts and 1-hexene initial concentrations were kept constant, respectively to 0.9×10^{-3} and 0.83 M. The conditions used correspond to Al/Ni ratios ranging from 50 to about 1000. Ln (R_{po}) versus ln [MAO] plot is given figure 2.

As could be expected from the variation of the catalytic activity with the Al/Ni ratio (Figure 1), for each catalytic system the initial polymerization rate follows successively two different rate laws. For Al/Ni ratios lower than r, Rp_0 increases sharply with the MAO concentration in agreement with an increase of the active species concentration. For Al/Ni ratios higher than r, the kinetic order with respect to MAO is equal to zero, in agreement both with the complete reaction of the nickel procatalyst with MAO and with a negligible role of MAO in excess on the active species formed.

The kinetic order with respect to the different nickel procatalysts was determined by the same approach. The MAO and 1-hexene initial concentrations were kept constant, respectively to 0.38 and 0.83 M, while the nickel procatalyst concentration was varied from 4.66×10^{-4} to 7.20×10^{-3} M which corresponds to Al/Ni ratios ranging from 50 to 800 (see Figure 3).

With the three Ni derivatives, the kinetic order with respect to nickel is fractional and close to 0.7. This suggests the formation of aggregated species, inactive towards 1-hexene polymerization, in equilibrium with active monomeric ones. Aggregation could be explained by the electrophilic character of the Ni active species and the non-polar character of the polymerization medium. Indeed, similar kinetic measurements undertaken with **1**/MAO in solvent of higher polarity or complexing ability, such as toluene or chlorobenzene, have shown an unitary nickel kinetic order [3].

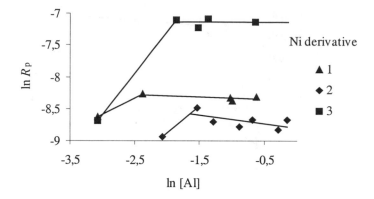

Fig. 2. Logarithmic variations of 1-hexene polymerization initial rate versus MAO concentration ; (R-Ph)$_2$ α-diimine NiBr$_2$ / MAO systems -(R=tBu ($\blacklozenge$), R=iPr ($\blacktriangle$), R=Me ($\blacksquare$)) in cyclohexane at 20°C. [Ni] = 0.9×10^{-3} M; [1-hexene] = 0.83 M ; 0.045 < [MAO] < 0.87 M ; 50 < Al/Ni <980

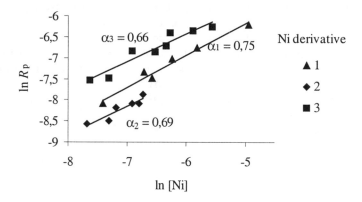

Fig. 3. Logarithmic variations of the 1-Hexene polymerization rate versus nickel procatalyst concentration ; (R-Ph)$_2$ α-diimine NiBr$_2$ / MAO systems -(R=tBu ($\blacklozenge$), R=iPr ($\blacktriangle$), R=Me ($\blacksquare$))in cyclohexane at 20°C. [MAO] = 0.38 M; [1-hexene] = 0.83 M; 4.66×10^{-4} < [Ni] < 7.20×10^{-3}M; 50 < Al/Ni < 980.

At last, the polymerization kinetic order with respect to 1-hexene was determined following the same procedure (see Figure 4). Nickel procatalysts and MAO concentrations were kept constant, respectively to 0.9×10^{-3} M and 0.23 M.

This corresponds to an Al/Ni ratio equal to 260 ensuring the complete formation of active species (plateau of activity). 1-Hexene concentration was varied from 0.3 to 2.28 M. As can be seen, the kinetic order with respect to 1-hexene is very dependent on the phenyl alkyl substituent of the diimine ligands. Indeed, the more active system, 3/MAO, leads to a negative monomer kinetic order whereas 1/MAO and 2/MAO catalytic systems exhibit a zero or slightly positive monomer kinetic order, respectively. An interpretation of these data will be proposed in the general discussion.

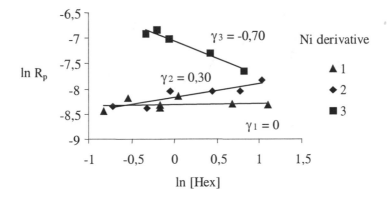

Fig. 4. Logarithmic variations of the 1-hexene polymerization rate versus 1-hexene concentration ; (R-Ph)$_2$ α-diimine NiBr$_2$ / MAO systems -(R=tBu (◆), R=iPr (▲), R=Me (■)) in cyclohexane at 20°C. [MAO] = 0.23 M; [Ni] = 0.9×10^3 M ; Al/Ni = 260; 0.3 < [1-hexene] < 2.28 M

Influence of the ligands on the UV/visible spectra

UV/visible spectroscopy is a very useful method to follow the activation process of zirconocenes by methylaluminoxane [4-7]. Recently, this spectroscopic method was successively applied to nickel/MAO systems. In toluene and chlorobenzene, it was shown that the monomer plays a determining stabilizing role on the 1/MAO catalytic species. This study allowed to discriminate between active species characterized by an absorption band centered around 520 nm and inactive ones, located at 710 nm [3].

In order to correlate with kinetic data, an UV/visible study of the three catalytic systems was performed in cyclohexane at 20°C. UV/visible spectra at the initial stage of the 1-hexene polymerization are given Figure 5. As may be seen, the relative intensity of the absorption bands centered at 520 and 710 nm respectively, is again very much dependent on the phenyl alkyl substituent of the diimine ligand.

In agreement with kinetics, the more active system (**3**/MAO) exhibits a main absorption band at 520 nm corresponding to active species whereas the 710 nm band attributed to inactive species is of much lower intensity. Contrarily, in the cases of **1**/MAO and **2**/MAO systems, the absorption band of inactive species (710 nm) is relatively much more intense than the one at 510-520 nm. Furthermore, the band corresponding to active species (510-520 nm) in these two systems is of comparable intensity in agreement with the close polymerization activity observed for the two catalytic systems **1** and **2**.

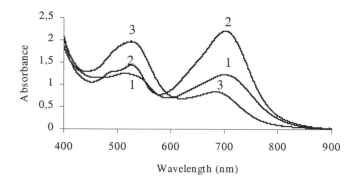

Fig. 5. UV/visible absorption spectrum of **1**, **2** and **3** activated by MAO after 5 min of polymerization. [Ni] = 0.9×10^3 M; [1-hexene] = 0.83 M and Al/Ni = 260.

Unlike **1**/MAO system, an additional absorption band centered at 490 nm is observed, in the cases of **2**/MAO and **3**/MAO catalytic systems. The band centered at 490 nm is not specific of the latter systems since UV/visible studies carried out with **1**/MAO in the presence of other monomer (styrene) was also observed [8]. The exact identification of the different species formed is currently under investigation.

The UV/visible spectra recorded at 20°C for the various systems remain unchanged during most of 1-hexene polymerization, in agreement with stable active species. Contrarily, a fast deactivation of species issued from **1**/MAO system was previously observed at 20°C in toluene and chlorobenzene. The stronger ability of active species to aggregate in cyclohexane could explain their better stability with time. When the polymerizations are nearly complete (after > 80% conversion), a deactivation of the three catalytic systems is however observed as indicated by the transformation, through an isosbestic point, of species located at 510-520 nm into the ones at 710 nm. This change with time of the UV-visible spectrum confirms the previous observations dealing with the stabilizing role of the monomer onto active species.

Influence of the ligands on polyhexene chains characteristics.

Brookhart and co-workers already reported the living character of α-olefins polymerization in the presence of similar catalytic systems at low temperature (-10°C).

In this study, the livingness of 1-hexene polymerization was checked by measuring the molar masses of the polyhexenes obtained at various $[Monomer]_0/[Ni]_0$ ratios. The experimental molar masses of polyhexenes were determined by size exclusion chromatography using both a polystyrene calibration (RI detection) and light scattering detection. The experimental molar masses of polyhexenes versus $[Hex]_{consumed}/[Ni]_0$ ratios are reported Figure 6 where they can be compared with theoretical values determined assuming one chain formed by initial Ni complex.

As already reported [3], for [Hex]/[Ni] lower than 1000, the experimental molar masses increase almost linearly with the [Hex]/[Ni] ratio whereas the molar mass distribution remains lower than 1.3. This suggests an almost controlled character of the 1-hexene polymerization. However, experimental molar masses higher than the theoretical ones indicate that only a fraction of the α-diimine[N,N] Ni derivatives yields active species. This might result either from an incomplete activation of the Ni compound or from their rapid deactivation at the very beginning of the reaction. A comparison of the experimental molar masses obtained with the three catalytic systems suggests that the efficiency of the 3/MAO system is higher than the two other catalytic systems. This observation is in agreement with UV-visible data, which show the higher proportion of active species formed in this system (ratio of 520 nm to 710 nm bands).

For high [Hex]/[Ni] ratio, however, experimental Mn values level off and molar mass distribution increases in agreement with the occurrence of chain transfer reactions, as already reported [3]. It is worth noting that chain transfer is higher with 3/MAO system, showing again the important role of the diimine ligand structure onto the stabilization of active species.

The polymerization of 1-hexene in the presence of α-diimine[N,N] nickel dibromide / MAO catalytic systems yields polyhexene chains with ethylene sequences due to migration of the metal through the so-called "walking" mechanism (see Scheme 2). We have previously shown that the monomer concentration is an important parameter that affects the number of butyl branches of the polyhexene chains (168 butyl branches/1000 C for a normal polyhexene).

We specifically analyzed the effect of diimine ligand structure onto the rate of metal migration along the chain, responsible to the formation of ethylene sequences in place of normal hexene units. The rate of butyl branches versus the monomer concentration, for each catalytic system, is given in figure 7.

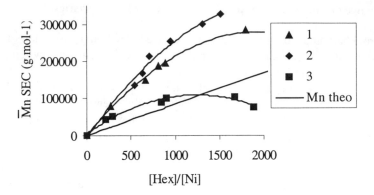

Fig. 6. Experimental polyhexenes $\overline{\text{M}}_n$ versus [Hex-1-ene] / [Ni] ratios.

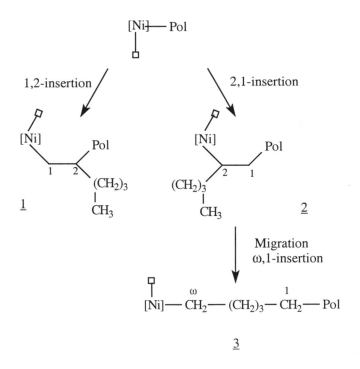

Scheme 2.

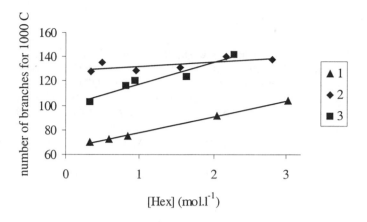

Fig. 7. Effect of the initial 1-hexene concentration on the number of branches in cyclohexane at 20°C.

As already described by Brookhart, the more hindered catalytic system, **1**, eases the metal migration and gives higher ratio of ethylene sequences, i.e. the smaller amount of butyl branches. Another important feature that can be seen on figure 7 is the influence of 1-hexene concentration on the number of branches. Whereas with **1** and **3**, the number of branches increases with hexene concentration, it remains constant with **2**. In a previous paper, the presence of two types of active species (species 1 and 2) having different polymerization rate was proposed. This proposal is in agreement with kinetic data reported in this study : in the case of catalytic systems **1** and **3**, the null or negative monomer kinetic order suggests that at lower hexene concentration, 2,1-insertion mode accompanied by metal migration along the butyl branch ("walking" mechanism) is favored, leading to Met-CH_2-R active species, 3 of higher reactivity. On the contrary, in the case of **2**, the slightly positive kinetic order with respect to hex-1-ene suggests that the metal migration process is much less sensitive to the monomer concentration as confirmed by data of Figure 7.

Conclusion

This study points out that substituents on phenyl rings of the diimine ligands play first a determining role on the activation processes and on the stabilization of active species. They also affect the frequency of metal migration, the more hindered substituent (t-butyl) yielding higher ethylene sequences.

Further work dealing with the polymerization of styrene in the presence of similar catalytic systems will be published in a forthcoming paper.

References

1   Ittel, S.D.; Johnson, L.K.; Brookhart, M. *Chem. Rev.* **2000**, *100*, 1169
2   Johnson, L.K.; Killian, C.M.; Brookhart, M. *J. Am. Chem. Soc.* **1995**, *117*, 6414
3   Peruch, F.; Cramail, H.; Deffieux, A. *Macromolecules* **1999**, *32*, 7977
4   Coevoet, D.; Cramail, H.; Deffieux, A. *Macromol. Chem. Phys.* **1998**, *199*, 1451
5   Coevoet, D.; Cramail, H.; Deffieux, A. *Macromol. Chem. Phys.* **1998**, *199*, 1459
6   Coevoet, D.; Cramail, H.; Deffieux, A. *Macromol. Chem. Phys.* **1998**, *200*, 1208
7   Pedeutour, J.N., Coevoet, D.; Cramail, H.; Deffieux, A. *Macromol. Chem. Phys.* **1999**, *200*, 1215.
8   Cramail, H. ; Deffieux, A.; Gomont, S.; Peruch, P. *unpublished results*

1,3,5-Triazacyclohexane Complexes of Chromium as Homogeneous Model Systems for the Phillips Catalyst

Randolf D. Köhn[a], Guido Seifert[a], Gabriele Kociok-Köhn[a], Shahram Mihan[b], Dieter Lilge[b], Heiko Maas[b]

[a]Dept. of Chemistry, University of Bath, Bath, BA2 7AY, UK, E-mail: r.d.kohn@bath.ac.uk
[b]BASF AG, Research Laboratories, Departments of Polyolefins and Oligomers, ZKP/E – M 505, 67056 Ludwigshafen, Germany, E-mail: shahram.mihan@basf-ag.de

Abstract. 1,3,5-triazacyclohexane complexes **2** prepared from trizazacyclohexanes **1** and $CrCl_3$ can be activated by MAO or $AlR_3/PhNMe_2H(BPh^F_4)$ to give solutions that can polymerise and/or trimerise ethylene depending on the N-substituents R with unprecedented high activities. α-olefins are selectively trimerised or co-polymerised with ethylene. We have varied these substituents R in symmetrical and asymmetrical R_2R'-triazacyclohexanes including some with different functional groups for a better understanding of this dependence. A detailed study of the activities and the polymer structures shows that these systems are very good models for the Phillips catalyst. The homogeneous reactions can be studied by several spectroscopic methods especially for the α-olefin trimerisation. The results show that the triazacyclohexane stays co-ordinated during the catalysis and that mono-nuclear metallacyclic complexes are likely involved.

Over the past 50 years, several transition metal systems have been found that can catalyse the polymerisation and oligomerisation of olefins. Heterogeneous Ziegler-Natta systems based on early transition metals are the most successful and produce a large variety of polyolefins. Homogeneous single site model systems, foremost the metallocenes, have been developed which are able to produce highly stereo regular polymers and have become useful industrial catalysts. The study of these homogeneous systems has tremendously improved the detailed understanding of the mechanism. Chain propagation is largely based on olefin insertion into metal-C or H bonds followed by chain transfer via β-H elimination. This general mechanism is often termed the hydride mechanism and seems to be common among other transition metal systems as well.

Late transition metal systems often have a higher tendency for chain transfer versus propagation that leads to good dimerisation or oligomerisation catalysts. One highly successful system is the Nickel based SHOP catalyst for the ethylene oligomerisation. Blocking the chain transfer pathway with sterically demanding groups in similar systems has lead to an exciting development of highly active homogeneous polymerisation catalysts based on late transition metals such as Fe, Co, Ni, and Pd. Again, a hydride mechanism appears to be the mechanism.

The heterogeneous Phillips catalysts [1] based on CrO_3/SiO_2 for the polymerisation of ethylene are known for nearly 50 years and still produce a large fraction of the world production of HDPE (>7 million t/a)[2]. This system has many unusual features compared to the other transition metal catalysts. First of all, it is a purely inorganic system that does not require any metal alkyl co-catalyst and can be activated by ethylene itself although activation can also be achieved with other reducing reagents such as CO or aluminium alkyls. Contrary to most catalysts based on the hydride mechanism, the molecular weight of the polymer is quite insensitive to hydrogen but can be regulated by the reaction temperature. End group analysis of the polymer shows not only the expected single methyl and vinyl end groups but additional methyl groups, vinylidene and some internal olefin end groups. Such end groups have been found in systems based on the hydride mechanism when additional isomerisation, mis-insertion or chain-walking steps are involved. However, these groups are found in Phillips systems in consistently similar ratios.

When activated with metal alkyls the Phillips catalyst can also oligomerise ethylene to α-olefins that can be in-situ co-polymerised giving polymer with side chains. However, these oligomers do not follow the Flory-Schultz distribution typical for systems based on the hydride mechanism. There is a high selectivity for the trimer of ethylene, 1-hexene, and subsequent co-polymers with butyl side chains. Co-polymers can also be obtained directly by adding α-olefins.

Contrary to the early and late transition metal systems, the nature of the active species and the mechanism in the chromium systems is still a matter of debate. This is largely due to the fact that the Phillips catalyst is very difficult to study. Highly active catalysts are obtained only at high dilution of chromium on the silica surface (< 1 w%) and the chemistry of the surface chromium is very rich with many different species at various oxidation states and nuclearities. However, only a small fraction of the total chromium is known to be active. Thus, surface analytical methods mostly detect inactive compounds. Generally, a mononuclear chromium compound in the oxidation state +III or maybe +II directly bound to the silica via two or three oxygen atoms is believed to be the active species (Figure 1).

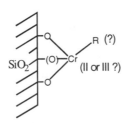

Fig. 1. Possible co-ordination environment of the active site in the Phillips catalyst

Many homogeneous model systems [3] show only limited activity [4] and most of them fail to reproduce the properties of the Phillips catalyst and a true model has yet to be found. As an alternative chromium system, the Union Carbide catalysts based on chromocene on silica has been introduced 30 years ago which is more accessible to studies. One cyclopentadienyl appears to stay attached to chromium and various mono-cyclopentadienyl chromium complexes show polymerisation activity. However, these cyclopentadienyl systems are quite different from the Phillips catalyst. They generally do not co-polymerise ethylene and α-olefins, the molecular weight of the polymer shows high hydrogen response, the polymers do not have the same end group distribution and are mostly linear and there is no selectivity for trimerisation.

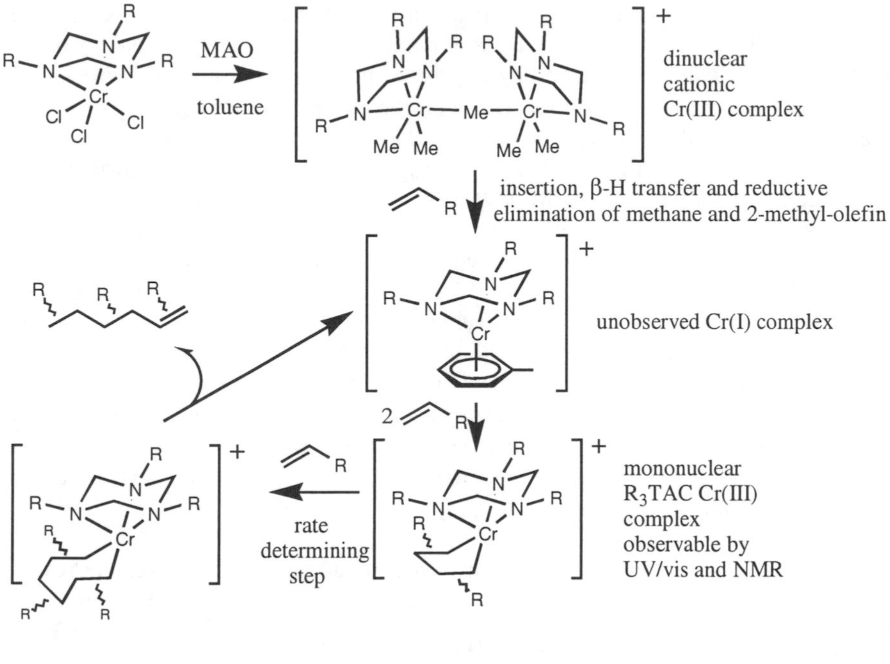

Scheme 1. Proposed mechanism for the selective trimerisation

In an interesting development over the last 20 years, homogeneous chromium systems have been found that can trimerise ethylene with >90% selectivity to 1-hexene. These systems generally consist of some soluble chromium complex, aluminium alkyls and some amine, mostly pyrroles [5]. However, little is known about the active species in these complex mixtures largely due to the difficulty of obtaining useful NMR spectra of these highly paramagnetic compounds. Briggs [6] and others have proposed a mechanism via metallacycles analogous to Scheme 1 which has gained some support by the results of Jolly [7] by showing that complex **3** (Scheme 4) can react with ethylene under reducing conditions (activated

Mg) to a metallacyclopentane complex and that the larger metallacycloheptane complex readily decomposes under reductive elimination to the trimer 1-hexene.

Apparently, a reduced chromium species and oxidative addition of two olefins to a metallacyclopentane is involved. Only after insertion of a third olefin, resulting in a metallacycloheptane, reductive elimination becomes possible. Thus, oligomerisation is selectively terminated after three olefins. This means that a new mechanism for the C-C bond formation between olefins is involved. A similar mechanism has recently been found for $Cp^*TiMe_3/B(C_6F_5)_3$ in the polymerisation of ethylene and in the co-polymerisation with styrene [8].

For several years we have been investigating the co-ordination chemistry of N-substituted 1,3,5-triazacyclohexanes 1 [9]. The three hard donor nitrogen atoms can facially co-ordinate metals and may model the two to three oxygen atoms believed to be bound to chromium in the Phillips catalyst. The few investigations on larger triazacycloalkanes, mainly triazacyclononanes, indicate that the N-substituted ligands are sterically too demanding to allow high activity for the polymerisation of ethylene or even α-olefins. Our results on the co-ordination chemistry of 1 show that complexes with low steric demand by the ligand due to the small N-metal-N angle of 60-65° can be formed. This small angle and the mis-directed nitrogen lone pairs lead to relatively weak M-N bonds. This can been observed by UV/vis spectroscopy and X-ray crystallography of the $CrCl_3$ complexes 2. The observed 10Dq value of 14,000 cm^{-1} is much lower than expected for (amine)$_3CrCl_3$ (17,000 cm^{-1}) and the Cr-N distance of 210-214 pm is longer than expected for Cr(III)-N(amine) (200 pm). However, these values indicate a comparable if not stronger bond than in the sterically weakened bonds in N-substituted triazacyclononane complexes. Table 1 lists some distances in various complexes 2 in comparison to an analogous triazacyclononane complex. The structural parameters of the co-ordination environment in 2 does not vary with the bulk of the N-substituent R whereas steric repulsion increases the Cr-N and Cr-Cl distances in the triazacyclononane complex.

Table 1. Structural parameter of 2 and an analogous triazacyclononane complex of $CrCl_3$

N-substituent R	angle N-Cr-N	distance Cr-N	distance Cr-Cl
4-pentynyl [10]	65.7(2)	210(1)	228(1)
n-octyl [11]	65.8(1)	210.5(11)	228.7(4)
$CH_2CMe_2CH_2CH=CH_2$ [12]	66.0(1)	211.1(4)	228.8(17)
S-CHMePh [12]	65.4(3)	213(2)	228.5(4)
CH_2Ph [12]	65.8(4)	211(2)	227.8(2)
Bu (triazacyclononane) [13]	82.4(4)	214.9(6)	232.4(6)

Results and Discussion

We have been able to characterise a few organometallic chromium(III) complexes with 1 and, in the case of zinc, even cationic alkyl complexes have been synthesised. Thus, it should be possible to generate cationic alkyl complexes of chromium which may be able to catalyse the ethylene polymerisation. Indeed, complex 2 (R=Me) reacts with methylaluminoxane (MAO) to give a highly active catalyst. However, good solubility of the complexes is crucial for achieving the high productivities. This solubility problem is solved by introducing longer alkyl chains as N-substituents in **2**. The solubility in toluene increases dramatically with octyl and dodecyl substituents [14]. Complexes **2** can be prepared by reacting **1** with $CrCl_3(THF)_3$ or more conveniently from $CrCl_3$ (stored under air), **1**, toluene and zinc powder by simple heating under a stream of argon.

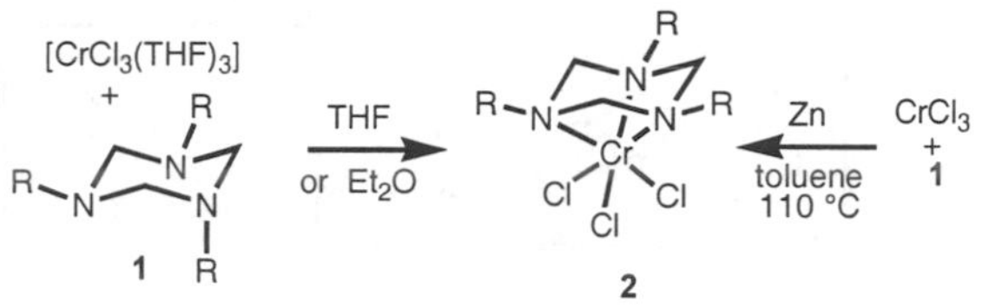

Scheme 2. Syntheses of the complexes **2**

Since the ligands **1** can also be prepared by heating the corresponding primary amine and paraformaldehyd in toluene, a simple one-pot synthesis of **2** is possible. The complexes **2** are generally inert towards water and air and can be cleaned by chromatography on silica.

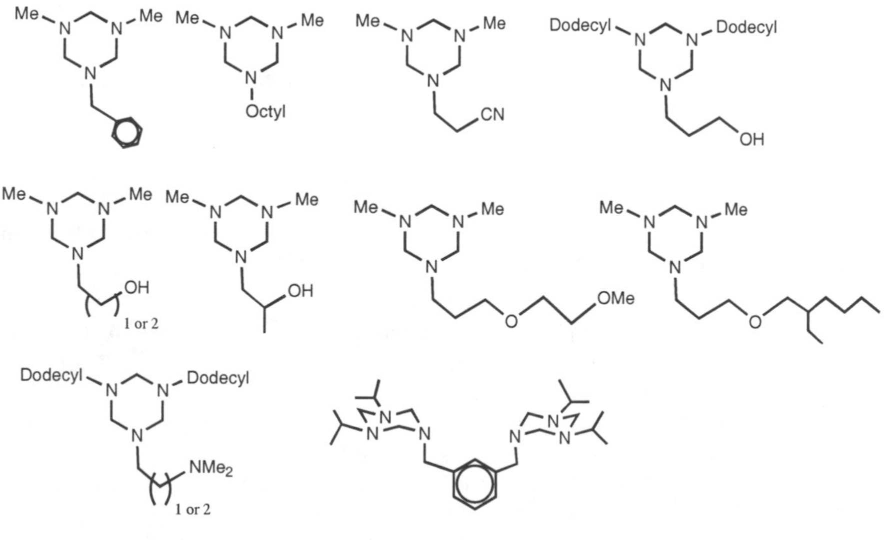

Scheme 3. Selected asymmetrically substituted triazacyclohexanes **1**.

By using two different amines in the synthesis of **1**, a mixture of asymmetrically substituted triazacyclohexanes can be obtained. If they are not easily separable by distillation or crystallisation they can be converted to a mixture of the chromium complexes and separated by chromatography afterwards. This allows not only the synthesis of symmetrically substituted triazacyclohexane complexes but also a large variety of asymmetrically substituted ones including some with functional groups such as ethers, nitriles, amines (Scheme 3).

The systems **2**/MAO have much higher activity than the best co-catalyst free complex [15] and the best non-Cp-system [16]. The activity is comparable to [(nBuCp)$_2$ZrCl$_2$]/MAO under the same condition (650 Kg molCr^{-1}h^{-1}) and the only more active chromium system is **3**/MAO from Jolly [4b] (Scheme 4).

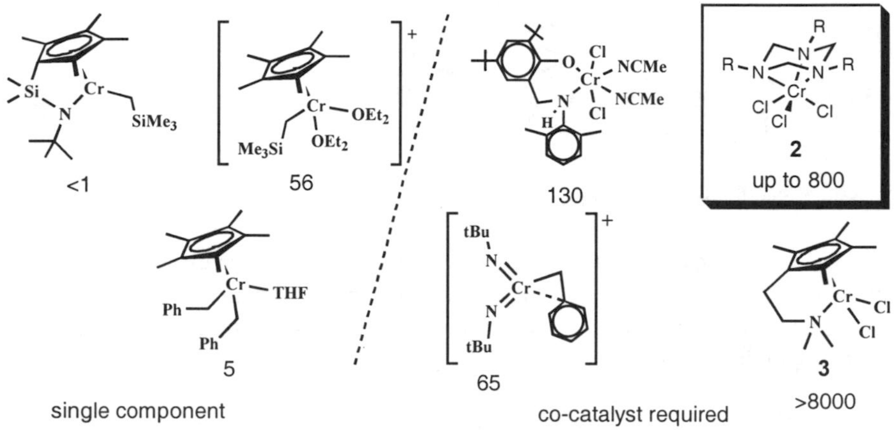

Scheme 4. Selected chromium complexes for ethylene polymerisation with their activity in [4a] in [kg molCr^{-1}h^{-1}] (for **2** at 40°C and 1 bar ethylene in toluene over 1 h)

Maximum activity for MAO as co-catalyst is reached at Al:Cr ≈ 300. However, activation of **2** can also be achieved with 1.6 eq. DMAB (dimethylanilinium tetrakis(pentafluorophenyl)borate) and 20-50 eq. Al(iBu)$_3$ giving similar activity. The activity depends strongly upon the N-substituents R. Table 2 gives the activities for selected complexes with different ligands **1**. Similar to the Phillips catalyst, molecular weights of polyethylene produced by activated **2** are highly dependent on the polymerisation temperature.

Generally, the activity improves with the higher solubility of the long chain substituents, decreases with increasing steric bulk or stronger donor functionalities. The molecular weights of the polymers under these conditions are around 40,000 (M_w) with M_w/M_n=2-4 which is typical for a single-site catalyst.

End group analysis of the polymers (IR, ^{13}C-NMR) shows more methyl groups than expected and additional vinylidene and some internal olefin besides the expected vinyl end groups in a distribution that is typical for end groups produced by the Phillips catalyst. Since the molecular weights of the polymers differ, the end groups are best compared relative to the total number of olefinic groups (set to 100%) (Table 3).

Table 2. Activities in [kgmolCr^{-1}h^{-1}] of complexes **2** with R´R$_2$-N-substituted triazacyclohexanes

N-substituent	Activity
R=R´=Me	455
R=R´=octyl	490
R=R´=dodecyl	717
R=R´=cyclohexyl	5
R=R´=t-butyl	39
R=R´=1,1-Me$_2$-dodecyl	3
R=R´=CH$_2$CH$_2$Ph	800
R=R´=CH$_2$Ph	220
R=R´=s-CHMePh	135
R´=(CH$_2$)$_2$NMe$_2$, R=dodecyl	185
R´=(CH$_2$)$_3$NMe$_2$, R=dodecyl	165
R´=dodecyl, R=(CH$_2$)$_3$NMe$_2$	17
R´=(CH$_2$)$_2$CN, R=ethyl	91
R´=(CH$_2$)$_2$OH, R=dodecyl (-HCl)	9

Table 3. End group distribution relative to the sum of olefinic groups set to 100%.

end group	CH$_2$=CHR	CH$_2$=CR$_2$	RHC=CHR	Me
PE (Phillips)[a]	84-92%	7-13%	1-4%	150-300%
PE (**2**, R=dodecyl)	82%	12%	6%	240%
decenes(**2**, R=dodecyl)	87%	8%	5%	200%

[a] Typical range from industrial polymers produced under various conditions.

Additionally, 1-hexene as the trimer of ethylene and some decenes as "co-trimers" of 1-hexene and ethylene can be found in the solution and butyl side chains in the polyethylene are indicative of some 1-hexene built into the polymer. α-Olefins react with the catalyst system nearly exclusively to the corresponding trimers [11]. The homogeneous catalysis of this α-olefin trimerisation can be observed by UV/vis and NMR spectroscopy. A single mononuclear chromium(III) complex is observed during the catalysis, probably a metallacyclopentane complex. ^{2}H NMR studies with a deuterated ligand proves that the triazacyclohexane is co-ordinated to the paramagnetic chromium in the active complex. The kinetic data can be fit to rate laws derived from the metallacyclic mechanism outlined in scheme 1.

Thus, our system is also able to reproduce the selectivity for trimerisation. Interestingly, analysis of the decene isomers by NMR shows an end group distribution similar to that in the polymers (Table 3). Addition of 1-hexene to the solution increases the content of butyl side chains substantially and a co-polymer can be obtained. All observed isomers of the "co-trimers" can be explained by the metallacyclic mechanism according to scheme 5. A similar scheme for the homo-trimerisation of α-olefins also yields all observed isomers and supports the formation of these products by the metallacyclic mechanism.

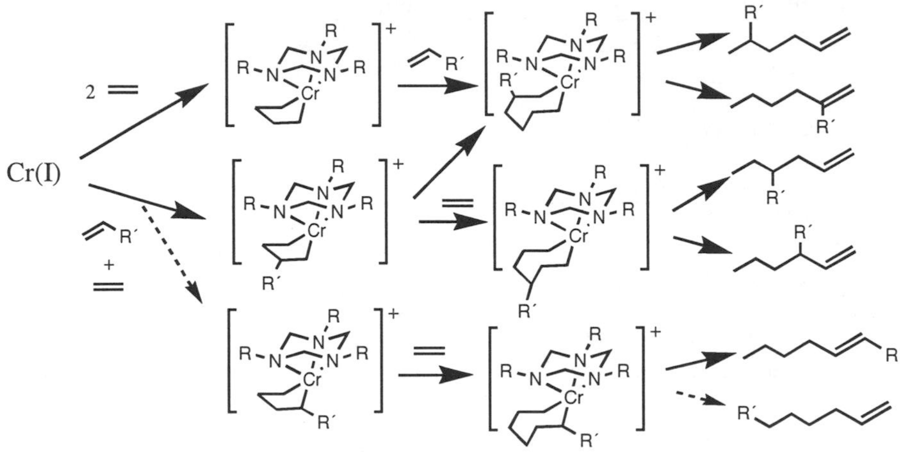

Scheme 5. Origin of the end groups in the product of "co-trimerisation" of RCH=CH₂ and two ethylene.

The observed end groups of the polymer could be due to several steps of insertions, eliminations and re-insertions in 1,2 and 2,1 fashion. However, the qualitative and quantitative similarity of the end group distribution between the polymer and the "co-trimer" suggests that the same mechanism via metallacycles is involved with R´ being the polymer chain – at least for the end group formation.

Thus, activated **2** represents the first true homogeneous model for the Phillips catalyst that can reproduce many important properties and the results indicate that the typical end group distribution may be closely linked to a required trimerisation activity.

R.D. Köhn thanks the Deutschen Forschungsgemeinschaft and the Fonds der Chemischen Industrie as well as Prof. H. Schumann (TU Berlin) for support. The authors thank BASF AG for encouragement and support.

References

1   J. P. Hogan, R. L. Banks, Phillips Petroleum, US 2,825,721 (1958); CA = 52:8621h; M. P. McDaniel, *Adv. Catal.*, 1985, **33**, 47; C. E. Marsden, *Plast. Rubber Compos. Process. Appl.*, 1994, **21**, 193; T. E. Nowlin, *Prog. Polym. Sci.*, 1985, **11**, 29; S. M. Augustine, J. P. Blitz, *J. of Catalysis*, 1996, **161**, 641.

2   M. Rätzsch, Polymerwerkstoffe '98 Merseburg, 23.-25. September 1998; *Kunststoffe*, 1996, **86**, 6; R. Messere, A.F. Noels, P. Dournel, N. Zandona, J. Breulet, *Proceedings of Metallocenes '96*, 6-7 März 1996, Düsseldorf, 309.

3   K. H. Theopold, *Eur. J. Inorg. Chem.*, 1998, 15.
4   a) G. J. P. Britovsek, V. C. Gibson, D. F. Wass, *Angew. Chem.*, 1999, **111**, 448; *Angew. Chem., Int. Ed. Engl.*, 1999, **38**, 428; b) A. Döhring, J. Göhre,1 P. W. Jolly, B. Kryger, J. Rust, and G. P. J. Verhovnik, *Organometallics*, 2000, **19**, 388.
5   M. E. Lashier (Phillips Petroleum Company) *EP 0780353* (1995).
6   J.R.Briggs, *J. Chem. Soc., Chem. Commun.*, 1989, 674.
7   R. Emrich, O. Heinemann, P. W. Jolly, C. Krüger, G. P. J. Verhovnik, *Organometallics*, 1997, **16**, 1511.
8   C. Pellecchia, D. Pappalardo, L. Oliva, M. Mazzeo, G.-J. Gruter, *Macromolcules*, 2000, **33**, 2807.
9   a) M. Haufe, R. D. Köhn, G. Kociok-Köhn, A. C. Filippou, *Inorg. Chem. Commun.*, 1998, **1**, 263; b) R. D. Köhn, G. Kociok-Köhn, *Angew. Chem.*, 1994, **106**, 1958; *Angew. Chem., Int. Ed. Engl.*, 1994, **33**, 1877; c) R. D. Köhn, G. Kociok-Köhn, M. Haufe, *J. Organomet. Chem.*, 1995, **501**, 303; d) M. Haufe, R. D. Köhn, R. Weimann, G. Seifert, D. Zeigan, *J. Organometal. Chem.*, 1996, **520**, 121.
10   M. V. Baker, D. H. Brown, B. W. Skelton, A. H. White, *J. Chem. Soc., Dalton Trans.*, 2000, 763.
11   R. D. Köhn, M. Haufe, G. Kociok-Köhn, S. Grimm, P. Wasserscheid, W. Keim, *J.Chem.Soc., Chem. Commun.*, submitted.
12   R. D. Köhn, M. Haufe, G. Kociok-Köhn, unpublished results.
13   S.-J. Wu, G. P. Stahly, F. R. Fronczek, S. F. Watkins, *Acta Crystallogr.*, 1995, **C51**, 18.
14   R. D. Köhn, M. Haufe, S. Mihan, D. Lilge, *J.Chem.Soc.,Chem. Commun.*, in press.
15   P. A. White, J. Calabrese, K. H. Theopold, *Organometallics*, 1996, **15**, 5473.
16   V. C. Gibson, C. Newton, C. Redshaw, G. A. Solan, A. J. P. White, D. J. Williams, *J. Chem. Soc., Dalton Trans.*, 1999, 827.

Rare Earth Half-Sandwich Catalysts for the Homo- and Copolymerization of Ethylene and Styrene

Jun Okuda*, Stefan Arndt, Klaus Beckerle, Kai C. Hultzsch, Peter Voth, and Thomas P. Spaniol

Institut für Anorganische Chemie und Analytische Chemie, Johannes Gutenberg-Universität Mainz, Duesbergweg 10-14, 55099 Mainz, Germany
E-mail: okuda@mail.uni-mainz.de

Abstract: The synthesis of rare earth metal half-sandwich hydrido complexes [Ln(η^5:η^1-C$_5$Me$_4$SiMe$_2$NCMe$_3$)(THF)(μ-H)]$_2$ (Ln = Y, Lu, Yb, Er, Tb) through σ-bond metathesis of the alkyl complexes [Ln(η^5:η^1-C$_5$Me$_4$SiMe$_2$NCMe$_3$)(CH$_2$SiMe$_3$)(THF)], easily accessible by the reaction of the amino-cyclopentadiene with [Ln(CH$_2$SiMe$_3$)$_3$(THF)$_2$], was developed. The dimeric lanthanide hydrido complexes are highly fluxional involving THF dissociation and cis–trans isomerization of the linked amido–cyclopentadienyl ligand. The presence of a monomer–dimer equilibrium is suggested by cross-over experiments. They were tested as single-site, single-component catalysts for the polymerization of ethylene, α-olefin, and styrene, as well as alkyl acrylate and acrylonitrile. The hydrido complexes polymerize ethylene slowly, whereas they form isolable mono(insertion) products with α-olefins and with styrene. The yttrium n-alkyl complexes [Y(η^5:η^1-C$_5$Me$_4$SiMe$_2$NCMe$_3$)(R)(THF)$_n$] (R = (CH$_2$)$_n$CH$_3$, n = 3–9), prepared by the mono(insertion) of α-olefins, initiate polymerization of styrene in a relatively controlled manner. Thus, styrene was polymerized by the in-situ formed n-hexyl complex to give atactic polystyrenes with narrow molecular weight distributions and somewhat enriched syndiotacticities. Sequential addition of *tert*-butyl acrylate allows the synthesis of poly(styrene-*block-tert*-butyl acrylate).

Introduction

Structurally well-characterized lanthanocene hydrides and alkyls of the type [Ln(η^5-C$_5$R$_5$)$_2$X]$_2$ (X = H, alkyl) have recently been attracting interest as cocatalyst-free, homogeneous polymerization catalysts for both nonpolar and polar monomers [1]. Exceedingly high activity for ethylene polymerization by [Ln(η^5-C$_5$Me$_5$)$_2$H]$_2$ (Ln = La, Nd, Lu) [2a] as well as living polymerization (Ln = Sc) were reported [2b], following earlier observation of the polymerization activity for some organolanthanide complexes [3]. More recently, the syndiospecific alkyl methacrylate polymerization by decamethylsamarocene derivatives [Sm(η^5-

C$_5$Me$_5$)$_2$X]$_2$ [4] began to offer an attractive possibility for the synthesis of new polymethacrylate materials, including novel AB diblock copolymers consisting, e.g., of nonpolar and polar segments [5]. The use of organolanthanides such as [Nd(η^5-C$_5$Me$_5$)$_2$Cl$_2$Li(Et$_2$O)$_2$] as effective chain-transfer agents for the inexpensive magnesium dialkyl MgR$_2$ also led to a more practical method of synthesizing related materials [6]. These lanthanocene derivatives active for polymerization may be considered as the isoelectronic analogs of the group 4 metallocene alkyl cations [M(η^5-C$_5$R$_5$)$_2$X]$^+$, with 14 valence electrons, that is, without the presence of weakly bonding counter-anions, thus avoiding complications caused by ion-pairing effects and ill-defined cocatalysts such as methylaluminoxane (MAO) [7].

Catalyst systems based on rare earth metal complexes with only one cyclopentadienyl ligand are expected to be even more active towards sterically demanding, functionalized monomers such as styrene or acrylates. Conventional synthesis of mono(cyclopentadienyl) rare earth complexes [Ln(η^5-C$_5$R$_5$)X$_2$(L)$_n$], where at least one X ligand is a hydride or an alkyl, is often hampered by ate-complex formation due to the high Lewis acidity and electrophilicity of the rare earth metal center [8]. A further complication lies in the uncontrolled formation of the thermodynamically more stable metallocene derivatives [9]. Thus, σ-bond metathesis of an appropriate cyclopentadiene derivative, in particular of a linked amino–cyclopentadiene (C$_5$R$_4$H)SiMe$_2$NHR' [10], and a lanthanide alkyl appeared to be a highly attractive synthetic pathway for the preparation of lanthanide complexes with only one supporting ring ligand [11].

Results and Discussion

Synthesis of Alkyl and Hydrido Complexes

To circumvent the problems associated with salt metathesis, we successfully employed the lanthanoid tris(alkyl) complex [Ln(CH$_2$SiMe$_3$)$_3$(THF)$_2$] (Ln = Y, Lu, Yb, Er, Tb), which allows the facile isolation of the linked amido–cyclopentadienyl complex [Ln(η^5:η^1-C$_5$Me$_4$SiMe$_2$NCMe$_3$)(CH$_2$SiMe$_3$)(THF)] as well as the bis(alkyl) complexes [Y(η^5-C$_5$Me$_4$SiMe$_2$X')(CH$_2$SiMe$_3$)$_2$(THF)] (X' = Me, Ph, C$_6$F$_5$) under very mild conditions (Scheme 1) [11].

These alkyl complexes are extremely sensitive and cannot be stored for prolonged periods of time [12a]. The yttrium complex [Y(η^5:η^1-C$_5$Me$_4$SiMe$_2$NCMe$_3$)(CH$_2$SiMe$_3$)(THF)] decomposes in hydrocarbon solution with a half-life of $t_{1/2}$ = 10 h. The choice of easily accessible Ln(CH$_2$SiMe$_3$)$_3$(THF)$_2$ [13] which is significantly more reactive than [Ln{CH(SiMe$_3$)$_2$}$_3$] is a trade-off against the problematic presence of THF [12b]. Analogous reactions of [Ln{CH(SiMe$_3$)$_2$}$_3$] (Ln = Yb, Lu) with (C$_5$Me$_4$H)SiMe$_2$NHCMe$_3$ to give Lewis-base-free complexes [Ln(η^5:η^1-C$_5$Me$_4$SiMe$_2$NCMe$_3$){CH(SiMe$_3$)$_2$}] require rather forcing conditions [12c]. The THF ligand in [Y(η^5:η^1-C$_5$Me$_4$SiMe$_2$NCMe$_3$)(CH$_2$SiMe$_3$)(THF)] is labile ($\Delta G^\ddagger$(223 K) = 40.5 kJ mol^{-1}) and is engaged in a dissociative equilibrium with the

12-electron species $[Y(\eta^5:\eta^1\text{-}C_5Me_4SiMe_2NCMe_3)(CH_2SiMe_3)]$. When this process is compared to that of the lutetium complex, the THF ligand is clearly less labile $(\Delta G^{\ddagger}(293\ K) = 62.3\ \text{kJ mol}^{-1})$ in the lutetium complex, because of its smaller size (*Shannon* ionic radii for coordination number 8: Lu, 0.977; Y, 1.019 Å).

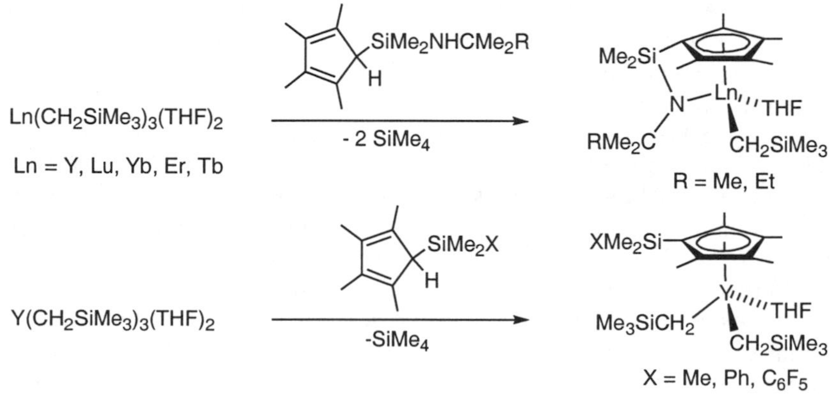

Scheme 1.

Hydrogenolysis of the linked amido–cyclopentadienyl alkyl complexes with dihydrogen or phenylsilane cleanly leads to the dimeric hydrido complex $[Ln(\eta^5:\eta^1\text{-}C_5Me_4SiMe_2NCMe_3)(THF)(\mu\text{-}H)]_2$ (Scheme 2).

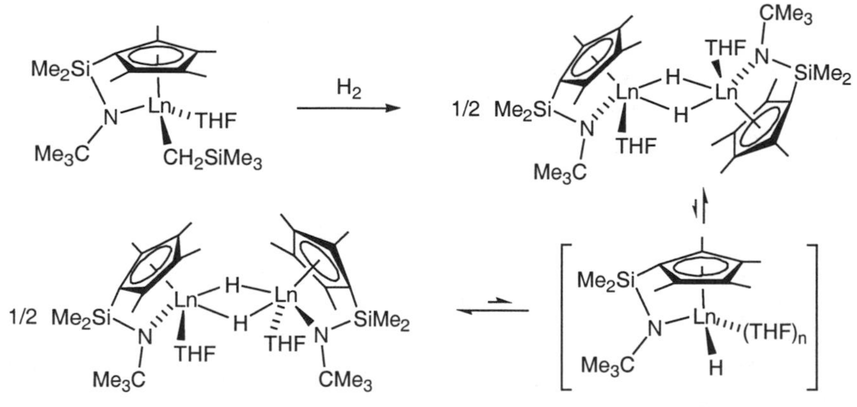

Scheme 2.

The diamagnetic hydrido complexes of yttrium and lutetium were fully characterized by X-ray diffraction in the solid state and by variable-temperature NMR spectroscopy in solution. They are highly fluxional with respect to THF dissociation, monomer–dimer equilibrium, and rotation about the metal–metal axis.

Two diastereomers, which differ in the relative configurations of the two metal centers within a square pyramidal coordination sphere, can be detected at lower temperatures. Since some of the exchange processes detectable by NMR spectroscopy do not necessarily require the dissociation of the dimeric unit, we were concerned to obtain evidence that dissociation into mononuclear units $[Ln(\eta^5:\eta^1-C_5Me_4SiMe_2NCMe_3)(H)(THF)_n]$ does occur in solution.

The presence of monomeric units can be inferred from various scrambling reactions [12b,14]. Thus, mixing a benzene solution of $[Y(\eta^5:\eta^1-C_5Me_4SiMe_2N-CMe_3)(THF)(\mu-H)]_2$ with that of the lutetium analog at room temperature instantaneously gave a 1:2:1 statistical mixture containing the homo- and hetero-metallic complexes $[Ln(\eta^5:\eta^1-C_5Me_4SiMe_2NCMe_3)(THF)(\mu-H)]_2$ (Ln = Y, Lu) and $[YLu(\eta^5:\eta^1-C_5Me_4SiMe_2NCMe_3)_2(THF)_2(\mu-H)_2]$. In a double cross-over experiment, $[Y(\eta^5:\eta^1-C_5Me_4SiMe_2NCMe_3)(THF)(\mu-D)]_2$ was mixed with $[Y(\eta^5:\eta^1-C_5Me_4SiMe_2NCMe_2Et)(THF)(\mu-H)]_2$ to give a statistical mixture with the deuterium distributed over all three complexes. Similar scrambling reactions between lanthanocenes are significantly slower [15].

Olefin Insertion and Polymerization

The hydrido complex $[Y(\eta^5:\eta^1-C_5Me_4SiMe_2NCMe_3)(THF)(\mu-H)]_2$ smoothly reacts with a variety of olefinic substrates (Scheme 3). Ethylene undergoes sufficiently slow sequential insertion at low temperatures ($<-30\,°C$) to give a mixture of n-alkyl complexes $[Y(\eta^5:\eta^1-C_5Me_4SiMe_2NCMe_3)(R)(THF)_n]$, some of which can be independently synthesized by the insertion reaction of α-olefin with the dimeric hydride. Initially it was somewhat surprising that the crystals isolated from the reactions of $[Y(\eta^5:\eta^1-C_5Me_4SiMe_2NCMe_3)(THF)(\mu-H)]_2$ and α-olefins were THF-free. However, single-crystal structural analyses of a series of n-alkyl complexes $[Y(\eta^5:\eta^1-C_5Me_4SiMe_2NCMe_3)(\mu-R)]_2$ (R = n-butyl, n-pentyl, n-hexyl) revealed the presence of β-agostic interactions which appears to alleviate the high Lewis acidity of the trivalent rare earth metal center. Initially, it was also puzzling that these THF-free n-alkyl complexes showed no activity toward polymerization of α-olefins or styrene. Few related structures of n-alkyl complexes have been reported in the literature [16]. In particular, it is interesting to note that the scandium n-propyl complex $[Sc(\eta^5:\eta^1-C_5Me_4SiMe_2NCMe_3)(\mu-CH_2CH_2CH_3)]_2$, described earlier by Bercaw et al [16a], is capable of initiating living oligomerization of propylene and of higher α-olefins, since it is dissociated into the monomeric 12-electron alkyl species $[Sc(\eta^5:\eta^1-C_5Me_4SiMe_2NCMe_3)(\mu-CH_2CH_2CH_3)]$. However, the THF-free dimeric alkyl complexes of yttrium dissolve in THF to give the extremely sensitive and reactive monomeric n-alkyl complex $[Y(\eta^5:\eta^1-C_5Me_4SiMe_2NCMe_3)(R)(THF)_n]$.

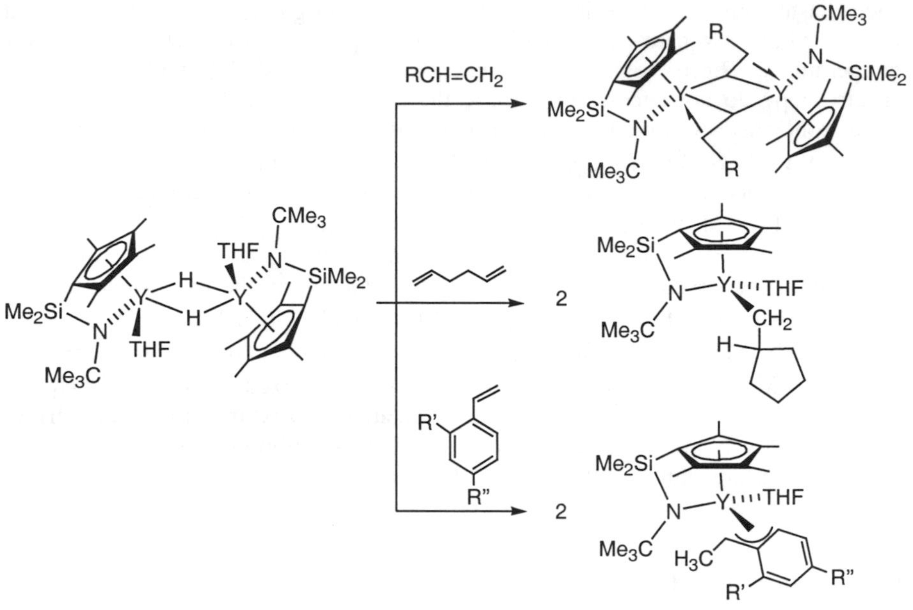

Scheme 3.

Whereas ethylene is slowly polymerized by the hydrido complex at room temperature to give linear polyethylene with $T_m = 136\ °C$ and an activity of <1 kg polymer/mol Y•bar•h, polymerization of α-olefins, dienes, or styrene does not occur, as stable mono(insertion) products form in these cases (Scheme 3). α-Olefins cleanly give the above-mentioned *n*-alkyl complexes, either as isolable dimers or monomeric THF adducts. 1,5-Hexadiene reacts with the hydride complex by cyclization to give the monomeric cyclopentylmethyl complex $[Y(\eta^5:\eta^1\text{-}C_5Me_4SiMe_2NCMe_3)\{CH_2CH(CH_2)_4\}(THF)]$.

Most interestingly, styrenes with not more than one *ortho*-substituent afford bright yellow 1-phenethyl complexes. No further reaction is observed with excess styrene. The color indicated charge transfer between the π-electrons of the phenyl ring and the yttrium center. An X-ray diffraction study of $[Y(\eta^5:\eta^1\text{-}C_5Me_4SiMe_2NCMe_2Et)(CHMeC_6H_4{}^tBu\text{-}4)(THF)]$ confirmed that the insertion occurred in a 2,1- or "secondary" fashion and that the phenyl ring is engaged in η^3-coordination. Variable-temperature NMR spectroscopy further revealed fluxional behavior that includes THF dissociation, phenyl ring coordination, and rotation about the *ipso*- and α-carbon bond. The yttrium atom becomes chiral at low temperatures ($T < -10\ °C$) as a result of the tight bonding of the THF ligand on the NMR timescale. However, even at −80 °C only one diastereomeric pair (R_Y,R_C and S_Y,S_C) was observed (Scheme 4). These results are relevant to the mechanism of syndiospecific styrene polymerization by cationic mono(cyclopentadienyl)-titanium(III) complexes, discovered by Ishihara et al. at Idemitsu [17]. Here an intermediate with 2,1-inserted last monomer and an η^4-coordinated styrene [Ti(η^5-C_5R_5){CH(CH$_2$P)C$_6$H$_5$}(η^5-H$_2$C=CHC$_6$H$_5$)]$^+$ (**P**: growing syndiotactic polystyrene chain) was postulated by Zambelli et al. [18].

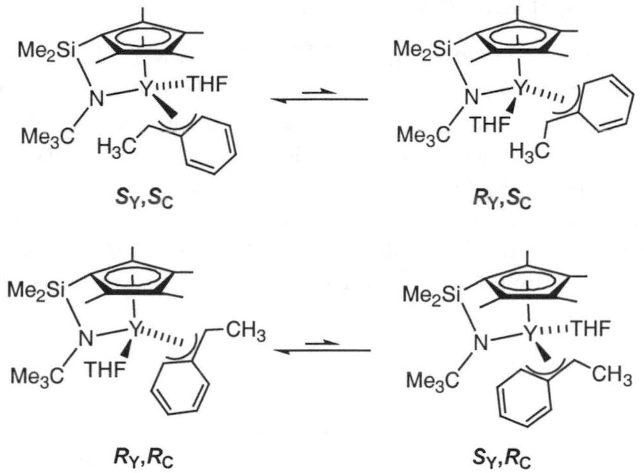

Scheme 4.

The polymerization of styrene, however, cannot be initiated efficiently by the hydrido complex or the styrene insertion product, probably because the THF present in the system cannot be displaced by styrene. The yttrium n-alkyl complex [Y(η^5:η^1-C$_5$Me$_4$SiMe$_2$NCMe$_3$)(R)(THF)$_n$] (but *not* the THF-free dimer) polymerizes styrene in a controlled manner to give atactic polystyrene with low polydispersity (Table 1). Thus, 50 equivalents of styrene give polystyrene with $M_n = 24,100$ and $M_w/M_n = 1.10$; $mm = 0$, $mr = 29.5$, $rr = 70.5\%$. The degree of syndiospecificity is significantly lower than that observed when styrene is polymerized by mono(cyclopentadienyl)titanium(III) complexes (typically $rr > 97\%$). The polystyrenes obtained are soluble in acetone and show no melting temperature according to DSC analysis.

Table 1. Polymerization of styrene with [Y(η^5:η^1-C$_5$Me$_4$SiMe$_2$NCMe$_3$)(n-hexyl)(THF)$_n$].[a]

run no.	t (h)	T (°C)	[M$_n$]/[Y]	yield (%)	M_n	M_w	M_w/M_n
1	24	25	50	100	24100	26600	1.10
2	40	25	115	99	34300	38400	1.12
3	72	25	240	90	61300	75400	1.23
4	24	50	130	90	30900	4300	1.39
5	24	75	130	100	8000	15300	1.91

[a] Polymerization conditions: 19 µmol of yttrium complex, 1.5 mL of benzene.

This polymerization is not strictly living, but sufficiently controlled (linear increase of molecular weights with conversion and monomer to yttrium ratio) to allow AB block copolymerization without considerable increase in molecular weight distributions. Thus, poly(styrene-*block*-*tert*-butyl acrylate) can be prepared by sequential addition of styrene and *tert*-butyl acrylate.

The yttrium hydrido complex $[Y(\eta^5:\eta^1\text{-}C_5Me_4SiMe_2NCMe_3)(THF)(\mu\text{-}H)]_2$ as well as the alkyl complex $[Y(\eta^5:\eta^1\text{-}C_5Me_4SiMe_2NCMe_3)(CH_2SiMe_3)(THF)]$ polymerize the polar monomers *tert*-butyl acrylate and acrylonitrile (Scheme 5). *tert*-Butyl acrylate is polymerized at temperatures as low as $-30\,°C$ (i.e., well below the decomposition temperature) to give poly(*tert*-butyl acrylate) in high yields and with molecular weights $M_n > 20,000$. The molecular weight distributions of the resulting polymers are in the range of $M_w/M_n = 1.5$–2.0 and the polymer microstructure as determined by ^{13}C NMR spectroscopy is predominantly atactic (Table 2).

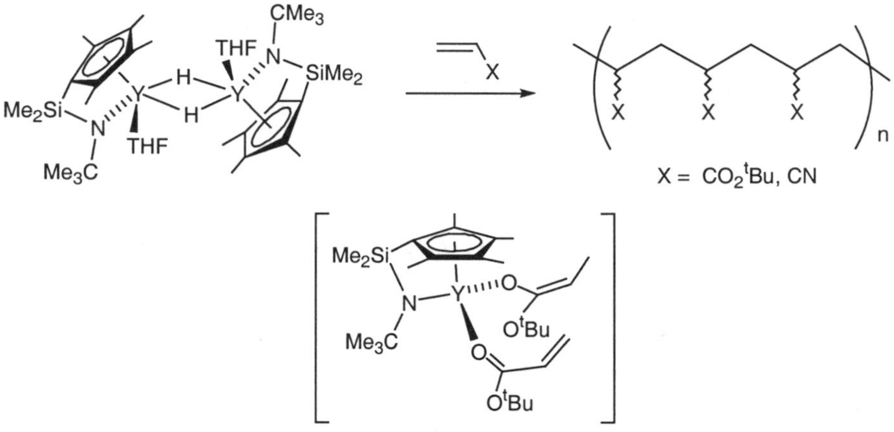

Scheme 5.

Table 2. Polymerization of *tert*-butyl acrylate by $[Y(\eta^5:\eta^1\text{-}C_5Me_4SiMe_2NCMe_3)\text{-}(CH_2SiMe_3)(THF)]$ (runs 1–3) and $[Y(\eta^5:\eta^1\text{-}C_5Me_4SiMe_2NCMe_3)(THF)(\mu\text{-}H)]_2$ (run 4).[a]

run no.	t (h)	T (°C)	$[M_0]/[Y]$	yield (%)	M_n	M_w	M_w/M_n
1	2	25	95	85	24900	38900	1.56
2	18	-30	79	98	30000	59000	1.97
3	18	-30	201	90	38100	61200	1.61
4	3	25	181	11	13200	22300	1.69

[a] Polymerization conditions: 25 μmol of yttrium complex, 5 mL of toluene.

More recently Collins et al. reported isospecific polymerization of methyl methacrylate with cationic zirconium enolate complex supported by the linked *tert*-butylamido–cyclopentadienyl ligand, $[Zr(\eta^5:\eta^1\text{-}C_5Me_4SiMe_2NCMe_3)\text{-}$

{OC(OR)=CMe$_2$}(L)]$^+$ which obviously is isolelectronic with the yttrium initiating system [19]. Therefore, for the mechanism of the acrylate polymerization, the group-transfer mechanism, already well-established for the lanthanocene systems [1,5], may be inferred.

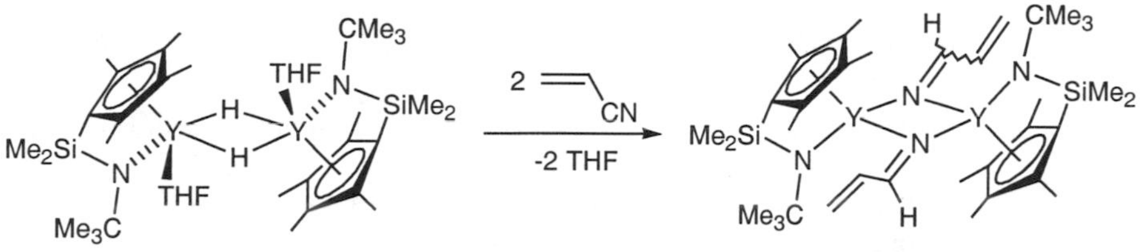

Scheme 6

An intense red solution forms as soon as acrylonitrile is added to a toluene solution of the hydride, and the precipitation of yellow atactic poly(acrylonitrile) soon follows. This unusual color is ascribed to an intramolecular charge-transfer band of an f^{0}d^0-complex to an electron acceptor. According to GPC results, the poly(acrylonitrile) samples exhibit molecular weights in the range of 10^5, but broad molecular distributions of $M_w/M_n > 5$. For the mechanism of the acrylonitrile polymerization, we propose a group-transfer-type polymerization related to that for the *tert*-butyl acrylate polymerization, involving a keteniminato intermediate [1,5]. However, when the yttrium hydrido complex [Y(η^5:η^1-C$_5$Me$_4$SiMe$_2$NCMe$_3$)-(THF)(μ-H)]$_2$ is treated with one equivalent of acrylonitrile at –78 °C, a mixture of what appears to be cis and trans isomers of dimeric μ-vinylimido complex is obtained (Scheme 6). Obviously, the methyl keteniminato complex, the product of 1,4-addition, is not detectable, but its tautomers resulting from 1,2-addition is formed. Nonetheless, this compound is capable of polymerizing acrylonitrile.

Conclusion

In conclusion, we have shown that half-sandwich alkyl and hydrido complexes of rare earth metals of the type [Ln(η^5:η^1-C$_5$Me$_4$SiMe$_2$NCMe$_3$)(X)(THF)] (X = H, alkyl) can be synthesized by the σ-bond metathesis methodology. The reactivity studies of the dimeric hydride revealed a rich insertion chemistry for α-olefins, to give structurally well-characterized mono(insertion) products. Efficient and controlled polymerization of styrene could be developed with monomeric *n*-alkyl yttrium complexes as initiators. Preliminary results indicate a significant increase in activity when the larger lanthanides Tb and Er are used, whereas the smaller elements Yb and Lu are less active or inactive.

Similar to the decamethylsamarocene initiators [1,5], rare earth metal complexes supported by a linked amido–cyclopentadienyl ligand, and possibly by a substituted cyclopentadienyl ligand [20], appear to constitute a versatile polymerization initiator system for both nonpolar and polar monomers. Recent work on ethylene–styrene copolymerization [21] as well as butadiene polymerization [22] by organolanthanide complexes clearly hint at the still unfathomed potential of rare earth metal initiated polymerization.

Acknowledgment

We thank the Deutsche Forschungsgemeinschaft, the Bundesministerium für Bildung und Forschung, BASF AG (Kunststofflaboratorium), and the Fonds der Chemischen Industrie for financial support. We also acknowledge the helpful collaboration with Prof. R. Mülhaupt, Freiburg and Prof. M. Schmidt, Mainz.

References

1   (a) H. Yasuda, E. Ihara, *Adv. Polym. Sci.* **133**, 53 (1997). (b) H. Yasuda in *Catalysis in Precision Polymerization* (S. Kobayashi, Ed.), Wiley, 1997, p. 189.
2   (a) G. Jeske, H. Lauke, H. Mauermann, P. N. Swepston, H. Schumann, T. J. Marks, *J. Am. Chem. Soc.* **107**, 8091 (1985). (b) B. J. Burger, M. E. Thomspon, W. D. Cotter, J. E. Bercaw, *J. Am. Chem. Soc.* **112**, 1566 (1991).
3   D. G. H. Ballard, A. Courtis, J. Holton, J. McMeeking, R. Pearce, *J. Chem. Soc., Chem. Commun.* 994 (1978).
4   H. Yasuda, H. Yamamoto, K. Yokota, S. Miyake, A. Nakamura, *J. Am. Chem. Soc.* **114**, 4908 (1992).
5   (a) H. Yasuda, M. Furo, H. Yamamoto, A. Nakamura, S. Miyake, N. Kibino, *Macromolecules* **25**, 5115 (1992). (b) H. Yasuda, H. Nakamura in *Progress and Development of Catalytic Olefin Polymerization* (T. Sano, T. Uozumi, H. Nakatani, M. Terano, Eds.), Technology and Education Publishers, Tokyo, 2000, p. 241.
6   J. F. Pelletier, A. Mortreux, X. Olonde, K. Bujadoux, *Angew. Chem. Int. Ed. Engl.* **35**, 1854 (1996).
7   Y.-X. Chen, T. J. Marks, *Chem. Rev.* **20**, 1391 (2000).
8   H. Schumann, J. A. Meese-Marktscheffel, L. Esser, *Chem. Rev.* **95**, 865 (1995).
9   (a) K. C. Hultzsch, T. P. Spaniol, J. Okuda, *Organometallics* **16**, 4845 (1997). (b) K. C. Hultzsch, T. P. Spaniol, J. Okuda, *Organometallics* **17**, 485 (1998). (c) J. Okuda, F. Amor, K. E. du Plooy, T. Eberle, K. C. Hultzsch, T. P. Spaniol, *Polyhedron* **17**, 1073 (1998).
10   (a) A. L. McKnight, R. M. Waymouth, *Chem. Rev.* **98**, 2587 (1998). (b) J. Okuda, T. Eberle in *Metallocenes* (R. L. Halterman, A. Togni, Eds.), Wiley-VCH, Weinheim, 1998, p. 418.
11   (a) M. Booij, N. H. Kiers, J. J. Heeres, J. H. Teuben, *J. Organomet. Chem.* **364**, 79 (1989). (b) Y. Mu, W. E. Piers, D. C. MacQuarrie, M. J. Zaworotko, V. G. Young. *Organometallics* **15**, 2720 (1996).
12   (a) K. C. Hultzsch, T. P. Spaniol, J. Okuda, *Angew. Chem. Int. Ed.* **38**, 227 (1999). (b) K. C. Hultzsch, P. Voth, K. Beckerle, T. P. Spaniol, J. Okuda, *Organometallics* **19**, 228 (2000). (c) S. Tian, V. M. Arrendondo, C. L. Stern, T. J. Marks, *Organometallics* **18**, 2568 (1999).

13 M. F. Lappert, R. Pearce, *J. Chem. Soc., Chem. Commun.* 126 (1973).
14 S. Arndt, P. Voth, T. P. Spaniol, J. Okuda. *Organometallics*, in press.
15 A. Z. Voskoboynikov, I. N. Parshina, A. K. Shestakova, K. P. Butin, I. P. Beletskaya, L. G. Kuz'mina, J. A. K. Howard, *Organometallics* **16**, 4041 (1997).
16 (a) P. J. Shapiro, W. D. Cotter, W. P. Schaefer, J. A. Labinger, J. E. Bercaw, *J. Am. Chem. Soc.* **116**, 4623 (1994). (b) C. J. Schaverien, *Organometallics* **13**, 69 (1994). (c) D. Stern, M. Sabat, T. J. Marks, *J. Am. Chem. Soc.* **112**, 9558 (1990).
17 N. Ishihara, T. Seimiya, M. Kuramoto, M. Uoi, *Macromolecules* **19**, 2465 (1986).
18 A. Zambelli, L. Oliva, C. Pellecchia, *Makromol. Chem., Macromol. Symp.* **48/49**, 297 (1991).
19 H. Nguyen, A. P. Jarvis, M. J. G. Lesley, W. M. Kelly, S. S. Reddy, N. J. Taylor, S. Collins, *Macromolecules* **33**, 1508 (2000).
20 J. N. Christopher, K. R. Squire, J. A. Canich, T. D. Shaffer (Exxon Chemical Co.), PCT WO 00/18808 A1.
21 (a) Z. Hou, H. Tezuka, Y. Zhang, H. Yamazaki, Y. Wakatsuki, *Macromolecules* **31**, 8650 (1998). (b) Y. Zhang, Z. Hou, Y. Wakatsuki, *Macromolecules* **32**, 939 (1999).
22 R. Taube in *Metalorganic Catalysts for Synthesis and Polymerization* (W. Kaminsky, Ed.), Springer, Heidelberg, 1999, p. 531.

3. Traditional Catalysts

The New Revolutionary Development Of Catalysis As The Driving Force For The Commercial Dynamic Success Of Polyolefins

P. Galli[*], G. Vecellio[**]

[*]Honorary Chairman Montell Technology,"G. Natta" R&D, P.le Donegani,12, Ferrara, Italy
E-mail: paolo.galli@eu.montell.com
[**]Montell Technology, "G. Natta" R&D, P.le Donegani, 12, Ferrara, Italy
E-mail: giancarlo.vecellio@eu.montell.com

Abstract. In the long history of materials, the development of the polyolefins represents a real unique adventure. The discovery of the new Ziegler-Natta catalyst families in the fifties activated a process that entered in a dynamic revolutionary development, still very much in progress in the present days, after having been kept quiet and at a very low profile for two decades. Their early development, in spite of the bright promises, was more difficult than expected and their commercial growth was very disappointing, at least during the first 20 years of their life. In the last 20 years we have assisted in the tremendous technology development and in an ever increasing explosive, commercial success, which is incomparable in the history of materials. Considering the past and looking at the future, strange and interesting similarities appear, being that in both cases there were, and there are, bright promises and heavy drawbacks for the polyolefin material family. We have to recognise that in both cases the commercial success has been, and will only be, the result of a continuous process technology innovation and consequent polymer property envelope expansion generated by an intensive commitment in basic research on catalysts, polymerisation processes and material technology. The situation at the beginning of the new millennium is totally different, but there are analogies at present that could perhaps be faced and solved by adopting the approach that has been the key to the solution of the problems of the sixties: the understanding, optimisation and management of the catalyst towards the new polymer properties.

1. INTRODUCTION

On re-considering the history of the Ziegler-Natta catalysts and their commercial exploitation, we may identify two periods in their development (Table 1):

1)   after the very early years of the discovery and the start up of the laboratory and pilot plant research activities in the second half of the 50s, a very slow, difficult, often disappointing, commercial development. We may consider them as the first, difficult "POOR" years.

2)   then, from the late 70s till the late 90s, we had a progressively increasing, dynamic, eventually explosive growth and commercial success.
     We may consider them as the second, aggressive, successful "RICH" years: the years of the boom of Polyolefin (PO) materials!

Table 1. Historic survey and critical analysis of the polyolefin technology development in the years 1955-2000.

YEARS	FACTS & SITUATIONS	IMPACT / RESULTS
1955-1960 **The "discovery" years**	The highest expectations and enthusiasm for a bright future.	Promises: new materials; new properties; promising, mild condition technology; low cost monomers.
1960-1980 **The first 20 "poor" years**	The hard reality: the unexpected technology issues; the unsatisfactory catalyst properties and gamut of property balances.	Drawbacks: rigid, non versatile, non reliable processes; slow and difficult technology commercialisation; poor capability of polymer property envelope expansion. Disappointing market growth.
1980-2000 **The following explosive 20 "rich" years**	New frontiers emerging catalysts: high expectation for a sensational polymer property expansion and the satisfaction of new market needs.	Promises: novel catalyst families; new properties and materials; low cost; environmental friendly technologies and materials.
1985-1995 **The new "promise" in these years**	New frontiers emerging catalysts: high expectation for a sensational polymer property expansion and the satisfaction of new market needs.	Promises: novel catalyst families; new properties and materials; low cost; environmental friendly technologies and materials.
1995-2000 **The "disappointment" years**	The hard reality: the unexpected technology issues; the unsatisfactory catalyst gamut of property balances; the missed achievement of an "ideal catalyst" model.	Drawbacks: too sophisticated, difficult to handle and manage catalysts; incomplete and unbalanced gamut of properties; expensive, slow, difficult to optimise technology commercialisation.

Today, at the beginning of the 3rd millennium, even after the last explosive 20 years, we may notice interesting analogies with the early situation of the polyolefin technology from the mid-50s and early 60s.

There are bright promises in terms of the possibility of the generation of new properties and even novel materials, capitalising on the high potential of new catalyst families and of revolutionary concept processes.

There are drawbacks in terms of difficulty to handle and manage the new sophisticated and delicate catalyst families and their problematic scale-up with still expensive and not optimised commercialisation processes.

The most successfully adopted key in the solution of the drawbacks of the sixties, has to become the guide line for the solution of the problems of the years 2000: the understanding, optimisation and management of catalysts towards an updated model of IDEAL CATALYST able to generate the new polymer properties desired in a practical way, without heavy limitations and constraints.

2. REVISITING THE PAST

2.1 The bright promises and the tough reality

The fifties were the years of the main discoveries in the Ziegler-Natta catalysis. There was a lot of enthusiasm for the novelty of the catalytic system able to operate in mild, easy conditions, a deep curiosity for the new materials and their set of properties, a strong expectation to have good and low cost materials, because of the easily available monomers and the expected low cost processes.

The enthusiasm and the push were on the scientists' part, and also the top management of several companies. In November 1955, in Montecatini, a large scale pilot plant for the polymerisation of ethylene and propylene started operating in the Ferrara plant in Italy.

In January 1957, a very first multipurpose BATCH polymerisation plant for the production of the first high density injection moulding polyethylene (PE) called "MOPLEN RO" and polypropylene (PP) started operating (Fig. 1)

It was mainly dedicated to the production of PE; a larger scale, more sophisticated plant fully dedicated to PP started operating in Ferrara in November 1957. (Fig. 2)

Those plants were projected, built and started up in an extremely short time at the cost of tremendous efforts, in terms of commitment, on the researchers' and management's part.

They all made great sacrifices, pushed and supported by the great enthusiasm based on the trust of the fluttering promises of the new family of catalysts and materials.

However, very soon reality turned out to be completely different! Analysing the complex situation of those years, we may conclude that it was the logical consequence of the total lack of understanding of the catalytic system.

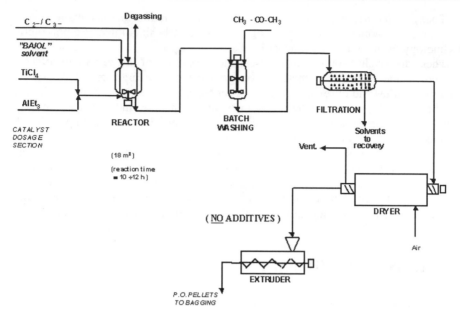

Fig. 1. The very first multipurpose polyolefin (PO) plant: January 1957; Ferrara (Italy)
Catalyst: $TiCl_4/AlEt_3$

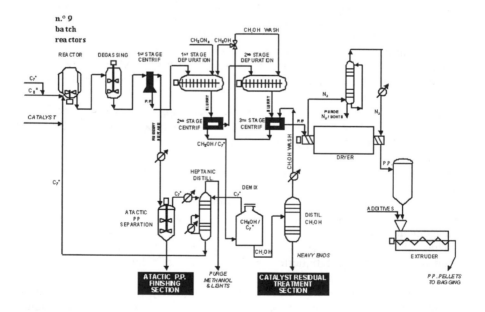

Fig. 2. The very first polypropylene (PP) plant: November 1957; Ferrara (Italy)
Catalyst: $3TiCl_3AlCl_3/AlEt_2Cl$

It was the main reason for its lack of management and of technological and commercial drawbacks.

The drawbacks were in the rigid, non-versatile processes, the low activity and poor selectivity of the catalysts, the total lack of understanding and capability of management of the reasons of polymer morphology, which all resulted in obstructing the possibility of any significant expansion of the polymer property envelope.

Those drawbacks affected both the PE and the PP families, mainly for the catalyst yield.

The poor catalyst selectivity was mainly the disadvantage of the PP, even if the high amount of low molecular weight-wax materials also affected the PE families for many years.

The total lack of understanding and controlling the polymer morphology affected all the plants, heavily spoiling all the possible efforts made in order to improve quality and cost.

The situation was at a "dead end" because the solution had to be the result of the difficult or impossible compromise between two conflicting needs: to make the polymer purification step effectual and to make the commercial plant operability reliable, fast and cost efficient. (Fig. 3)

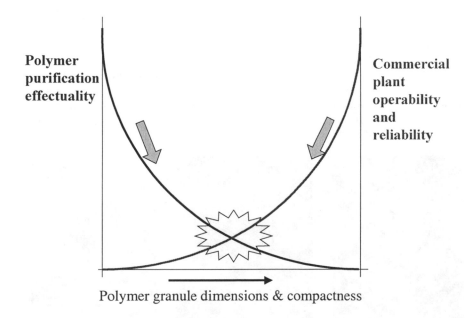

Polymer purification effectuality

Commercial plant operability and reliability

Polymer granule dimensions & compactness

Fig. 3. The two conflicting needs

1. To make the polymer purification step effectual, manageable and easy:
   a. it meant to extract and remove the catalyst residuals and the polymer by-products in a practical and not too time and money consuming industrial operations. By-products were waxes and low molecular weight polymers, and specifically, in the case of polypropylene, the atactic and sindiotactic polymers.
   b. it needed to produce polymer granules of very small dimensions and /or very high porosity in the polymerisation step, in order to allow an easy and manageable mass transfer process in and out of the polymer granules, as specified in the previous point a).

2. To make the overall operability of the commercial plant manageable, reliable and cost efficient:
   a. it meant, and means, to obtain medium or large size and compact morphology polymer granules.
       The granule porosity was a big issue because it meant high polymer fragility during the plant operations, which generated large amounts of very fine and irregular polymers, creating big problems.
   b. it meant to perform the extraction in the same way as all the other operations in an economically and sustainable short time.

As a matter of fact, the main difficulties encountered in the commercialisation of the process were directly connected to the low catalyst activity and poor selectivity, in the same way as the inconstant, unpredictable and unmanageable morphology of the polymer generated during the polymerisation.

This totally unpredictable, random, polymer morphology (such as fine particles, extra fines, or vice-versa, very big, porous, fibre like or popcorn like etc.,) (Fig. 4) created insurmountable obstacles to the smooth running of industrial operations, such as discharge from the reactors, slurry or powder, transportation, filtration, purification, centrifugation, drying etc.

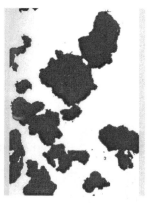

Fig. 4. Examples of bad polymer morphology

The life of the industrial plant was an adventure, or better, a nightmare, in a continuous precarious balance between conflicting needs, such as product quality or plant operability. The plants always had to live under "Damocle's sword", regarding the alternative to have to shift in between a total plugging of the lines or not to be able to achieve a decent quality of the materials in terms of purification from the catalyst residual and undesired polymerisation by- products.

As a matter of fact, small morphology was very favourable, for example in the polymer depuration step, it made the extraction easier for both the catalyst residuals and for the polymeric by-products, waxes, atactic polypropylene etc.

However it was a nightmare in all polymer transfers, from one step to another, in the filtration, centrifugation, drying, etc. The big polymer particle morphology acted in the opposite way. The worst and not infrequent case was to have to deal with the irregular random mix of very small fines and big particles together.

But indeed, the worst in such an already difficult situation, was the total unpredictability of the phenomenon in the commercial plant and the lack of any possibility of reacting properly to the issue!

There was no simple and elegant solution for the issue if we tried to cope with it in a direct way. The real, unique, definitive and only solution was, and is, to increase the catalyst activity and selectivity in order to make all the operations of chemical material removal from the granule unnecessary. Easier said than done!

It meant to be able to achieve a breakthrough in the Ziegler-Natta catalysis.

2.2   The catalyst system development

To cope with, and to solve such a difficult and   "dead end" situation, a revolutionary jump in catalyst understanding had to be achieved in order to obtain a major breakthrough in catalyst science. The heavy commitment on the part of many companies all over the world in that field, obtained the result.

The discovery of the high activity "δ" MgCl2-supported catalyst for the ethylene polymerisation in 1968, became the first of the two major breakthroughs which have revolutionised the development of the PO technologies in the following 30 years.

It brought about the elimination of the need of the removal of the catalyst residues with a significant simplification of the polymerisation processes.

The elimination of these constraints allowed a remarkable reduction in investment costs, but more important, it disclosed the way of the achievement of several new degrees of freedom in the concept of the new generation of catalysts.

For the time being, the main and only practical impact has been the one on the PE technology, the first one to enjoy the benefits of the new catalysts.

The possibility of eliminating the polymer purification section, with all its costs and constraints had two important impacts:

1)   the slurry plants became simpler, easier to be managed, with consequent production cost reduction and improved product quality. In 1971, Montecatini had already completed the construction of the first high yield plant in the world, in Brindisi, Italy. At the beginning of 1972, it was already

commercialising the first injection moulding, blow moulding and roto-
moulding P.E grades.

2) the gas phase process became an achievable reality: the elimination of the
   purification step made this feasible. In a few years the gas phase process
   became an industrially affirmed reality mainly thanks to the efforts of the
   U.C.C.
   In addition to the simpler, larger size, reliable new plants, another very
   important result was achieved: a novel PE material was invented and
   developed: the linear low density PE (LLDPE)!
   All that, gave an impressive contribution to the commercial growth of P.E in
   the following decades.

However the impact was higher than that.

This breakthrough really disclosed a new way with a great potential for the
catalyst, the PO technology in general and P.P in particular.

The polypropylene was initially unable to take advantage of that discovery. The
new MgCl2-based catalyst was not at all able to offer a decent degree of
selectivity, guaranteeing an acceptable level of polymer, stereoregularity.

On the contrary, the results in terms of stereospecificity, of the first "δ" MgCl2-
based catalyst were a disaster! The other key catalyst properties were poor or
missing, making any technology improvement unachievable:

- the need of extracting the atactic PP remained or even increased;
- the constraints BARRING the freedom to produce large compact granules,
  remained;
- the poor plant operability and reliability remained;
- the high production cost remained;
- the constraint to the generation of new polymer properties remained.

Clearly, it appeared that any further progress was unthinkable without another
dramatic improvement in the catalytic system. The catalyst for the propylene
polymerisation was, and without doubt remains, the most complicated and
difficult. However, its development has been the key and the driver for the
dramatic development of all the polyolefin technology.

Since the beginning of the 70s, the study, understanding and development of the
PP catalyst has been the main driving force for the development of the catalyst and
technologies for the entire families of polyolefins.

Let us then take the case of the PP catalyst, its historical development, its
breakthrough and how it has been extended for the benefit of the entire polyolefins
family. PE has been, to a very large extent, carried ahead benefiting from the
understanding and development achieved in the area of the most demanding and
difficult "travelling companion" polypropylene.

The needs of its technology, the tremendous commitment in the understanding
of the catalyst and the developments of its technology, became the new and main
driving force for the entire family of polyolefins.

Let us consequently concentrate from now on the development of the catalyst for the polymerisation of propylene.

It was realised that it was mandatory to look for a catalyst showing an entire, complete, all embracing set of properties. Just one, even if an outstanding property, such as activity or selectivity, was not at all enough ! It appeared that it was not possible to achieve a complete success just by operating empirically, looking at the new catalyst properties, one by one.

It became necessary to elaborate and to develop the research in line with a MODEL of the catalyst to achieve.

The model for the "new dream", the all embracing ideal catalyst, was imagined and called "the extreme target: the ideal catalyst", a catalyst, that in addition to processing a very high catalytic activity and great selectivity, would have imparted the desired proper morphological structure to the appropriate product, whether this was a polymer, a copolymer or a polymer alloy. (Fig. 5)

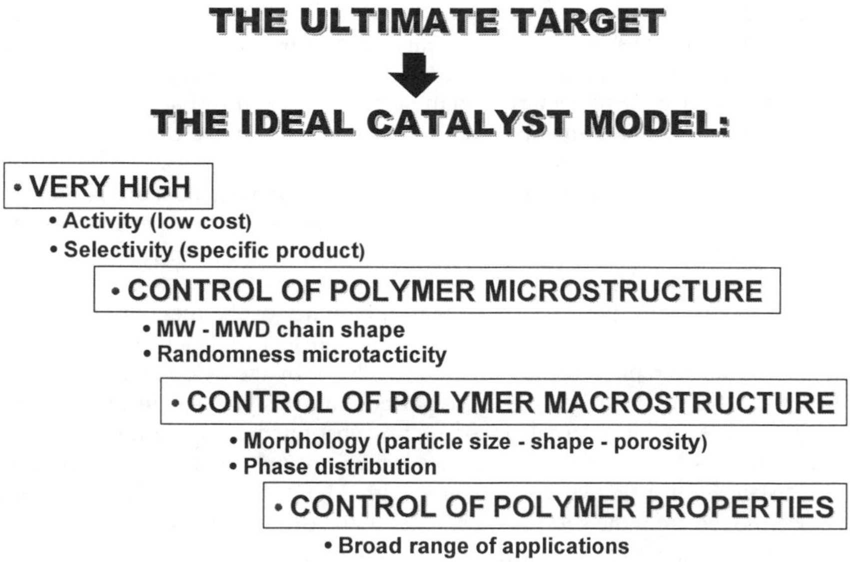

THE ULTIMATE TARGET
⬇
THE IDEAL CATALYST MODEL:

- **• VERY HIGH**
  - • Activity (low cost)
  - • Selectivity (specific product)
  - **• CONTROL OF POLYMER MICROSTRUCTURE**
    - • MW - MWD chain shape
    - • Randomness microtacticity
    - **• CONTROL OF POLYMER MACROSTRUCTURE**
      - • Morphology (particle size - shape - porosity)
      - • Phase distribution
      - **• CONTROL OF POLYMER PROPERTIES**
        - • Broad range of applications

Fig. 5. The "ideal catalyst " model

Starting from Natta's discovery, it is possible to see how the studies on the catalyst made a continuous and still very much in progress, improvement in terms of yields achievable.

The discovery of the active support "δ" $MgCl_2$ disclosed the possibility of simplifying the process with the elimination of the expensive, polluting and even more limiting and constraining catalyst residual elimination steps.

It allowed the revolution of the PE technology, but it was nevertheless not enough for PP.

To enjoy the same simplification the PE technology had in the 70s, it became essential to achieve both of the first two prerequisites of the ideal catalyst:

VERY HIGH — - **ACTIVITY**
and
- **SELECTIVITY**

This target became the main objective of the researchers operating in the area of PP in the years 1968-75.

In 1975, the target was achieved with the discovery of the so-called 3rd generation catalyst, giving for the first time high yield and high selectivity.

These high-activity Ziegler-Natta catalysts comprise $MgCl_2$, $TiCl_4$ and an "internal" electron donor and are typically used in combination with an aluminium alkyl cocatalyst such as $AlEt_3$ and an "external" electron donor added in polymerisation.

The first catalyst systems contained ethyl benzoate as internal donor and a second aromatic ester as external donor, but nowadays the catalysts most widely used in polypropylene manufacture contain a diester (e.g. diisobutyl phthalate) as internal donor and are used in combination with an alkoxysilane external donor of $RR'Si(OMe)_2$ or $RSi(OMe)_3$ type.

The requirement for an external donor when using catalysts containing an ester as internal donor is due to the fact that, when the catalyst is brought into contact with the cocatalyst, a large proportion of the internal donor is lost as a result of alkylation and/or complexation reactions. In the absence of an external donor, this leads to poor stereospecificity due to increased mobility of the titanium species on the catalyst surface. When the external donor is present, contact of the catalyst components leads to replacement of the internal donor by the external donor.

Most recently, research on $MgCl_2$-supported catalysts has led to systems not requiring the use of an external donor. 2,2-disubstituted-1,3-dimethoxypropanes were identified as bidentate internal donors which not only had the right oxygen-oxygen distance for effective coordination with $MgCl_2$ but which, unlike esters, were not removed from the support on contact with $AlEt_3$ and which, in contrast to alkoxysilanes, were not-reactive with $TiCl_4$ during catalyst preparation.

The best performance was obtained with bulky substituents in the 2-position.

The successive "generations" of high-activity and high-selectivity $MgCl_2$-supported catalyst systems for polypropylene are summarised as follows as far as the yield increase is concerned (Fig. 6):

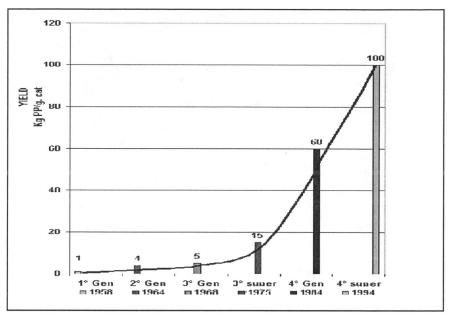

Fig. 6. The increase of activity of catalyst systems

The benefits achieved for the PE technology were extended in the 2nd half of the 70s, as for the PP technology, eventually bringing about the elimination of the need of the removal of the atactic fraction and residues with a significant simplification of the polymerisation processes; the elimination of these constraints allowed a remarkable reduction in investment costs and brought about several new degrees of freedom. (Fig. 7)

From these discoveries, the "dream" of the "Ideal Catalyst" as the extreme target in the heterogeneous catalysis started appearing as an achievable reality.

The elimination of the need of the removal of the catalyst residues and to a large extent, of the atactic polymer fractions, were not such important facts in themselves, but mostly because they became the starting point and the base for all the future revolutionary developments.

As a matter of fact, the most important event for all the future scientific, technological and commercial development was not just the increase in the catalyst activity and selectivity. It was the green light to conceive, with a total freedom from any process constraints, the last approach to the "IDEAL CATALYST", having reduced the process constraints to zero.

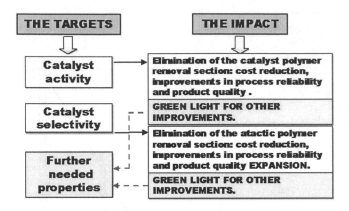

Fig. 7. The first step in the way towards the achievement of the ideal catalyst.

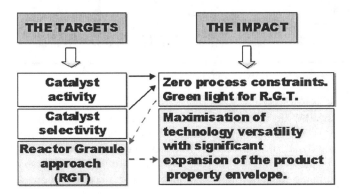

Fig. 8. The second step in the way towards the achievement of the ideal catalyst.

It enabled us to maximise the technology versatility so as to have a significant expansion of the product property envelope. (Fig. 8)

The fundamental research led to the second breakthrough: discovery and development of a new dimension [1] in heterogeneous catalysis and of the "reactor granule technology" (RGT).

Its discovery and understanding brought about the complete control of the catalyst-polymer granule genesis and growth: the RGT represents the highest result in the Science of structural versatility in the polymerisation technology.

The Reactor Granule Technology (RGT) approach led to:

- understanding of the catalyst/polymer replication;
- total management of polymer granule shape growth, dimension, compactness;
- achievement of a new dimension in heterogeneous catalysis;
- total control of internal and external polymer morphology;
- control of both dimensions and locations of the different compositions and

phases;
* expansion of the polymer property envelope.

The role of the CATALYST ARCHITECTURE [2,3] in the controlled expansion of the polymer granule was deeply explored, so as to reach an optimum equilibrium between the mechanical strength of the growing granule and the catalyst polymerisation activity [4].

Under appropriate polymerisation conditions, polymer particles with an internal morphology ranging from compact [5] to porous can be obtained (Fig.9).

Fig. 9. Different polymer morphologies from RGT

The polymer particle becomes the reactor itself in which polymerisation occurs, and by changing monomer it is possible to obtain another polymer intimately dispersed within the mass of the solid granule of the matrix. It became the basis to the "Reactor Granule Technology" (RGT) [6], fully exploiting the potential of the spherical form catalyst to generate alloys, thanks to the capability of building different polymers components inside the same granule.

The most important fact is that now it is possible to achieve an IDEAL mixing of different and even very different components inside the same granule, overcoming the difficulties of their mixing via the blending technology, and, in many cases, even the difficulty, or impossibility, of handling incompatible products, that cannot be processed alone because of their M.W. (either too low or too high) sticky, oily products (Fig. 10).

In this way, we have now generated a novel possibility of overcoming the difficulties of miscibility of the different polymer phases, with a degree of freedom clearly superior to the one we may have via the mechanical blending of the different polymer components and phases.

We are now approaching this new frontier, just entering the exciting future of the full exploitation of such a tremendous potential.

In the development of the polyolefin technology, the discovery and implementation of the "RGT" is today considered, together with the discovery of the high yield catalyst, based on active $MgCl_2$, a fundamental milestone in the polyolefin technology development.

**EXAMPLE OF POLYMERIZATION WITH PERFECT REPLICA OF A
SPHERICAL FORM CATALYST:
<u>FOUR PHASES</u> HETEROPHASIC TETRAPOLYMERIZATION
- COMPOSITE MATERIAL**

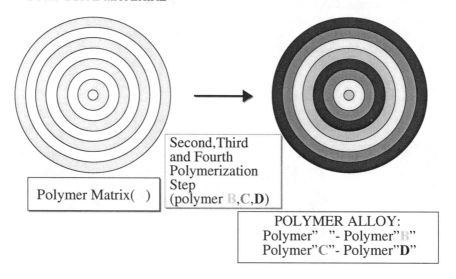

Polymer Matrix( )

Second, Third
and Fourth
Polymerization
Step
(polymer B,C,D)

POLYMER ALLOY:
Polymer" "- Polymer"B"
Polymer"C"- Polymer"D"

Fig. 10. Four phases polymer alloy achievement

3. Pondering over the present.

3.1 The process development

The different catalyst generations developed from the 1st to the 4th have allowed, or better still, have driven the development of a family of new and novel processes taking full advantage of the new virtuosity made available by the various generations of catalyst families with a complete revolution in the process concept, the polymer properties and the market development.

New, revolutionary generations of polyolefin production processes like *Spheripol, Hypol, Lip-Shac, Unipol, Novolen* (1982), *Catalloy* (1990), *Spherilene* (1994) and *Borstar* (1995), just to mention the most important, started in fact to revolutionise the market.

The processes, based on the use of the fourth generation catalyst, like *Spheripol*, *Catalloy* and *Spherilene*, make it possible to generate multipolymers, multiphase alloys and blends directly in synthesis, thus achieving materials not achievable with traditional technologies.

All that, made it possible to expand the polypropylene and polyolefin properties obtainable in polymerisation from the straight homopolymers or slightly modified copolymers, to highly modified copolymers, and to real polymer alloys without boundaries in the ratios between the different comonomer introduced, number and type of phases and final compositions.

The progressive simplifications of the process technology brought about a dramatic reduction of both: the CAPITAL COSTS and the PRODUCTION COSTS connected, first of all, to the dramatic energy savings.

But, most important, is the dynamic expansion of the PROCESS VERSATILITY with the sharp and endless broadening of the POLYMER PROPERTY ENVELOPE.

3.2 The product development

We would like to introduce this chapter by quoting a third party: Peter Bins and Ken Sinclair, who say in their last interesting multi-client study on PP "ULTIMATE PP MARKET POTENTIAL" by Bins & Associates: "Underpinning the success of these new technologies (i.e. the 1980s technology), however, was a new approach to serving existing markets and developing new ones. The new technologies were not of themselves valuable, their value had to be developed through and from the market. Leaders of the 1979/83 revolution such as Himont (now Montell) therefore devoted particular attention to the identification and satisfaction of *real* market needs. These needs were not for PP as such, but for functionalities or properties. Himont focused on selling *properties*, applying their new capabilities at the technology/market interface to design new PP grades having the processing behaviour and end-use performance to meet *real* market needs."

As a matter of fact the possibility to obtain polymer particles having:
- perfect spherical shape
- narrow particle size distribution
- high bulk density
- controlled degree of porosity
- controlled internal composition and morphology
- high process flowability

has allowed in particular to maximise the operability of production plants reducing investment and running costs dramatically, but, most important, to increase the process versatility, broadening the product property envelope.

The continuous development in the product properties has been, and will remain, the fundamental pre-requisite to the endless commercial success of PO in general and PP in particular.

In the case of PP it has brought about the continuous development of the PP property envelope shown in Fig 11.

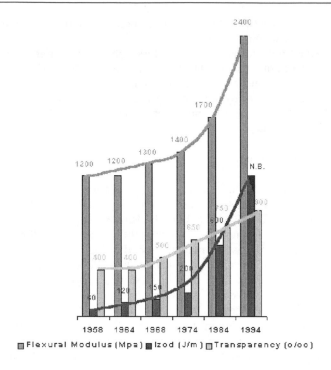

Fig. 11. Variation of properties for polymers by different catalyst generations

3.3 The recent advances: the polyolefin alloys (the definitive overcoming of the border line among the difficult families)

The *Catalloy* process

The *Catalloy* process has been planned to make maximum use of the Reactor Granule Technology allowing the repeated introduction of different monomers during propylene polymerisation thus generating a multiphase multipolymer alloy directly in the reactor. The spherical free-flowing resins obtained are fit for the subsequent conversion and are provided with an extremely wide range of properties which cannot be obtained with any other process for polypropylene.

The *Catalloy* process is a multistage, highly flexible, mainly gas-phase technology. The Reactor Granule based on the catalytic system used, permits the incorporation of multiple polymer structures within a single particle moving through the multistage process.

Also in this case the growing particle itself becomes the polymerisation reaction medium thus eliminating all the previous process constraints and allowing the production of unique reactor-made resins with properties no longer limited from mechanical considerations of the process.

The synthesis of polypropylene alloys directly in the reactor has demonstrated the capability of giving resins with unexpected new properties which could expand the property envelope of polyolefins and enable them to compete in properties with other non-olefinic plastic resins, such as Nylon, PET, TPU, PS, PVC, etc.(Fig. 12).

PROPERTY OVERLAP OF POLYMERS FROM *Catalloy* PROCESS
APPLIES TO SEVERAL SUBSEGMENTS
WITHIN THE ADDRESSABLE MARKETS IDENTIFIED

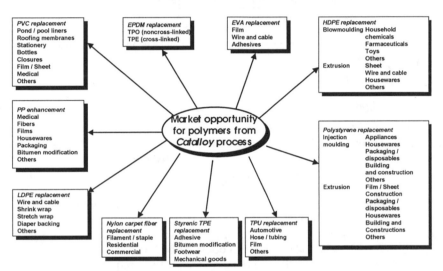

Fig. 12. New application fields for polymers obtained from the *Catalloy* process

The physical-mechanical properties of products made via *Catalloy* process cover the field of elastomeric and very rigid materials significantly broadening the already rich polymer property envelope of *Spheripol* process technology.

In particular, with this process, materials which have been obtained so far only via mechanical blends of different pre-formed polymers, can be better obtained directly in synthesis (Fig. 13).

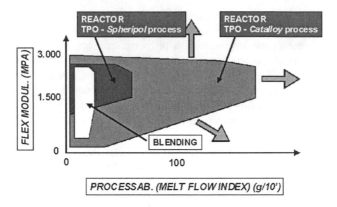

Fig. 13. Impact copolymer technologies: history of thermoplastic materials property envelope development

The *Spherilene* Process

The extension of the Reactor Granule Technology to polyethylene polymerisation allowed the production of a wide range of linear polyethylenes in the same plant the characteristics of which, in terms of safety, versatility, economics, efficiency and environmental advantages are similar to those of the *Catalloy* process for the polypropylene-based alloys.

The *Spherilene* process is an innovative and real "swing" technology developed to optimise the combination of the low-cost gas-phase technology and the flexibility in the product mix allowed by the catalytic system (from VLDPE with a density below 0.900 g/ml to HDPE with a density of 0.965 g/ml).

Currently, the use of polyethylene in many applications requires the blending of different grades or their blending with other resins in order to modify the processability or product properties.

By extending the Reactor Granule Technology to ethylene polymerisation it has been possible to produce the said polyethylene blends directly in synthesis.

It may be considered another "*Catalloy* like" approach to the PO alloys, starting from the polyethylene side.

3.4 Analysing the year 2000 situation

There is no doubt that the revolutionary development of the Ziegler-Natta, which reached its apex at the beginning of the 80s, has had an enormous positive impact on the growth of the POs.

We completely share and would like to once more quote Peter Bins and Ken Sinclair, saying:

"advances in catalyst technology in the late 1970s precipitated an industry revolution that more than doubled the markets accessible to PP, resulting in double-digit demand for more than a decade. The new catalysts introduced in the early 1980s were the key to making new high performance goods that greatly expand the substitution potential of PP, as well as its versatility in new market creation".

Quite rightly they also added:
"While PP demand growth remains strong, in the 6-8%/yr. range, how long can growth continue at these high levels? Is PP approaching its ultimate potential to substitute for more expensive plastics & other materials? How much further can the PP property envelope be expanded and who will likely capture the value which these new materials bring? Has the 1978-1982 technological revolution now run its course? .....Or is further hidden market potential yet to be revealed?"

The answer they stress has to be sought for in the TECHNOLOGY side:
"The growth of polypropylene (PP) from its commercial introduction in 1957 provides one of the best examples of market response to the introduction of new technology. Demand growth expressed in kg/capita for the first 25 years of the industry from 1957/58 to about 1982/83 could be fit quite well to a logistics curve moving toward a fixed target of ultimate per-capita consumption. As illustrated in the chart below, (Fig. 14) indications in 1983 were that the market was maturing -- approaching an upper limit of per-capita consumption -- and that future growth was trending toward growth in the economy in general. Expectations of medium term growth were in the region of 4 -5% per year.

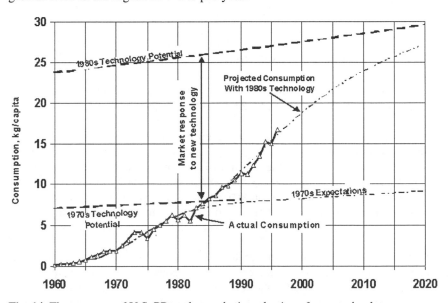

Fig. 14. The response of U.S. PP market to the introduction of new technology

These expectations, however, failed to take into account the structural changes in PPs life cycle brought about by technological innovations, particularly the development in the late 1970s of high-yield, highly stereoselective (HY/HS) catalysts. These boosted PP demand in virtually all markets world-wide, leading to global consumption growth of more than 10% per year through the 1980s. It became apparent that growth in all categories was following a new logistics curve with an ultimate per-capita consumption two to three times the previous target".

Now, at the beginning of the years 2000, we are noticing that the "projected consumption with the 1980s technology" is showing a trend towards a market that again, in the same way as it was doing in 1983, is maturing, approaching a new upper limit of per-capita consumption with a growth that is trending towards growth in the economy in general.

So the market growth wave, generated by the 1980s technology, is exhausting its impetus!

If we want to guarantee the continuation of the market growth at the present speed, we should successfully and rapidly make a 3rd phase of technology revolutionary innovations available, capable of boosting the PP demand, continuing to lead the global consumption growth of more than 10% per year through the next two decades. Are we moving ahead in the right way? Let us see what today's PROMISES for the future are!

4. The future: The promises for the years 2000

4.1 The further potential of the Z.N family: the mixed catalysis (the expansion of the RGT model to non Ziegler-Natta based polymerisation)

In recent years, the RGT has successfully been extended beyond the limits of the classical Ziegler-Natta catalysts expanding the technology potential to include and cover with the widest versatility the families of the homogeneous catalytic systems.

Understanding the replication phenomenon and the capability of engineering the catalyst architecture has made it possible to create and develop a new generation of Reactor Granule Technology based on mixed heterogeneous and non-heterogeneous catalysts: the Multicatalyst Reactor Granule Technology (MRGT) [7].

This discovery disclosed a novel way to most conveniently and widely exploit at an industrial level the potential of the best homogeneous polymerisation catalysts.

Indeed, the spherical particles can act as microreactors for further olefins or different polar functional monomer polymerisation once the homogeneous catalysts are introduced into their pores.

Thus, polymeric alloys are obtained in synthesis which synergically join the properties of the polymers obtained with titanium catalysts with those of new and completely different polymer families obtainable with the second catalytic species applied.

4.2 The *Hivalloy* engineering polymers

The first successful application of this mixed catalysis process was the *Interloy* process born from the need of making polyolefins compatible with polymers of different nature, mainly with polar polymers. In this process the catalyst is of the radical generator type.

The porous polyolefin granule containing still active Ti-sites supplies a highly reactive substrate within which an easy grafting and polymerisation reaction can occur with non-olefinic or polar monomers up to 50% by weight.

The target of the technology used to make *Hivalloy* products is to generate a new family of materials retaining the best properties of polyolefins, and adding the desired properties typical of the engineering thermoplastic materials (ETP) case by case.

Thanks to the characteristics of intimately alloying polyolefins and of non-olefinic materials together, *Hivalloy* products combine the most desirable properties of polypropylene (processability, chemical resistance, low density) with many desirable characteristics of engineering resins (stiffness/impact balance, mar and scratch resistance, reduced cycle time in moulding applications, improved creep resistance, high HDT) [8] (Fig. 15).

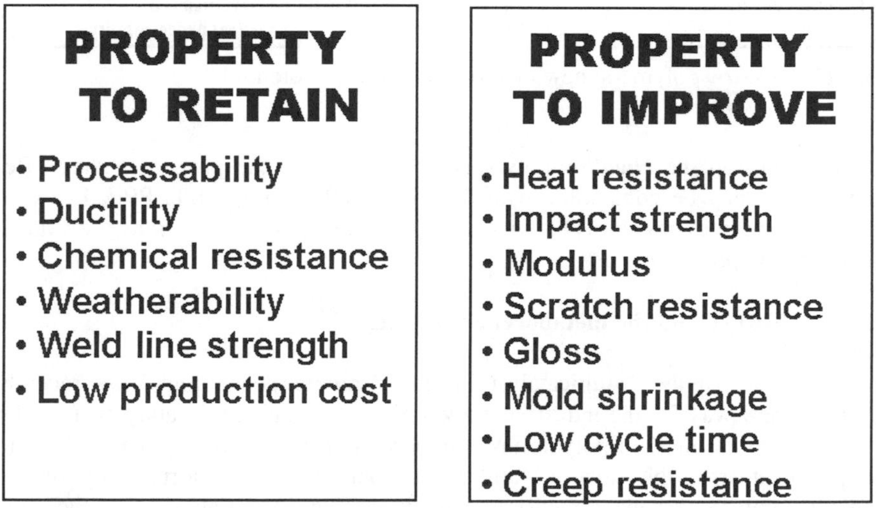

PROPERTY TO RETAIN	**PROPERTY TO IMPROVE**
• Processability • Ductility • Chemical resistance • Weatherability • Weld line strength • Low production cost	• Heat resistance • Impact strength • Modulus • Scratch resistance • Gloss • Mold shrinkage • Low cycle time • Creep resistance

Fig. 15. *Hivalloy* polymers property balance

The technology shows extreme flexibility allowing the polymerisation of the most diversified range of comonomers, such as styrenics, acrylics, SAN, alogenated monomers (Fig. 16).

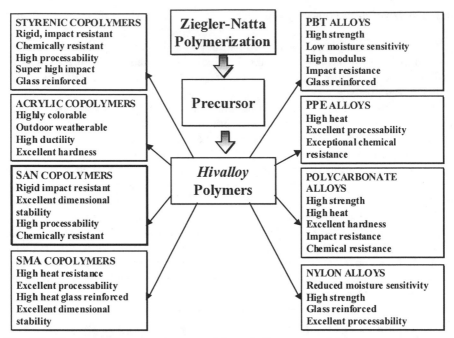

Fig. 16. *Hivalloy* polymers: new materials and alloys potential

In conclusion the *Hivalloy* polymers, in addition to the properties they offer themselves, may be compatible with different ETP such as PBT, PPE, PC, etc., thus offering the possibility of generating an extremely wide range of novel materials and properties.

4.3 The MRGT and the metallocene catalysts

The discovery of the "Multicatalyst Reactor Granule Technology" has recently opened a new way for the industrial exploitation of metallocene catalysts. In order to be used in the modern polyolefins commercial processes, these catalysts had to be provided with the same capability of controlling the morphology of the generated polymer (shape, dimension, density, mechanical resistance) and the same processability of 4th generation Ti-catalysts.

The most promising way to obtain such catalysts, turned out to be the supportation on granules of a porous olefinic polymer according to the Reactor Granule Technology. According to this original catalysis technology, activated metallocenes are introduced in the granules of the porous spherical polymer which behave as microreactors for the further polymerisation of olefins with the second catalyst.

The morphological characterisation of the polymers obtained from the MRGT applied to metallocenes shows an intimate blending of the two different polyolefins in terms of structure and molecular weight, having grown on the same granule [9].

In this mixed catalysis process, the advantages of multi-site heterogeneous catalysts, such as polymer processability, are combined with those of single-site catalysts, such as elastic properties and an optimised heterophasic copolymer structure with improved properties may be achieved.

Thus, the possibility exists to produce an extremely wide range of new properties in synthesis not obtainable with any other existing polymerisation or compounding technology.

4.4 The ultimate frontier of the Ziegler-Natta and the MRGT catalysis

a) The further potential of the new donors.

We have seen how the issue of the lack of stereospecificity of the first "δ" MgCl2-based catalyst has been solved via the introduction of proper "internal/external" donors, more recently just of "internal" donors.

It is amazing what has been discovered as their side effect and further potential: the capability of deeply changing not only the stereospecificity, but also the capability of the catalyst to generate the most convenient MMD.

In the case of PP we have now envisioned the possibility to shift from MMDs below 3 to the values higher than 12. In addition to that, limited to the broad MMDs, we have now at our disposal an even more powerful and elegant technology: the "Multizone Circulating Reactor" (MZCR).

b) The Multizone Circulating Reactor (MZCR) technology

In order to more widely exploit the potential of MRGT, a new process technology has recently been conceived for the polymerisation of alfa-olefins in the gas-phase [10](Fig. 17).

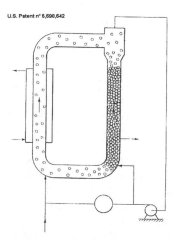

U.S. Patent n° 5,698,642

Fig. 17. The "Multizone Circulating Reactor" (MZCR)

Through this technology, the growing polymeric granule is kept continuously circulating between two interrelated zones, where two distinct and different fluodynamic regimes are realised. In the first zone, the polymer is kept in a "fast fluidisation"; leaving said zone, the transport gas is separated and the polymer crosses the second zone in the "packed bed mode" and is then reintroduced in the first zone. A complete and massive circulation is obtained between the two zones, managed by the balance of pressures between among them.

The fluodynamic peculiar regime of the second zone, where the polymer enters as dense phase in "plug flow", offers the opportunity of altering, through simple but substantial means, the monomeric composition with respect to the chain terminator (hydrogen).

Whenever required, by the same route and on the same growing granule, a multiple, alternate and cyclic as well as continuous polymerisation can be obtained, which attains the most complete and intimate mixing of different polymers with a substantial "homogeneity" of the final polymer.

Through this new technology, it has been easy to obtain very homogeneous PE and PP homopolymers with very broad molecular weight distribution and high stiffness and copolymers with enhanced flexural modulus/izod impact balance.

4.5 The new promises: the single-site catalysts

Single-site catalysts, and metallocenes in particular, have represented the most meaningful innovation in the recent history of polyolefins as far as their potentialities for generating new properties are concerned.

The single-site catalysts are a new, beautiful family of catalysts with a unique capability of creating tailor made polymer structures, controlling the key polymer structure parameters and the consequent properties, at a molecular level

Mostly because of such a unique and beautiful capability of planning catalyst structures and related polymer structure and properties, never seen before, they have a huge potential that has not yet been fully explored and still to be exploited.

New and novel polymer properties may and will be created; the experience we have achieved in the last decade allows us to predict that they have the potential to play a determinant role in the polymer developments in the years 2000, once that and provided that we will be able to overcome the critical issues that still affect, with several constraints, their technology development, heavily penalising their full commercialisation.

A few years ago the headlines claimed that metallocene catalysts would revolutionise the polyolefin industry. Companies certainly believed in this new technology and poured billions of dollars into research and development.

Commercial forecasts were particularly optimistic; so far "metallocenes" are failing to pay dividends for the companies that put faith in the new technology.

No doubt that in spite of the bright promises, the general picture is significantly below the expectations. What is the reason? We quote Peter Bins and Ken Sinclair again:

"Metallocene PP (mPP) has property combinations that differ from those of conventional Ziegler-Natta PP, and seem to be better in many respects. However, markets for mPP have been surprisingly slow to develop. Why is this? Why does Exxon, the leader in mPP commercialization, believe that metallocenes will have a greater impact on the PP industry than on the PE industry? Will mPP expand potential markets for PP only marginally, or is the industry facing another revolutionary change and a major boost in demand?"

In our opinion to make an analysis, and bearing the long past in mind, the situation appears singularly and surprisingly overlapping with the one of the Ziegler-Natta catalysts and products of the early 60s.

5. Conclusions

The key role played by the catalyst in developing the PO technology and market, is today universally accepted.

We have seen how the revolutionary progresses in the catalyst technology have generated a tremendous wave of innovation in the process and product developments, creating low cost, versatile processes, new products and applications with a huge, determinant impact on the market growth.

However we have to be aware that the wave of technology innovation of the 80s and 90s will soon start losing vigour.

If we want to guarantee the PO materials the same trend in the market growth as in the last 20 years, we have to be able to activate a new wave of technology innovation very soon.

Again, as the past teaches: it has to be generated by the catalyst.

Taking in due consideration the "teaching" of the past, we must keep in mind that in order to transform the new catalysts bright promises in successful commercial realities, the catalyst has to submit to some key prerequisites.

Nowadays, there is the need of a "new"-"old", all embracing approach, paying due attention to ALL the key properties that a successful catalyst should have. Its target should be the possible identification of an optimised model of catalyst, showing all the "VIRTUOSITIES" typical of the "IDEAL CATALIST MODEL" we proposed in the early 70s.

With an addition: today, we suggest an updated version, taking into better consideration a need nowadays certainly deserving more attention than in the past, i.e. the elimination of any toxicity risks.

Luckily, nowadays there are several interesting and promising developments in progress; we may identify and select the two main, broadest scope and more promising research fields amongst them:

1) to further understand, fine-tune, and specialise the Ziegler-Natta catalysts making them able to generate more specific, tailor-made polymer structures.

2) to re-direct the single-site catalyst development towards the achievement of a
    more balanced complete and convenient set of properties on better
    "virtuosity".(Fig. 18)

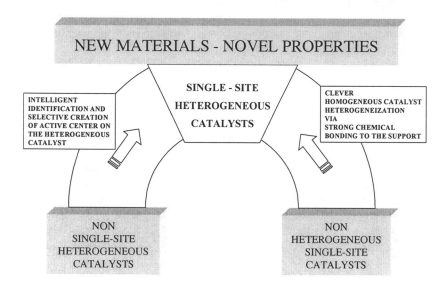

Fig. 18. The next fascinating steps for a catalyst's bright future

–   In the classical Ziegler-Natta area we definitively see the potential of a
    significant progress.
    We are already on the way to understanding, selecting and specialising the
    catalytic active centres in order to direct them towards the generation of the
    specific polymer molecular structures and products we need.

–   In the area of the single-site catalysts we certainly have the greatest capacities
    today and the largest amount of expectations because of the broad spectrum of
    bright promises they show in many fields.
    However, there are really bright promises that have nevertheless remained in
    the "PROMISES" field for much too long. The main drawback, penalising
    their full commercialisation, is that even if showing a set of some outstanding
    properties, they show a lack of other equally needed properties.
    By overwhelming that gap, endowing the single-site catalysts with the key
    "VIRTUOSITIES" typical of an ideal catalyst model, this will disclose them
    the way to play the major role in boosting the PO technology innovation wave
    during the next 20 years.
    It is the challenge of the beginning of the new millennium.

Acknowledgements:

The authors wish to thank A. Bolognesi for his contribution and support in the preparation of this work.

Note: *Spheripol, Catalloy, Spherilene, Hivalloy,* are trademarks of Montell.

References

1.   Galli P., *Proceedings of the II International Conference "The future of science has begun - chemical, biochemical and cellular topology",* Milan, 1992, p. 43.
2.   Galli P., Cecchin G., Simonazzi T., 32nd IUPAC Int. Symp. "Frontiers of macromolecular science", Kyoto, (1988) p. 91.
3.   Galli P., *J. Macromol. Sci.-Phys.* 1996, **35**, 427.
4.   Galli P., Barbè P.C., Noristi L. *28th IUPAC Macromol. Symposium* July 12-16, Amherst, Massachussets, USA, 1982.
5.   Galli P., *Progr. Poly. Sci.* 1994, **19**, 959.
6.   Galli P., *Macromol. Symp.* 1994, **78**, 269 and 1996, **112**, 1.
7.   Collina G., Pelliconi A., Sgarzi P., Sartori F., Baruzzi G., *Polym. Bull.* 1997, **39**, 241.
8.   Galli P., Haylock J.C., Albizzati E., De Nicola A.J., *35th Int. Symp. on Macromolecules,* IUPAC ACRON, OH, 1994.
9.   Galli P., Collina G., Sgarzi P., Baruzzi G., Marchetti E., *Jour. Appl. Pol. Sci.* 1977, **66**, 1831.
10. Govoni G., Rinaldi R., Covezzi M., Galli P., *U.S. Pat.* 5,698,642, 1997.

Basic Approaches for the Design of Active Sites on the Traditional Ziegler Catalysts

Hiroyuki Kono, Mikio Tomisaka, Hitoshi Matsuoka[1], Boping Liu, Minoru Terano[*]

School of Materials Science, Japan Advanced Institute of Science and Technology,
1-1 Asahidai, Tatsunokuchi, Ishikawa 923-1292, Japan, E-mail: terano@jaist.ac.jp
[1] Delegated from Tokuyama Corp.

Abstract. In this paper, three basic approaches for the design of active sites on the traditional Ziegler catalysts are discussed for the future breakthrough of this type of catalysts. Firstly, the variation of isospecific active sites formed on a $MgCl_2$-supported $TiCl_4$ catalyst in the absence or presence of external electron donor was specified by the stopped-flow method based on the isotacticity distribution of the polypropene. Besides the depression of the formation of some aspecific active sites, the effect of addition of external electron donor is to transfer some of the aspecific titanium species into isospecific active site. Secondly, hydrogen dissociation sites were found to be derived from the over-reduced Ti species due to deactivation during propene polymerization. Finally, the enantioselectivity and diastereoselectivity of the active sites for copolymerization of propene and 1, 5-hexadiene were studied. It was made clear that the steric hindrance in the vicinity of active sites increases with the isospecificity of the active sites.

1. Introduction

As one of the most typical examples of the Ziegler catalyst system, $MgCl_2$-supported Ti-based catalysts have achieved a spectacular success in improving the activity and stereospecificity of the catalyst for olefin polymerization. In spite of several decades of research efforts, many aspects concerning the active sites and polymerization mechanism in Ziegler catalysis still remain controversial.

The stopped-flow method for olefin polymerization with Ziegler catalysts was first developed by Keii and Terano in 1987 to evaluate specific kinetic parameters in the polymerization of propene with a $MgCl_2$-supported Ziegler catalyst [1-5]. The most predominant features of this technique is its quasi-living polymerization characteristic within a very short period of polymerization (ca. 0.2 s) and its ability to observe the polymer produced in the initial stage of polymerization, which reflects directly the nature of the active sites just after their formation.

In this paper, three basic approaches for the design of active sites on the traditional $MgCl_2$-supported Ziegler catalysts utilizing the stopped-flow techniques are discussed for the future breakthrough of this type of catalysts. The first approach is based on the analysis of variation of isospecific active sites induced by the addition of external electron donor for propene polymerization. The second approach is through the basic research concerning the hydrogen dissociation sites which are closely related with the state of active sites for propene polymerization. The third approach is established upon the study of the enantioselectivity and diastereoselectivity of the active sites for copolymerization of propene and 1, 5-hexadiene.

2. Experimental

2.1. Materials

Propene of research grade was used without further purification. $MgCl_2$ and Cyclohexylmethyldimethoxysilane (CMDMS) were kindly supplied by Toho Titanium Co., Ltd. Electron donor (ED) ethyl benzoate (EB) and CMDMS were dried over the molecular sieves 13X. Triethylaluminum (TEA, Tosoh Akzo Corp.) , EB and CMDMS were used as toluene solution. Heptane and toluene were purified by passing through the molecular sieves 13X column. 1, 5-hexadiene (purchased from Tokyo Chemical Industry) was refluxed and distilled over CaH_2.

2.2. Variation of isospecific active sites induced by external ED

The $MgCl_2$-supported $TiCl_4$ catalyst was prepared by a similar procedure that has been described elsewhere [6]. The stopped-flow polymerization of propene and estimation of kinetic parameters were carried out according to the method reported previously [1-5] using a basic-type stopped-flow system with or without EB or CMDMS as external ED. The polymer samples obtained were characterized by TREF, GPC and ^{13}C NMR.

2.3. Dissociation sites of hydrogen on the catalyst

The highly active $MgCl_2$-supported $TiCl_4$ catalyst with EB as internal ED used is the same as reported in reference [7]. A high-pressure-type modified stopped-flow system with three vessels was used for the polymerization. Vessel (A), (B) and (C) are equipped with water jackets. For the polymerization in the presence of hydrogen, the catalyst slurry and the $Al(C_2H_5)_3$ solution in heptane saturated by hydrogen were placed in vessel (A) and (B), respectively. Propene was placed in vessel (C). In the case of polymerization without hydrogen, vessel (A) and (B) were saturated with nitrogen, respectively. Pretreatment time was controlled around 0.05 to 0.2 second and polymerization time was 0.1 second. The stereoregularity and molecular weight distribution of the produced polymer were analyzed by the ^{13}C NMR and GPC, respectively. H_2-D_2 exchange reactions were also conducted by using the stopped-flow method [8]. The evolution of HD was checked by gas chromatography using special column at low temperature.

2.4. Study of the enantioselectivity and diastereoselectivity of the active sites.

The synthesis of the novel block copolymers (3a,3b) using the low-pressure-type modified stopped-flow method was carried out with $TiCl_4$/EB/$MgCl_2$ and $Al(C_2H_5)_3$ according to the method reported previously [9, 10]. For comparison, the PP homopolymer (1) and poly(methene-1,3-cyclopentane)-co-PP (PMCP-co-PP) copolymer (2) were also prepared using the same polymerization process. A PP/PMCP-co-PP blend (4) was prepared in a 50/50 weight ratio by solution blending was used as a reference. The resulting polymers were analyzed by GPC, ^1H NMR, TREF, ^{13}C NMR, DSC, etc.. Furthermore, the block copolymer was fractionated on the basis of crystallinity of PP part by TREF. The copolymer parts of each fraction were also characterized by GPC and ^{13}C NMR.

3. Results and discussions

3.1. Variation of isospecific active sites induced by external ED.

The variation of isospecificity of active sites on a $MgCl_2$-supported Ziegler catalyst was investigated for a better understanding of the nature of the active sites. Polypropene was prepared by stopped-flow method in the absence or presence of external electron donor and its isospecificity distribution was analyzed by TREF method. Thus, the variation of isospecific active sites formed on the catalyst can be detected. Fig. 1 shows the TREF diagrams of each polymer, and Tab. 1 shows the characterization of the highest isotactic fraction (elution temperature by TREF is over 112°C) of resulting polymer and kinetic parameters. It can be seen that highly isospecific active sites derived from the highest isotactic fraction also exist in the electron donor-free catalyst system. Furthermore, all the values of the highest isotactic fraction, such as Yield, $\overline{M_n}$, $\overline{M_w}$ over M_n, meso pentad fraction, k_p and [C*], are constant (Tab. 1) showing no effect on the highest isospecific active sites by the addition of electron donor. However, the [C*] derived from the second highest isotactic fraction (elution temperature by TREF is between 90°C and 112°C) of resulting polymers increased from 0.63 mol % to 0.69 mol % for EB and to 0.80 mol% for CMDMS. Simultaneously, the [C*] derived from the atactic fraction (elution temperature by TREF is below 20°C) of resulting polymers decreased from 7.58 mol % to 3.54 mol % for EB and to 2.56 mol % for CMDMS. It was made clear that some aspecific active sites were converted to isospecific active sites by the addition of external electron donors.

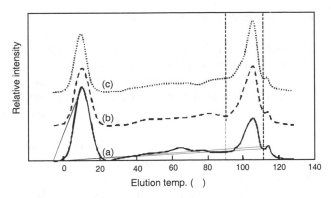

Fig. 1. TREF diagrams of PPs. The polymerization was carried out with MgCl$_2$-supported Ziegler catalyst (a) in the absence of electron donor. (b) in the presence of ethylbenzoate as external electron donor.   (c) in the presence of cyclohexylmethyldimethoxysilane as external electron donor.

Table 1. Characterization of the highest isotactic fraction of PPs[a] and kinetic parameters

External Donor	Yield (g/mol-Ti)	$\overline{M}_n$[b]	$\overline{M}_w/\overline{M}_n$[b]	$mmmm$[c] (mol%)	$k_{p,iso}$ (L/mol·s)	$[C^*_{iso}]$ (mol%)
-	16.7	41600	1.8	98.5	9300	0.041
EB	16.4	41700	1.7	98.4	9300	0.039
CMDMS	16.2	41500	1.7	98.6	9270	0.038

a) High isotactic fraction (>112   ) of PP was obtained by TREF analysis system.
b) Determined by GPC
c) Determined by ^{13}C NMR

Based on the above results, it can be postulated that there exist mainly three kinds of active sites in the MgCl$_2$-supported Ziegler catalyst: the first is isospecific titanium species which produce isotactic polypropene, the second is aspecific titanium species which creates atactic polypropene, the third is a sterically hindered aspecific titanium species due to the strong interaction with its surroundings and preferably make polymer with moderate isotacticity. The isospecific titanium species can be further classified into two groups: the highest and the second highest isospecific active sites. Besides the depression of the formation of some aspecific active sites, the effect of addition of external electron donor is to occupy one of the vacancy of some of the aspecific titanium species by coordination and consequently the later is transferred into isospecific active site with the second highest isospecificity without any effect on the highest isospecific active sites.

3.2. Dissociation sites of hydrogen on the catalyst

Hydrogen is very important in the industrial scale of the olefin polymerization process to control the molecular weight of the product. But the mechanism of the chain transfer reaction by hydrogen has not been clarified yet, though various techniques and approaches are used to solve this problem. Stopped-flow method has the advantage that the chain transfer reaction can be disregarded and the state of the active sites is constant.

Our group has studied the mechanism of the chain transfer reaction using the stopped-flow method. In the standard stopped-flow method, that means no pretreatment of the catalyst with cocatalyst, hydrogen has no effect on the chain transfer reaction. This is the first experimental data showing that hydrogen does not affect molecular weight of the polymer. But with pretreatment (for several minutes), hydrogen was effective even at the lower hydrogen pressure, though the polymer yield was decreased by the pretreatment. It seems, based on these results, that the active sites deactivation has some important relation with the chain transfer reaction by hydrogen.

In this study, the pretreatment time of the catalyst was changed into very short period (ca. $\leq 0.2s$) for investigating the relationship between the catalyst activity, the chain transfer reaction and the dissociation of hydrogen.

The results of propene polymerizations are showed in Figures 2 and 3. Fig.2 shows that hydrogen had no effect on the yield and that the yield decreased linearly as the pretreatment time increased irrespective of presence or absence of hydrogen. The decrease in the polymer yield indicates the increase of the amount of deactivated sites. Fig.3 shows that hydrogen acts as the transfer agent under these conditions, which is reflected by the decrease in molecular weight of the polymer. It was found, in addition, that the deactivation of the sites occurred within the same extent for both isospecific and aspecific sites. The stereoregularity of the polymer was not changed by the pretreatment. Hydrogen has no effect on the stereospecificity of the catalyst. The decrease of the yield (catalytic activity) indicates the increase of the amount of over-reduced Ti species due to catalyst deactivation.

Generally, chain transfer with hydrogen takes place through atomic hydrogen just as previously proposed by some researchers from kinetic method [11-14]. Therefore, H_2-D_2 exchange reactions were conducted to confirm the dissociation of hydrogen using the stopped-flow method. The evolution of HD was found exclusively in the case using pretreated catalyst. This reaction conditions were corresponded to the condition in reference [8]. Furthermore, the relationship between the active sites deactivation and the amount of HD produced was investigated using the modified stopped-flow method. The result shows HD increased linearly with the increase of the pretreatment time, which may be corresponding to the amount of the dead sites (Fig.4).

From polymerizations and H_2-D_2 exchange reactions, hydrogen dissociation is supposed to occur on the dead Ti species and then the atomic hydrogen acts as

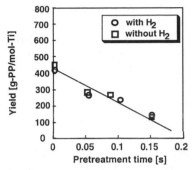

Fig.2. Dependence of yield on pretreatment time

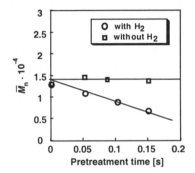

Fig.3. Dependence of $\overline{M}_n$ on pretreatment time

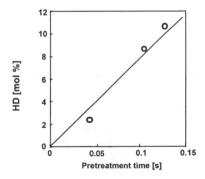

Fig.4. Results of H_2-D_2 exchange reaction

chain transfer agent. This means the Ti species have two important roles not only for the polymerization but also for hydrogen dissociation in the same system.

3.3. Study of the enantioselectivity and diastereoselectivity of the active sites

The stopped-flow polymerization of 1,5-hexadiene, which is a typical α,ω-diolefin, catalyzed by a $MgCl_2$-supported Ziegler catalyst was investigated to demonstrate that the cyclopolymerization proceeded under a quasi-living stage [9, 10]. The cyclopolymerization of 1,5-hexadiene gave unique poly(methylene-1,3-cyclopentane) (PMCP). There are two distinct stereochemical events in the cyclopolymerization: the enantioselectivity of the olefin insertion determines the tacticity, while the diastereoselectivity of the cyclization step determines whether *cis* or *trans* rings are formed. The study of the cyclopolymerization was useful to obtain additional insights into the active sites, which could not be obtained from a typical propene polymerization.

In this section, the novel block copolymer having a chemical link between the PP part and the copolymer part containing the cyclic unit of 1,5-hexadiene was synthesized by stopped-flow polymerization with the $MgCl_2$-supported Ziegler catalyst. In order to obtain a real block copolymer product, the polymerization was performed by means of the modified stopped-flow method, in which the first step ($\leq 0.1s$) of propene homopolymerization followed by the second step ($\leq 0.1s$) of 1,5-hexadiene and propene copolymerization can be finished within the lifetime of the growing polymer chains. In particular, the analysis of the chain segments in the block copolymers by several methods was applied to gain additional insight into the active sites.

The modified stopped-flow polymerization afforded a block copolymer containing a cyclic unit based on the cycliation of 1,5-hexadiene. The results of the block copolymerization are shown in Tab. 2. The cyclization ratio of 1,5-hexadiene in the polymers obtained by the stopped-flow polymerization method was measured by ^1H NMR analysis. The ^1H NMR spectra of the polymers showed no proton resonance of the vinyl group ($CH_2=CH-$) of the side chain. It indicates that the cyclization is faster than the insertions for propene and 1,5-hexadiene and all of the 1,5-hexadiene units existed in the structure of methylene-1,3-cyclopentane in the polymers. To clarify the existence of the block formation in the block copolymer, the copolymer and the corresponding PP/PMCP-*co*-PP blend were extracted with boiling heptane to examine the existence of free PMCP-*co*-PP in these systems. The polymers extracted were examined by ^{13}C NMR (Fig.5).

In the case of the block copolymer, the peaks due to PMCP-*co*-PP existed after extraction, as shown in Fig. 5 (b). On the other hand, the signals from PMCP-*co*-PP disappear after extraction of the PP/PMCP-*co*-PP blend. Therefore, these results suggest the formation of PP-*b*-(PMCP-*co*-PP) in the stopped-flow polymerization method, where the PMCP-*co*-PP is chemically linked to the PP.

Tab. 2. Synt hesis of PP-*b*-(PMCP-*co*-PP) by stopped-flow polymerization   [a]

Code	Polymerization time in s		Yield in g/mol·Ti	$\overline{M}_n$ $(\overline{M}_w/\overline{M}_n)$ [b]	Hexadiene [c] content in mol.-%	$^2 H_m$ [d] in J/g
	PP	PMCP-*co*-PP				
1[e]	0.1	—	470	9400 (3.0)	—	78
2[f]	—	0.1	290	7500 (3.3)	24	—
3a	0.1	0.1	850	19000 (3.0)	15	38
3b	0.15	0.05	970	22000 (3.2)	11	55

a) Synthesis of block copolymers was conduct ed with the catalyst with Al(C₂H₅)₃ (Al/Ti = 30) at 30 °C for 0.1-0.2 s.

b) Molecular weight and its distribution were det ermined by G PC. The molecular weight was calculat ed based on the univers al calibrat ion curve of polypropene.

c) determined by ¹³C NMR.

d) determined by DSC. The melt ing enthalpy of the polymer produced in code 2 could not be det ermined exact ly because of the amorphos polymer.

e) Homopolymerizat ion of propene.

f) Copolymerizat ion of 1,5-hexadiene and propene.

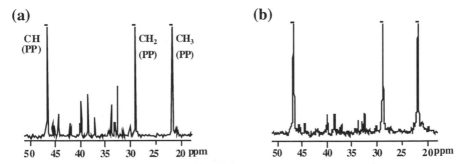

Fig. 5. ¹³C NMR spectra of (a) PP-b-(PMCP-co-PP) (3a) and (b) the unsolved part of the block copolymer after extraction with boiling heptane.

TREF diagrams (differential-type) of polymers except the PMCP-*co*-PP are shown in Fig. 6.   Two different elution peaks were observed for PP: a main peak at 112 °C and a small peak at 107 °C, suggesting the existence of two types of isospecific active sites.   The PP/PMCP blend shows bimodal elution patterns resulting in PP and PMCP existing at quite different regions of elution temperature (PP, 80~120 °C; PMCP-*co*-PP, 20~60 °C).   The elution peak of PP-*b*-(PMCP-*co*-PP) (3a) was apparently lower than those of PP.   The phenomenon is thought to be governed by the fact that the crystalline state is disturbed by the segregation process of incompatible parts in the block copolymer.   Fig.6 clearly shows that PP-*b*-(PMCP-*co*-PP) eluting at a temperature region between 20 °C to 120 °C is mainly composed of a unified component.   On the other hand, the TREF diagram of 3b was different from that of 3a because of the different crystallinity of 3a and 3b. In particular, PP and 3b show the same elution peaks of non-uniform active sites with different selectivities. Thus, the information of the stereospecific active sites can be indirectly obtained from the analysis of the copolymer part in the block copolymer. 3b was fractionated on basis of the tacticity of the PP part in the block copolymer. Each fraction was analyzed by GPC and ^{13}C NMR (Table 3). Fraction D and E contained isotactic PP had the lower 1,5-hexadiene content and the high *cis* value in copolymer part. The insertion of 1,5-hexadiene and the conformation of cis- and trans-ring probably is correlated to steric hindrance in the vicinity of active site. Therefore, it is suggested that the high steric hindrance exists in the vicinity of isospecific active sites, and the low steric hindrance exists in the vicinity of aspecific active sites.

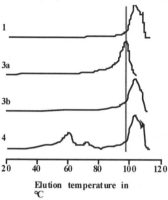

Fig. 6. Temperature rising elution fractionation diagrams of PP (1) , PP-b-(PMCP-co-PP) (3a and 3b) and PP/PMCP blend (4).

Tab. 3.   Results of fractionation by TREF

Fraction	Temp. (°C)	1,5-Hexadiene [a] content (mol%)	*cis* [a] (%)
A [b]	10	28	53
B [b]	10-50	13	56
C [b]	50-100	7	57
D [c]	100-107	5	57
E [c]	107-140	4	65

a) determined by ^{13}C NMR
b) ata-PP-*block*-(PMCP-*co*-PP)
c) iso-PP-*block*-(PMCP-*co*-PP)

4. Conclusions

In this paper, three basic approaches for the design of active sites on the traditional Ziegler catalysts are discussed. Firstly, The variation of isospecific active sites formed on $MgCl_2$-supported $TiCl_4$ catalysts in the absence or presence of external electron donor was specified by the stopped-flow method. Besides the depression of the formation of some aspecific active sites, the effect of addition of external electron donor is to transfer some of the aspecific titanium species into isospecific active site. Secondly, hydrogen dissociation sites were found to be derived from the over-reduced Ti species due to deactivation during propene polymerization induced by TEA pretreatment. Finally, the enantioselectivity and diastereoselectivity of the active sites for the synthesis of a novel block copolymer PP-b-(PMCP-co-PP) were studied. The insertion of 1,5-hexadiene and the conformation of cis- and trans-ring is correlated with the steric hindrance in the vicinity of active sites. It was made clear that the steric hindrance in the vicinity of active sites increases with the isospecificity of the active sites.

References

1. T. Keii, M. Terano, K. Kimura and K. Ishii, *Makromol. Chem., Rapid Commun.* **8**, 583 (1987)
2. M. Terano, T. Kataoka, T. Keii, *J. Mol. Catal.* **56**, 203 (1989)
3. T. Keii, in : *Catalyst Design for Tailor-made Polyolefins*, K. Soga, and M. Terano (Eds.), Kodansha-Elsevier, Tokyo 1994, p.1
4. T. Keii, *Macromol. Theory Simul.* **4**, 947 (1995)
5. T. Keii, M. Terano, K. Kimura, K. Ishii, in : *Transition Metals and Organometallics as Catalysts for Olefin Polymerization*, W. Kaminsky, H. Sinn (Eds.), Springer-Verlag, 1988, pp.1
6. H. Mori, M. Endo, K. Tashino, M. Terano, *J. Mole. Catal. A: Chem.* **145**, 153
7. (1999)
8. H. Mori, M. Yamahiro, M. Terano, M.Takahashi, T. Matsukawa, *Makromol.* Chem Phys. **201**, 289 (2000)
9. T. Keii, Y. Doi, E. Suzuki, M. Tamura, M. Murata, K. Soga, Makromol. Chem. **185**, 1537 (1984)
10. H. Mori, H. Yamada, H. Kono, M. Terano, *J. Mol. Catal. A, Chem.* **1997**, 125, 81
11. H. Mori, H. Kono, M. Terano, *Macromol. Chem. Phys.* **2000**, 201, 543
12. H. Mori, K. Tashino, M. Terano, *Macromol. Rapid Commun.* **16**, 651 (1995)
13. G. Natta, Chem. Ind. (Milan) **41**, 519 (1959)
14. K. Soga, T. Shiono, Polym. Bull. (Berlin) **8**, 261 (1982)
15. J. C. W. Chien, C. Kuo, *J. Polym. Sci., Part A: Polym. Chem.* **24**, 2707 (1986)

Surface Analytical Approaches for the Phillips Catalyst by EPMA and XPS

Boping Liu, Takami Shimizu, Minoru Terano

School of Materials Science, Japan Advanced Institute of Science and Technology, 1-1 Asahidai, Tatsunokuchi, Ishikawa 923-1292, Japan
E-mail: terano@jaist.ac.jp

Abstract. Electron probe microanalysis (EPMA) and X-ray photoelectron spectroscopy (XPS) were jointly applied to study the distribution state of Cr species on a Philllips catalyst with 1wt.% Cr loading. The EPMA images revealed the existence of aggregates of Cr species in sizes of 200-300 nm on the surface of the catalyst, which were supposed to be crystallized aggregates of chromia (Cr_2O_3). The variation of distribution state of Cr for the catalyst after being treated in various moisture-containing atmospheres had also been studied. It was found that aggregation of Cr species occurred in all treatments. Correspondingly, the variation of binding energy and full width of half maximum intensity (FWHM) of the Cr2p (3/2) spectrum from the XPS analysis gave further evidence for identification of all Cr species formed after the treatments. The formation mechanism of crystallized aggregates of chromia during the calcination of the Phillips catalyst was postulated to be related with a surface-stabilized trivalent Cr species.

1. Introduction

As one of the most important polyolefin catalysts, Phillips catalysts are responsible for several million tons of commercial production of polyolefins. In spite of more than 40 years of research since their discovery by Hogan and Banks in the early 1950s, many aspects concerning the state of Cr species as well as the polymerization mechanism still remain controversial [1, 2]. The distribution state of Cr species, which is closely related to the molecular structure and oxidation state of the chromium oxide on the catalyst surface, seems to play a crucial role for a basic understanding of the Phillips catalysts. In the literature, many different methods, such as oxygen chemisorption [3], magnetic susceptibility measurement [4], Uv-vis DRS [5], XRD [6], EPR [7], SIMS [8], Raman [9], EXAFS-XANES [10], PIXE [11] and SEM/EDS [12], etc., had been used to characterize the distribution state of Cr species. It was generally accepted that the CrO_3 could be highly dispersed and stabilized as surface monochromate, dichromate (sometimes even polychromate), through the interaction with the surface groups on silica gel during the calcination process at 300-900 °C. An important problem is concerning the state of aggregation of Cr species, that is the formation of chromia crystallites, which affects to a great extent the performance of the Phillips catalyst. Aggregated crystallites of chromia, which was thought to deactivate the catalyst for ethylene polymerization, was usually formed during the calcination process at high temperature (ca. over 500 °C). The higher Cr loading the more serious aggregation of Cr species and the lower activity for ethylene polymerization. That's maybe the main reason why the Cr loading for the typical industrial Phillips catalysts is limited at a low level ca. 1 wt.% to get the best catalytic performance. When the Cr loading is higher than the amount for a saturated monolayer supporting (ca. 5 wt% Cr for a support with $300m^2/g$ surface area), the aggregated chromia is formed through the thermal decomposition of the residue bulky chromium trioxide which can't be stabilized on the silica during calcination due to the lack of enough surface groups.   The most interesting topic is concerning the formation of aggregated chromia on a Phillips catalyst with a relatively low Cr loading ca. 1wt% or lower, where the Cr loading of an industrial catalyst usually locates and it is usually beyond the measurement sensitivity for detecting the occurrence of aggregated microcrystals by using the above-mentioned methods. Therefore, Some disputes concerning this topic are still in existence due to the lack of direct and conclusive evidence. For example, according to the report of Cimino et al. [13], when a Phillips catalyst with 1wt.% Cr was calcined over 600°C the formation of agglomerated Cr_2O_3 was detectable by XRD method. Whereas, Zaki [9] reported that the α-Cr_2O_3 detection was beyond the sensitivity of the XRD technique when the Cr loading was lower than 3wt.%, and this is supported by the recent results of Wang and Murata et al. [14]. Some Raman studies by Wachs et al. [15-17] on calcined Phillips catalysts with Cr loading from 0.1wt.% to 2wt.% also didn't show the presence of crystalline Cr_2O_3.

The EPMA method make it possible to directly map the distribution state of the Cr species on the Phillips catalyst surface with even an industrial level of Cr loading, and thus to confirm the formation of aggregated Cr species. XPS method is a powerful method for measuring the oxidation state of Cr in the Phillips catalyst [13, 18-29]. The binding energy (BE) of the Cr2p (3/2) spectrum, which will slightly shift with adsorption of various gases, is frequently used as a criterion for identifying the valence state of Cr species. On the other hand, the FWHM value of the Cr2p (3/2) spectrum is a reflection of the distribution state of the Cr species in its corresponding oxidation state, which also give valuable information especially for discriminating bulky chromium oxide phase from surface-stabilized Cr species. The partially thermal-reduced catalyst implemented in this study was found to be only slightly affected by the X-ray irradiation during the XPS measurement within 2hours. This agrees well with the report of Best et al. [19].   In this work, our approach was based on EPMA combined with XPS technique for study of aggregation mechanism of Cr species on a Phillips catalyst with 1wt.% of Cr.

2. Experimental

2.1. Catalyst and treatment

Nitrogen (99.9995%) were directly used. A Phillips catalyst was donated by Japan Polyolefin Co. Ltd., which was prepared from impregnation of silica gel with aqueous solution of CrO_3 followed by calcination in air up to 800°C. Treatment to this catalyst was carried out in open air under ambient condition for 24hours before EPMA characterization. Another two catalyst samples were treated at 800°C for 2 hours in moisture-saturated air and moisture-saturated N_2 flow, respectively, in a quartz microreactor.

2.2. Catalyst characterization

2.2.1. EPMA measurements

The EPMA measurements were carried out on a JEOL JXA-8900L system. Each sample was embedded on a carbon tape and fixed on a sample holder in the glove bag under nitrogen atmosphere. Then, the sample holder was put into the transfer chamber, by which the sample could be introduced into the main chamber without atmospheric exposure. The vaccum in the EPMA chamber was above 1×10^{-5} Torr. The EPMA was used with an accelerating voltage of 20 kV, a probe current of 2×10^{-8} A, a probe electron beam diameter of 1μm and a dispersive crystal of PETJ for wavelength dispersive X-ray analysis (WDX). Both map and line analysis of the Cr distribution were carried out with a depth sensitivity of ca. 1μm.

2.2.2. XPS instrumentation

XPS data were obtained on a Physical Electronics Perkin-Elmer Model Phi-5600 ESCA spectrometer with monochromated Al Kα radiation (1486.6eV) operated at 300W. Each catalyst sample was embedded on a conductive copper tape fixed on a sample holder. The sample holder was then put into a transfer vessel (PHI Model 04-110, Perkin-Elmer Co., Ltd.), which can be connected to the sample introduction chamber on the XPS instrument for sample transfer without atmospheric exposure. The prepared sample was degassed in the introduction chamber to 10^{-7} Torr before entering the main chamber, in which the vaccum was kept above 3×10^{-9} Torr during the 2hr XPS data acquisition. A neutralizer was used to reduce the charging effect to obtain a better signal to noise ratio. All binding energies were referenced to the Si 2p peak of silica gel at 103.3eV to correct for charging. The values of binding energy for all the surface-stabilized Cr species reported by McDaniel et al.[20] were used as reference data for the multiplet fitting of the Cr 2p XPS curves by the Gaussian–Lorentzian method.

3. Results and discussions

3.1. Mapping the distribution state of chromium on the original catalyst

Fig.1 shows the typical results of EPMA map and line curves of the Cr distribution on the Phillips catalyst. As it can be seen, the Cr species mostly dispersed uniformly on the surface of each particle. Whereas, the heterogeneity of Cr distribution on each individual catalyst particle was revealed in all the EPMA images, that is the existence of small amount of local aggregates of Cr species in sizes of 200-300 nm on the catalyst surface. These aggregates corresponding to the red patches in the map image and sharp peak in the line curves in Fig.1 were supposed to be microcrystals of chromia. This was supported by the XPS results shown in Tab.1. The coexistence of trivalent and hexavalent Cr species in the catalyst reveals the high possibility of formation of chromia due to partial reduction of Cr^{+6} to Cr^{+3} during thermal treatment at 800°C in the preparation process. The hexavalent Cr species with a binding energy of 581.81eV (Cr 2p (3/2)) exist in the forms of monochromate, dichromate and polychromate shown as fomulas Cr-A (+6), Cr-B (+6) and Cr-C (+6) (n ≥ 1) in Scheme 1. The CrO_3 (bulk) usually has a much lower binding energy of Cr 2p (3/2) at around 579eV to 580eV [20]. During the calcination, the bulk CrO_3 was stabilized on the support surface and the binding energy of the Cr 2p (3/2) simultaneously shifted to higher value.

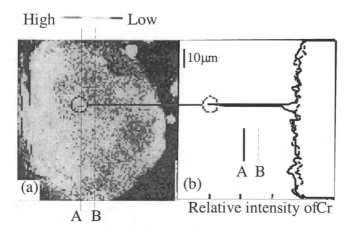

Fig.1. EPMA map(a) and line curves(b) of the chromium distribution on the original Phillips catalyst particles

Tab.1. XPS analysis of the Phillips type catalysts treated in various atmosphere

Gas media	Treatment conditions	Cr 2p (3/2)		Valent states	Atomic percentage (%)	Chemical state
		BE(eV)	FWHM(eV)			
Air	CrO₃[a]	579.24	2.53	+6	100	Bulky
Air	Cr₂O3[a]	576.59	3.03	+3	100	Bulky
N₂	No-treatment	577.21	4.43	+3	29.63	Stabilized
		581.81	9.62	+6	70.37	Stabilized
Open air	r.t., 1kg/cm² 24 hours	577.26	3.48	+3	27.96	Bulky
		579.95	6.97	+6	25.35	Bulky
		581.86	8.28	+6	46.7	Stabilized
Air-H₂O (g) flow	800 , 1kg/cm² 2 hours	577.58	3.28	+3	40.59	Bulky
		581.64	8.3	+6	59.41	Stabilized
N₂-H₂O (g) flow	800 , 1kg/cm² 2 hours	577.64	3.26	+3	39.22	Bulky
		581.74	8.3	+6	60.78	Stabilized

[a] CrO₃ and Cr₂O₃ in bulky phase were measured as standard samples for comparison

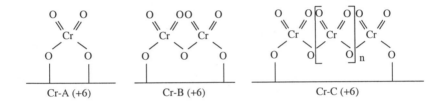

Cr-A (+6)   Cr-B (+6)   Cr-C (+6)

Scheme 1. Plausible structures of surface-stabilized hexavalent Cr species on silica.

The binding energy of 581.81eV (Cr 2p (3/2)) for hexavalent Cr species in this study agrees well with that of 581.60eV measured by M.McDaniel et al. for their hexavalent Cr/SiO₂ catalyst [20]. The Cr^{+3} species occurred on the catalyst usually has a binding energy of Cr 2p (3/2) at around 577eV to 578eV [20]. The XPS data of the catalyst studied in this work show that almost 30 percent of stabilized Cr^{+6} atoms was reduced to Cr^{+3} duaring the calcination process. Whereas, according to the EPMA results only a small part of these Cr^{+3} presents as aggregated chromia on the catalyst. The main part of the Cr^{+3} atoms may present as stabilized species which are chemically bonded to the silica surface with formulas Cr-D(+3) [30], Cr-E(+3) [31], Cr-F(+3) and Cr-G(+3) [5] as illustrated in Scheme 2. This sounds reasonable when judging from the FWHM value (4.43eV) of the Cr 2p (3/2) peak corresponding to the Cr^{+3} species. The FWHM value of the Cr 2p (3/2) peak for chromia in bulk phase usually lies well below 4.0eV. M.McDaniel et al. [20] reported the preparation of a standard stabilized Cr^{+3} sample with a FWHM value of 4.9eV for the Cr 2p (3/2) peak. The thermal-induced reduction of Cr^{+6} to Cr^{+3} and the consequent agglomeration of chromia when the Cr loading was much lower than the amount for a saturated monolayer supporting had been reported frequently in the literature, but the mechanism in detail still remained ambiguous. First of all, the role of moisture in the formation of aggregated chromia is the first consideration due to the inevitable formation of traces of moisture from the dehydroxylation of residue silanol groups on the catalyst surface during the whole calcination process.

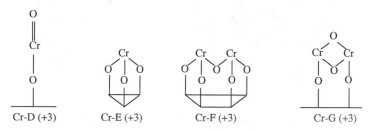

Scheme 2. Plausible structures of surface-stabilized trivalent Cr species on silica.

3.2. Variation of Cr distribution state after treatments in various atmospheres

3.2.1. In open air at r.t.

The catalyst was treated in the open air for 24 hours before EPMA and XPS measurements to check the effect of moisture on the aggregation of Cr species at r.t.. EPMA map image and line analysis curves shown in Fig.2 indicate serious aggregation of Cr species after the treatment. The aggregation of Cr species results in the formation of many microcrystal particles in different sizes ranging from several hundred nm to several microns. The concentration of Cr in the areas besides those particles decreased to a lower extent due to the aggregation. The XPS results of the treated sample (Tab. 1) show that more than one fourth of the stabilized Cr^{+6} species was converted into CrO_3(bulk) with a binding energy of 579.95eV. The significant decrease of FWHM value of the XPS Cr2p (3/2) spectra for the Cr^{+3} from 4.43eV to 3.48eV also indicates the simultaneous agglomeration of Cr^{+3} species induced by the interaction with moisture in the open air. Obviously, the chemical bonds between the stabilized Cr^{+3} species and the support were cleft by the moisture. Moreover, the moisture also acted as a catalyst to accelerate the aggregation process of the bulky chromia formed. The slight decrease of the atomic percentage of Cr^{+3} species is also due to its serious aggregation. As some of the Cr^{+3} species in the aggregated particles was beyond the detection of XPS methods. The surface atomic concentration of Cr atom measured by XPS (Tab. 2) also decreased from 1.15% to 0.98% due to the same reason.

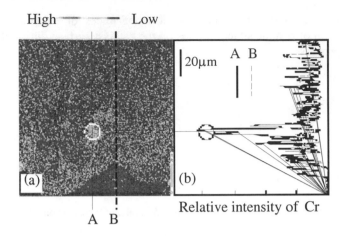

High ——— — Low

A B

|20µm

(a)   (b)

A B

Relative intensity of Cr

Fig. 2. EPMA map(a) and line curves(b) of the Cr distributin on the Phillips catalyst after being treated in the open air at r.t. for 24 hours.

Tab.2. The surface atomic concentration of all the elements except hydrogen on the Phillips type catalysts treated in various atmospheres by XPS

Atmosphere/temperature	Atomic concentration (%)			
	Cr	O	Si	C
N_2/r.t.	1.15	65.65	30.15	3.06
Open air/r.t.	0.98	65.54	29.98	3.5
Air-H_2O(g) flow/800	0.88	64.69	33.88	0.54
N_2-H_2O(g) flow/800	0.88	65.12	33.05	0.94

3.2.2. In moisture saturated air and nitrogen flow at 800°C

In order to investigate the effect of moisture on the aggregation of Cr species under high temperature, the catalyst was further treated at 800°C for 2 hours in moisture-saturated air and moisture-saturated nitrogen flow, respectively, before EPMA and XPS measurements. The EPMA map images and line analysis results reveal the highly aggregated Cr species on the catalyst after each treatment and the aggregation degree is similar for the two samples. Fig. 3 shows the EPMA map image and line analysis of the catalyst sample treated in moisture-saturated air flow. Due to the formation of significantly large size of aggregated Cr area (above 10×10 µm), the surface density of the dispersed small aggregate particles decreased to a lower extent. The XPS data of this two samples in Tab. 1 show that all the stabilized Cr^{+3} was converted to Cr^{+3} species in bulky phase. Moreover, about 10%

of stabilized Cr^{+6} species was transferred into Cr^{+3} species in bulky phase. The EPMA and XPS results show that the presence of moisture rather than the using of nonoxidizing gas (N_2) seems to be the crucial factor stimulating the reduction and aggregation process under high temperature.   The conversion of stabilized Cr species into bulky chromia by water vapor and serious aggregation of chromia were corresponding to the significant decrease of the FWHM values of the $Cr_{2p}(3/2)$ XPS spectra. As illustrated in Tab. 2, the significant decrease of surface atomic concentration of Cr atom from 1.15% to 0.88% also give clue to the serious aggregation of chromia in the two samples treated by moisture-saturated gas flows.

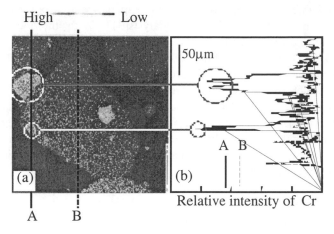

Fig. 3. EPMA map (a) and line curves (b) of the Cr distribution on the Phillips catalyst after being treated in moisture-saturated air flow at 800°C for 2 hours.

3.3. Mechanism speculation of the formation and aggregation of chromia

Based on the information obtained so far, a mechanism concerning the formation of α-chromia microcrystals during the calcination of a Phillips catalyst with 1 wt.% Cr loading is speculated as following.□After all the CrO_3 had been stabilized on the silica during the calcination process, further calcination only facilitates the dehydroxylation of the residue hydroxyl groups, and emits traces of water vapor. The thermal reduction of chromate into bulky chromia through the intermediate of CrO_3 (bulky) by this traces of water vapor seems have the lowest possibility. According to a mechanism proposed by Dalla Lana et al. that traces of water act as a catalyst for the stabilization of CrO_3 during the calcination process at high temperature [32]. As reported by McDaniel et al.[6]. the surface siloxane groups formed by dehydroxylation are also reactive with the CrO_3 especially at over 400°C. So even if the cleavage occurs the very small amount of CrO_3 formed will be instantly re-esterified with the neighboring hydroxyl or siloxane groups. As the calcination proceeds further at 800°C, another factor, that is the strain in all the surface Si-O bonds developed by the successive dehydroxylation of hydroxyl groups, will become more and more dominant in affecting the chemical stability of the chromate species. Lower Cr loading means more hydroxyl groups are removed by condensation rather than by reaction with CrO_3 so as to create much stronger strain in all the surface Si-O bonds. At a certain critical point of Si-O strain, the high temperature induced partial reduction of chromate will occur leading to the formation of stabilized Cr^{+3} species which is even resistant to oxidization by air. B.M.Weckhuysen and I.E.Wachs had also demonstrated the partial reduction of chromate into a stabilized Cr^{+5} species: a mono-oxo species with one terminal Cr=O bond and three Cr-O-support bonds, by Raman spectroscopy of zirconia, alumina and titania supported Phillips catalysts[31]. With the increasing of stabilized Cr^{+3} species, some of them may be cleft from the support surface and converted into Cr_2O_3 by traces of moisture. The high aggregation rate of the amorphous chromia at 800°C will produce a small amount of chromia microcrystals which can't be either re-oxidized or re-esterified.   Higher temperature, longer duration and higher content of moisture in the calcination process are sure to bring about more serious partial reduction and aggregation of stabilized Cr species through the above-mentioned mechanism as demonstrated in Scheme 3. The XPS evidence of the existence of the stabilized Cr^{+3} species on the catalyst also strongly supports this mechanism.

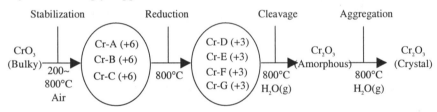

Scheme 3. Plausible mechanism of formation of Cr_2O_3 microcrystals during calcination for Phillips catalysts with low level of Cr loading.

4. Conclusions

The distribution state of Cr species on a Philllips catalyst with a typical industrial level of Cr loading 1wt.% was investigated by EPMA and XPS methods. The Cr species mostly dispersed uniformly on the surface of each catalyst particle. Whereas, three kinds of Cr species namely stabilized Cr^{+6} species, stabilized Cr^{+3} species and a few chromia microcrystals in sizes of 200-300 nm were identified on the catalyst. Based on the evidences concerning the variation of the states of Cr for the same catalyst after being treated in the open air at r.t., moisture-saturated air or moisture-saturated N_2 flow at 800°C, a plausible mechanism concerning the thermal-induced reduction of stabilized Cr^{+6} species and the formation of α-chromia microcrystals during the calcination in the preparation process of a Phillips catalyst with a low Cr loading ca. 1wt.% was speculated. The increasing strain in the surface Si-O bonds developed by the successive dehydroxylation of residue hydroxyl groups at high temperature is envisaged as the main factor to initiate this reduction under even an oxidizing gas media. The trace of moisture from the successive dehydroxylation is presumed to cleave the stabilized Cr^{+3} species into bulky chromia and subsequently to accelerate the latter's crystallization process leading to the formation of α-chromia microcrystals.

References

1. M. McDaniel, *Adv. Catal.* **33** (1985) 47.
2. B.M.Weckhuysen and R.A. Schoonheydt, *Catal. Today* **51** (1999) 215.
3. H. Charcosset, A. Revillon and A. Guyot, *J. Catal.* **8** (1967) 326.
4. H. Charcosset, A. Revillon and A. Guyot, *J. Catal.* **9** (1967) 295.
5. C. Groeneveld, P.P.M.M. Wittgen, A.M. Van Kersbergen, P.L.M. Mestrom, C.E. Nuijten and G.C.A. Schuit, *J. Catal.* **59** (1979) 153.
6. M.P. McDaniel, *J. Catal.* **67** (1981) 71.
7. A. Ellison, T.L. Overton and L. Bencze, *J.Chem.Soc.– Faraday Trans.* **89** (1993) 843.
8. A. Ellison, *J. Chem. Soc. – Faraday Trans.* **80** (1984) 2567.
9. M.I. Zaki, N.E. Fouad, J. Leyrer and H. Knozinger, *Appl. Catal.* **21** (1986) 359.
10. A. Ellison, G. Diakun and P. Worthington, *J. Mol. Catal.* **46** (1988) 131.
11. A. Rahman, M.H. Mohamed, M. Ahmed and A.M. Aitani, *Appl. Catal. A: General* **121** (1995) 203.
12. H. Schmidt, W. Riederer and H.L. Krauss, *J. Prakt. Chem.* **338** (1996) 627.
13. A. Cimino, B.A. De Angelis, A. Luchetti and G. Minelli, *J. Catal.* **45** (1976) 316.
14. S. Wang, K. Murata, T. Hayakawa, S. Hamakawa and K. Suzuki, *Appl. Catal. A: General* **196** (2000) 1.
15. D.L. Kim, J.M. Tatibouet and I.E. Wachs, *J. Catal.* **136** (1992) 209.
16. M.A. Vuurman, I.E. Wachs, D.J. Stufkens and A. Oskam, *J. Mol. Catal.* **80** (1993) 209.
17. D.L. Kim and I.E. Wachs, *J. Catal.* **142** (1993) 166.
18. Y. Okamoto, M. Fujii, T. Imanaka and S. Teranishi, *Bulletin of Chemical Society of Japan* **49** (1976) 859.
19. S.A. Best, R.G. Squires and R.A. Walton, *J. Catal.* **47** (1977) 292.
20. R. Merryfield, M. McDaniel and G. Parks, *J. catal.* **77** (1982) 348.
21. A. Cimino, D. Cordischi, S.D. Rossi, G. Ferraris, D. Gazzoli, V. Indovina, G. Minelli, M, Occhiuzzi and M. Valigi, *J. Catal.* **127** (1991) 744.
22. J.R. Sohn and S.G. Ryu, *Langmuir* **9** (1993) 126.
23. D. Gazzoli, M. Occhiuzzi, A. Cimino, G. Minelli and M. Valigi, *Surf. Interface Anal.* **18** (1992) 315.
24. D. M. Hercules, A. Proctor and M. Houalla, *Accounts of Chemical Research* **27** (1994) 387.
25. B.M. Weckhuysen, A.A. Verberckmoes, A.L. Buttiens and R.A. Schoonheydt, *J. Phys. Chem.* **98** (1994) 579.

26. I. Hemmerich, F. Rohr, O. Seiferth, B. Dillmann and H.J. Freund, *Zeitschrift Fur Physikalische Chemie - International Journal of Research in Physical Chemistry & Chemical Physics* **202** (1997) 31.

27. P.C. Thune, C.P.J. Verhagen, M.J.G. van den Boer and J.W. Niemantsverdriet, *J. Phys. Chem.* **101** (1997) 8559.

28. D.H. Cho, S.D. Yim, G.H. Cha, J.S. Lee, Y.G. Kim, J.S. Chung and I.S. Nam, *J. Phys. Chem.* **102** (1998) 7913.

29. Q. Xing, W. Milius and H.L. Krauss, *Z. Anorg. Allg. Chem.* **625** (1999) 521.

30. K.G. Miesserov, *J. Polym. Sci.: Part A-1* **4** (1966) 3047.

31. B.M. Weckhuysen and I.E. Wachs, *J. Phys. Chem. B* **101** (1997) 2793.

32. W.K. Jozwiak and I.G.D. Lana, *J. Chem. Soc. – Faraday Trans.* **93** (1997) 2583.

Main Kinetic Features of Ethylene Polymerization Reactions with Heterogeneous Ziegler-Natta Catalysts in the Light of Multi-Center Reaction Mechanism

Yury V. Kissin

Edison Research Laboratory, Exxon Mobil Chemical Co., Edison, NJ, USA
Current address: Rutgers University, Department of Chemistry, 610 Taylor Rd., Piscataway, NJ 08854-8087, USA, E-mail: ykissin@rutchem.rutgers.edu

Abstract   The previously developed kinetic scheme of ethylene polymerization reactions with heterogeneous Ziegler-Natta catalysts (refs 1-3) states that the catalysts have several types of active centers which have different activities, different stabilities, produce different types of polymers, and respond differently to reaction conditions. Each type of center produces a single polymer component (Flory component), a material with the same structure (copolymer composition, isotacticity, etc.) and a narrow molecular weight distribution with M_w/M_n=2.0. This paper examines several features of ethylene polymerization reactions in the view of this mechanism. They include temperature and cocatalyst effects on molecular weight distribution, as well as the effect of reaction parameters (temperature, ethylene and hydrogen partial pressure, α-olefin and cocatalyst concentration) on molecular weights of Flory components.

1 Introduction

At the previous symposium on olefin polymerization with Ziegler-Natta catalysts (Hamburg, 1998), we presented a new kinetic scheme for ethylene polymerization reactions with heterogeneous Ti-based Ziegler-Natta catalysts [1]. According to this scheme, the catalysts have several types of active centers which have different activities, different stabilities, produce different types of polymers, and respond differently to reaction conditions. Each type of center produces a material (a Flory component) with a uniform structure (copolymer composition) and with a narrow molecular weight distribution characterized by the M_w/M_n ratio of 2.0.

The following differences between the centers were determined:

1. Molecular weights of Flory components differ quite significantly from center to center. For example, when an ethylene homopolymerization reaction is carried out at 80°C in the absence of hydrogen, the catalyst has five different types of centers which produce polymer molecules with M_w values ranging from 10,000 to over 1,500,000.

2. Some centers copolymerize α-olefins with ethylene quite well (reactivity ratios r_1 are 40 to 50) whereas other centers copolymerize α-olefins poorly, $r_1 \sim$ 1000.

3. Each type of center responds differently to catalyst poisons.

4. Different active centers decay at different rates. As a result, polymer properties (average molecular weight, molecular weight distribution, average copolymer composition, etc.) noticeably change with reaction time.

2 New Kinetic Scheme

Earlier, a large volume of kinetic data [1-3] showed that ethylene/α-olefin copolymerization reactions exhibit several unusual kinetic features (discussed below). These features can be explained under one assumption, that a growing polymer chain containing one ethylene unit, the $Ti-C_2H_5$ group, inserts ethylene molecules relatively slowly. The stability of the $Ti-C$ bond in this group can be the result of a relatively strong β-agostic interaction between H atoms of its CH_3 group and the Ti atom. The proposed kinetic mechanism differs from the standard mechanism of ethylene polymerization reactions in two main features: (a) it assumes the existence of the stable $Ti-C_2H_5$ bond in equilibrium with an noncomplexed $Ti-C_2H_5$ bond capable of ethylene insertion, and (b) it assumes that when an α-olefin is present in the reaction, it inserts into the $Ti-H$ bond and bypasses the stage of the stabilized $Ti-C_2H_5$ group.

Addition of these two features provided plausible explanations for peculiarities of ethylene polymerization reactions with Ti-based catalysts:

1. Ethylene never exhibits a level of reactivity in homopolymerization reactions expected from its reactivity in copolymerization reactions with α-olefins [4,5] because a significant fraction of active centers is incapable of ethylene insertion due to the stable nature of the $Ti-C_2H_5$ group.

2. Introduction of any α-olefin to an ethylene polymerization reaction invariably results in a markedly higher reaction rate. This effect is the consequence of reactivation of polymerization centers due to the insertion of an α-olefin molecule into the $Ti-H$ bond.

3. The rate of ethylene homopolymerization with Ti-based catalysts has the reaction order with respect to C_E exceeding one, usually in the 1.7-1.9 range. This effect is related to the interplay of two reaction steps which depend on C_E, insertion of an ethylene molecule into the $Ti-H$ bond (resulting in the stabilized $Ti-C_2H_5$ group) and insertion of an ethylene molecule into the $Ti-C_2H_5$ bond.

4. Introduction of hydrogen causes a significant reversible depression of the ethylene polymerization rate due to a more frequent generation of $Ti-H$ bonds which leads to the formation of stabilized $Ti-C_2H_5$ bonds.

Any modification of the standard reaction scheme requires reevaluation and, in some cases, a new explanation of numerous previously known features of catalytic polymerization reactions. Many years of combined experience in using Ziegler-Natta catalysts for polymerization of various olefins and their copolymerization produced a number of empirical rules about the catalyst behavior. They include temperature effects, olefin concentration effects, the chemistry of chain initiation and chain termination reactions, etc. [4,5]. Several examples below demonstrate how the introduction of the multi-center catalytic scheme involving the stabilized Ti–C_2H_5 group affects interpretation of these long-known effects.

2 Experimental Part

Polymerization experiments were carried out in 0.5- and 3.8-liter stainless-steel autoclaves equipped with magnet-driven propeller stirrers, manometers, and external steam or electric jackets. Clean reactors were dried in a nitrogen flow at 100°C for 60 minutes and then cooled to 30°C. Solvent (n-heptane) and 1-hexene were added to a reactor under nitrogen flow, then the nitrogen atmosphere was replaced with ethylene. $AlEt_3$ (Akzo Nobel, 25-wt.% solution in hexane) and $AlMe_3$ (Witco, 25-wt.% solution in heptane) were added as cocatalysts. After that the reactor was heated to a desired temperature and hydrogen was introduced followed by a catalyst. Most experiments were carried out with a silica-supported catalyst with a Ti content of 3 wt.% [6]. Finally, ethylene was admitted to maintain a desired reaction pressure. It was continuously supplied to compensate for its consumption in the course of a polymerization reaction. The polymer slurry was dried at 25°C for 15-20 hours, then the polymer was additionally dried under vacuum at 75°C for two hours and, in the case of ethylene/α-olefin copolymers, roll-milled to achieve product uniformity.

The molecular weight and the molecular weight distribution were determined by the GPC method at 145°C with a Waters 150C Liquid Chromatograph (two columns 10^6, 10^4 and 10^3 Å). Resolution of GPC curves into Flory components [7] was carried out with the Scientist program (MicroMath Scientific Software). Copolymer compositions (reported as C_{Hex}^{copol}, mol.%) were measured by IR using a Perkin-Elmer Paragon 1000 FTIR spectro-photometer. Compositional uniformity of copolymers was analyzed with the Crystaf method. The melt index, I_2, and the melt flow ratio, I_{21}/I_2, were measured according to the ASTM method D-1238.

4 Interpretation of Kinetic Features of Ethylene Polymerization

4.1 Temperature Effect on Reaction Kinetics

An increase of reaction temperature affects Ti-based catalysts in four ways [4,5]:
1. The catalysts become more active.
2. The catalysts become less stable, especially above 90°C.
3. Average molecular weights decrease, presumably due to a higher probability of chain termination reactions, both in chain transfer with a monomer and via β-hydrogen elimination.
4. The ability of a catalyst to copolymerize α-olefins with ethylene increases.

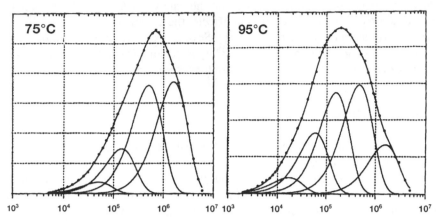

Fig. 1. GPC curves and their resolution into Flory components for ethylene/1-hexene copolymers prepared at 75 and 95°C in the absence of hydrogen, C_E=0.51-0.54 M, 2-h runs

The use of the multi-center model provides a much mode detailed and varied picture of the catalyst behavior. Figure 1 shows GPC curves of ethylene/1-hexene copolymers prepared at 75°C and 95°C under the same monomer concentration. The figure also gives the resolution of the GPC curves into Flory components Their parameters are listed in Table 1. As Figure 1 shows, the Flory peaks do not shift much with temperature. The biggest change is observed in the contents of Flory components: the contents of the two components with the highest molecular weights, IV and V, significantly decrease at a higher temperature (Table 1). This change, rather than an increase in the probability of chain transfer reactions, explains the decrease of M_w(av.) with temperature (Table 1). If one takes into account that the centers producing components IV and V have a poor ability to copolymerize α-olefins with ethylene, one should expect that copolymers produced at higher temperatures will have higher average contents of α-olefins, the prediction borne out by the experimental data (Table 1).

The second example of the temperature effect was studied for the same copolymerization reaction in the presence of hydrogen under conditions usually employed for the LLDPE synthesis. The average results (typical for the literature on Ziegler-Natta catalysis) are as expected: as the temperature increases, the catalyst productivity increases, the molecular weight of the copolymers decreases and the 1-hexene content in them increases. However, the Flory component analysis showed that most of these changes again could be interpreted as due to a change in the proportion of low molecular vs. high molecular weight components in the product mixture. Crystaf analysis of the copolymers confirmed that the

Table 1. Temperature effect on active centers of Ti-based catalyst in ethylene/1-hexene copolymerization the absence of hydrogen. Conditions: C_{Hex}^{mon}=2.2 M, C_E~0.52 M

Temp.	M_w (av.)	M_w/M_n	C_{Hex}^{copol}	Component	M_w	Fraction
75°C	857,100	4.88	0.5 mol. %	I	15,510	0.3%
				II	49,250	4.2%
				III	144,700	16.2%
				IV	504,500	39.0%
				V	1,590,000	40.3%
85°C	704,000	5.40	~1.0 mol. %*	I	17,200	1.0%
				II	54,500	7.6%
				III	140,500	22.2%
				IV	453,800	38.0%
				V	1,630,300	31.3%
95°C	433,000	6.84	2.6 mol. %	I	17,000	4.8%
				II	58,200	18.1%
				III	155,200	30.2%
				IV	467,700	32.5%
				V	1,560,000	14.4%

* Corrected for different reaction times, 4 h instead of 2 h in other experiments

materials prepared at higher temperatures have lower contents of highly crystalline components IV and V (components with a low 1-hexene content), similarly to that shown in Table 1.

4.2 Chain Transfer Reactions in Ethylene Polymerization Reactions

It is customary to discuss four chain transfer reactions in olefin polymerization reactions with Ziegler-Natta catalysts [4,5]. They are shown below using ethylene polymerization as an example:
Chain transfer with a monomer:

$$C_p^*(n) + E \ \longrightarrow(k_t^E)\rightarrow \ C_E^* + \text{Polymer}(n) \tag{1}$$

Spontaneous chain transfer (β-hydrogen elimination):

$$C_p^*(n) \ \longrightarrow(k_t^\beta)\rightarrow \ C_H^* + \text{Polymer}(n) \tag{2}$$

Chain transfer with an organoaluminum compound (cocatalyst):

$$C_p^*(n) + AlR_3 \ \longrightarrow(k_t^{Al})\rightarrow \ C_R^* + R_2Al-\text{Polymer}(n) \tag{3}$$

Chain transfer with hydrogen:

$$C_p*(n) + H_2 \quad —(k_t^H)\rightarrow \quad C_H* + \text{Polymer}(n) \tag{4}$$

According to the general steady-state kinetic approach [5], the average polymerization degree v is: v = Chain growth rate/Σ(Chain transfer rates). Its reciprocal value is used to determine kinetic parameters of chain transfer reactions:

$$1/v = (k_t^E \cdot C_E + k_t^{Al} \cdot C_{Al} + k_t^\beta + k_t^H \cdot C_H)/k_p \cdot C_E \tag{5}$$

The multi-center approach to the polymerization kinetics adds a number of nuances to this standard treatment because it allows separation of two different major effects: (a) the variation in chain transfer rates for a given type of center with reaction conditions, and, (b) effects of chain transfer agents on reactivities of different types of centers. In the latter case, α-olefins, cocatalysts and hydrogen are viewed as catalyst modifiers, either poisons or promoters.

Hydrogen is the main chain transfer agent used for the control of molecular weights of all polyolefins produced with heterogeneous Ziegler-Natta catalysts [4]. In the case of ethylene polymerization reactions, hydrogen effects are the most straightforward and easiest to interpret. Hydrogen suppresses the overall activity of Ziegler-Natta catalysts in ethylene polymerization reactions. However, the kinetic behavior of the catalysts does not change in any appreciable manner [2,3]. For example, in the case of ethylene homopolymerization at 80-90°C, reaction rates reach a maximum after 30-40 minutes and then maintain a relatively steady level, with or without hydrogen. This means that hydrogen does not affect the activation chemistry and stability of polymerization centers.

Table 2 compares the results of GPC peak resolution analysis for two ethylene homopolymers produced at 80°C, one prepared at $P_H=0$ and the another at high a P_H. In a first approximation, hydrogen does not affect contributions of different Flory components: hydrogen poisons all centers to approximately the same degree. In the case of a high P_H, Eq. 5 can be reduced to two terms, one for chain transfer with ethylene (Reaction 1) and another with H_2 (Reaction 4). As a result, linear correlations exist between the $1/v$ values for each Flory component (their M_w values are listed in Table 2) and the C_H/C_E ratio, with the k_t^H/k_p value as the slope and the k_t^E/k_p value as the intercept. These rate constant ratios for each type of active center are given in Table 3. They demonstrate the great efficiency of hydrogen as the chain transfer agent.

An important prerequisite for the applicability of the reaction scheme including Reactions 1-4 is a linear dependence between $1/v$ and the C_H/C_E ratio (or the P_H/P_E ratio) provided that C_{Al} is kept constant [5]. The validity of this assumption in the case of ethylene homopolymerization reactions was indeed shown for components III, IV and V, based on GPC data for polymers produced at several C_H/C_E. Because the presence of hydrogen does not affect contributions of different Flory components (Table 2), average molecular weights can also be used for these $1/v$ vs. P_H/P_E correlations, even their approximate estimations from viscosity parameters such as the melt index, as Fig. 2 demonstrates. It should be stressed that the linearity of the plot up to very high P_H/P_E ratios apparent from the figure is typical for ethylene homopolymerization reactions only. Deviations from the linearity occur even in the case of ethylene/α-olefin copolymerization because hydrogen poisons different active centers in these reactions to a different degree [1-3]. Propylene polymerization reactions are also different: at high P_H/P_{Pr} ratios, the molecular weight of polypropylene is not affected by hydrogen anymore [8,9].

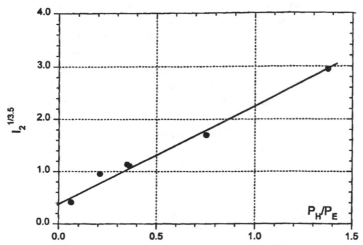

Fig. 2. Hydrogen effect in ethylene homopolymerization reactions in coordinates of Eq. 5. $1/\nu$ values (from average molecular weights, estimated from the melt index) vs. P_H/P_E ratios in ethylene homopolymerization reactions at 85°C

Table 2. Hydrogen effect on Flory components in ethylene homopolymerization with Ti-based catalyst, 1-hour experiments at 80°C

P_H	P_E	Component	M_w	Fraction
0	0.38 MPa	I	10,700	3.2%
		II	47,500	10.6%
		III	146,600	27.0%
		IV	462,800	38.0%
		V	1,577,000	21.1%
0.69 MPa	0.38 MPa	I	1,550	2.6%
		II	6,200	9.2%
		III	18,500	24.9%
		IV	50,300	45.2%
		V	145,800	18.0%

Table 3. Hydrogen effect on kinetic parameters of active centers in ethylene polymerization

Component	I	II	III	IV	V
k_t^E/k_p	0.0028	0.0012	0.0004	0.00012	0.00035
k_t^H/k_p	~0.20	0.049	0.016	0.0063	0.0021 (80°C)

The choice of a cocatalyst influences the performance of Ti-based catalysts in many respects. It affects their activity, polymerization kinetics, molecular weights and molecular weight distributions of polyolefins, etc. [4,5]. Here we address only the issue of the chain transfer with organoaluminum compounds (Reaction 3) and its significance compared to other chain transfer reactions. We carried out two ethylene/1-hexene copolymerization reactions with the same catalyst activated with AlEt₃ at concentrations which differed 5 times. The overall results are consistent with the existing notion: because AlEt₃ is a chain transfer agent, the M_w(av.) value of the polymer prepared at a high C_{Al} is lower (103,000 vs. 127,200) whereas the ability of the catalyst to copolymerize ethylene with 1-hexene and the molecular weight distribution remain approximately the same.

When the molecular weight parameters of each Flory component in the two copolymers are compared, a more detailed picture of the cocatalyst concentration effect emerges. AlEt₃, in a high concentration, has two effects: it suppresses activities of all types of active centers approximately to the same degree (fractions of Flory components do not change) and it acts as a weak chain transfer agent. However, it affects center I (the center which produces macromolecules with the highest α-olefin content and the lowest molecular weight) much stronger than other centers. Calculations using Eq. 5 give the following k_t^{Al}/k_p values:

Center:	I	II	III	IV	V
k_t^{Al}/k_p:	0.304	0.051	0.012	0.006	0.003

If one takes into account that cocatalysts are usually used at much lower concentrations than monomers, the absolute k_t^{Al}/k_p values signify that, in practical terms, one can neglect the chain transfer with a cocatalyst in comparison with two major chain transfer reactions, those with hydrogen and with a monomer.

The β-hydrogen elimination reaction (Reaction 2) is usually regarded an important factor controlling molecular weights of polyolefins [4,5]. Detailed kinetic analysis shows, however, that quantitative evaluation of the role of this reaction is elusive. Only experiments in the absence of hydrogen can be used for this purpose. Because one expects that the β-hydrogen elimination reaction is much more pronounced when the last monomer unit in a polymer chain has the tertiary C–H bond, we compared molecular weights of Flory components in an ethylene homopolymer and ethylene/1-hexene copolymers prepared at 85°C in the absence of hydrogen.

Introduction of any α-olefin in an ethylene polymerization reaction results in significant kinetic changes [1-3], the most pronounced being an increase of the fractions of Flory components I, II and III. However, a comparison of molecular weights of Flory components provides information about the significance of the β-hydrogen elimination reaction. Centers IV and V react with 1-hexene very poorly and one cannot expect to see any changes in their molecular weights, the effect that was indeed found. Centers II and III, however, are sensitive to the 1-hexene presence. Our data show that doubling of the 1-hexene concentration (and the resulting doubling of the C_{Hex}^{copol} value) does not affect molecular weights of these components in any systematic way. This result suggests that the β-hydrogen elimination reaction is a relatively rare event even at this high temperature.

4.3 Reaction Order with Respect to Monomer Concentration

Ethylene homopolymerization reactions exhibit the reaction order with respect to C_E (or P_E) greatly exceeding one and usually in the 1.7-2.0 range [3,10-12]. One example of this dependence is shown in Fig. 3-A. It gives the results of a multi-step slurry experiment in which P_E was increased step-wise. In this experiment, the effective reaction order with respect to P_E is ca. 1.8.

The multi-center model of ethylene polymerization reactions explains both the high reaction order and the activating effect of α-olefins as a result of stability of the Ti–C₂H₅ group [1-3]. This model implies also that there may be other types of catalysts which, in the absence of significant stabilization of the M–C₂H₅ group, may be free of both these effects. In such cases, the reaction order with respect to C_E should be close to one, and introduction of an α-olefin in an ethylene polymerization reaction should not affect the reaction rate or may even decrease it, depending on α-olefin concentration [3].

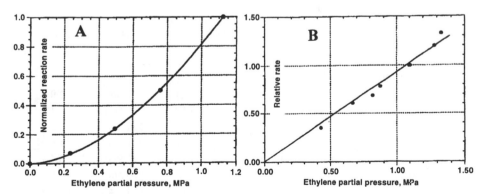

Fig. 3. Relative reaction rates as a function of P_E in multi-step ethylene polymerization reactions with Ti-based catalyst at 80°C (A) and with Cr oxide-based catalyst at 90°C (B)

This hypothesis was tested by studying kinetics of ethylene polymerization reactions with a supported chromium oxide-based catalyst in several multi-step slurry and gas-phase experiments. Figure 3-B shows the plot of the relative reaction rates in one of the experiments vs. P_E. In this case, the rate of the ethylene polymerization reaction has the first-order dependence on P_E. Several experiments of a similar type in heptane slurry at 75 and 90°C confirmed the first-order dependence. The same catalyst was used at 90°C to test the α-olefin effect. In agreement with the model, the reaction kinetics is not affected at a low 1-hexene concentration, 0.23 M, but the catalyst activity is noticeably decreased at a high 1-hexene concentration, 1.12 M.

4.4 Cocatalyst Type Effect on Molecular Weight Distribution of Polyethylene

Replacement of AlEt$_3$ with AlEt$_2$Cl usually depresses activity of Ti-based catalysts and results in an increase of molecular weight of polymers [4,5]. We examined this effect in ethylene polymerization reactions using a Ti-based catalyst activated with AlEt$_3$ and AlEt$_2$Cl. On a superficial level, differences between AlEt$_3$ and AlEt$_2$Cl as cocatalysts are as follows (Table 4): the AlEt$_2$Cl-activated catalyst has a lower productivity, it produces a polymer with a higher molecular weight and a broader molecular weight distribution.

Comparison of GPC curves of the two copolymers reveals two features:

1. The position of the GPC curve shifts to higher molecular weights in the copolymer prepared with AlEt$_2$Cl. This means that molecular weights of all Flory components are increased in this copolymer.

2. A noticeable high molecular weight shoulder appears in the GPC curve of the copolymer prepared with AlEt$_2$Cl.

Table 4. GPC data for ethylene/1-hexene copolymers produced with Ti-based supported catalyst activated with AlEt$_3$ and AlEt$_2$Cl in the presence of hydrogen. Conditions: 85°C, $C_E = 0.54$ M, $P_H = 0.276$ MPa, 2-h runs

Cocat.	Product., G/g cat h	C_{Hex}^{copol}, mol %	M_w (av.)	M_w/M_n	Comp.	M_w	Fraction
AlEt$_3$	2820	2.2	103,500	3.8	I	8,500	2.2%
					II	23,000	12.9%
					III	54,000	38.1%
					IV	124,400	35.2%
					V	311,100	11.6%
AlEt$_2$Cl	2400	2.0	160,000	4.3	II	21,900	8.4%
					III	60,500	41.8%
					IV	154,700	36.5%
					V	577,300	13.2%

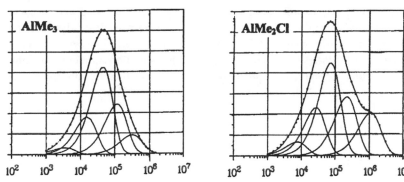

Fig. 4. Comparison of GPC curves of ethylene/1-hexene copolymers prepared at 85°C under the same conditions with Ti-based $MgCl_2$-supported catalyst activated with $AlMe_3$ and $AlMe_2Cl$

A more detailed analysis of the GPC data is given in Table 4. If one neglects the results for low molecular weight components I and II (their M_w estimation is not sufficiently precise), the cocatalyst replacement obviously brings about two types of changes: molecular weights of components III and IV increase somewhat (by 12 and 24%, respectively), whereas the molecular weight of component V nearly doubles. As a result, this component appears as a prominent shoulder on the GPC curve.

In the case of other types of Ti-based catalysts, particularly those using $MgCl_2$ as a support, replacement of AlR_3 with AlR_2Cl not only increases the molecular weight of component V over 3 times (Fig. 4) but increases its fraction nearly two times as well. As a result, the molecular weight distribution of such polymers increases dramatically. For example, the M_w/M_n value for the copolymers in Fig. 4 increases from 5.5 to 10.9, although Flory components I-IV in both products are similarly spaced and their average M_w/M_n values remain similar, 4.2 vs. 4.9.

5 Conclusions

The formulation of the multi-center kinetic scheme of ethylene polymerization reactions with heterogeneous catalysts [1-3] and the development of kinetic tools necessary for the application of the scheme significantly expands our understanding of working of the catalysts. Previous kinetic investigations of the effects of reaction variables (temperature, type of cocatalyst, concentrations of α-olefins, hydrogen, etc.), as well as the studies of catalyst modifiers and poisons, were mostly limited to the analysis of their effects on the overall catalyst activity and the average molecular weight [4,5]. Now it becomes possible to evaluate each of these effects on activities of different types of centers and on molecular weights of respective Flory components. Such an analysis significantly improves our understanding of catalyst behavior and provides additional tools for catalyst and process control.

6 Acknowledgments

The experimental work of J. L. Garlick and D. A. Brown (polymer synthesis), J. J. Orvos (GPC measurements) and Dr. H. Fruitwala (Crystaf analysis) is greatly appreciated. The author thanks Drs. R. I. Mink and T. E. Nowlin (Edison Research Center, Exxon Mobil Chemical Company) for helpful discussions.

7 References

1   Kissin YV, Mink, RI, Nowlin TE, Brandolini AJ (**1999**) In Kaminsky W (ed) Metalorganic Catalysts for Synthesis and Polymerization, Springer, Berlin, p 60
2   Kissin YV, Mink RI, Nowlin TE, Brandolini AJ (**1999**) *Topics in Catalysis* 7: 69
3   Kissin YV, Mink RI, Nowlin TE, Brandolini AJ (**1999**) *J Polym Sci, Part A: Polym Chem* 37: 4255, 4273, 4281
4   Boor J Ziegler-Natta Catalysts and Polymerizations, Academic Press, New York, **1979**
5   Kissin YV Isospecific Olefin Polymerization with Heterogeneous Ziegler-Natta Catalysts, Springer, New York, 1985
6   Mink RI, Nowlin, TE (**1995**) US Patent 5,470,812
7   Kissin YV (**1995**) *J Polym Sci, Part A: Polym Chem* 33: 227
8   Vizen EI, Rishina LA, Sosnovskaya LN, Dyachkovsky FS, Dubnikova IL, Ladygina TA (**1994**) *Eur Polym J* 30: 1315
9   Spitz R, Bobichon C, Guyot A (**1989**) *Makromol Chem* 190: 707, 717
10   Kissin YV (**1989**) *J Molec Catal* 46:220
11   Kryzhanovskii AV, Gapon II, Ivanchev SS (**1990**) *Kinetics Catalysis* 31: 90
12   Han-Adebkun GC, Ray WH (**1997**) *J Appl Polym Sci* 65: 1037

Hydrogen Effects in Propylene Polymerization with Ti-Based Ziegler-Natta Catalysts. Chemical Mechanism

Yury V. Kissin[*a)] and Laura A. Rishina[b)]

[a)] Rutgers University, Department of Chemistry, 610 Taylor Rd., Piscataway, NJ 08854-8087, USA, E-mail: ykissin@rutchem.rutgers.edu
[b)] Semenov Institute of Chemical Physics, 117977 Moscow, Kosygin st. 4, Russia, E-mail: rishina@online.ru

Abstract The hydrogen activation effect in propylene polymerization reactions with Ti-based catalysts is usually explained in the literature as a consequence of the secondary insertion of a propylene molecule into the $Ti-CH_2CH(CH_3)$-Polymer bond resulting in the formation of a sleeping active center. This article proposes a different mechanism of the hydrogen activation effect. It is based on the assumption that propylene insertion into the Ti−H bond has poor regioselectivity, in contrast to its insertion into the Ti−C bond, which is highly regioselective. $Ti-CH(CH_3)_2$ species formed after the secondary insertion of propylene into the Ti−H bond is stable due to the β-agostic interaction between the H atom of its CH_3 group and the Ti atom. However, the Ti−H bond is restored in a reaction with hydrogen. Validity of this mechanism is demonstrated in the study of ethylene/α-olefin copolymerization reactions with two Ti-based catalysts. GC analysis of ethylene-propylene co-oligomers prepared in the presence of hydrogen showed that the probability of the secondary insertion of a propylene molecule into the Ti−H bond is only 3-4 times lower than the probability of its primary insertion.

1 Introduction

Propylene polymerization reactions with $MgCl_2$-supported Ziegler-Natta catalysts are strongly activated in the presence of hydrogen [1-11]. The activation effect is reversible: removal of hydrogen reduces the polymerization rate and addition of hydrogen in the course on an established polymerization reaction results in immediate activation of the catalysts [1,8,11]. Proposed explanations of this effect include hydrogen effects on the number of active centers [6,8], oxidation of Ti^{2+} to Ti^{3+} [10], hydrogenation of inactive Ti allylic species [1-3,11], and hydrogenation of unsaturated chain ends that poison active sites [12]. Currently, the most widely accepted hypothesis for the effect focuses on regioselectivity of the catalysts [9,13-

16]. Active centers of all Ziegler-Natta and metallocene catalysts are occasionally capable of the secondary insertion (2,1-insertion) of a propylene molecule into the M–C bond resulting in the formation of a "sleeping" active center $M-CH(CH_3)R'$:

$$M-CH_2CH(CH_3)R + CH_3CH=CH_2 \rightarrow M-CH(CH_3)CH_2CH_2CH(CH_3)R \qquad (1)$$

This center is inactive in the propylene insertion reaction due to a very low reactivity of the $M-CHR'R''$ bond. In the absence of other chain transfer reactions, the center may remain inactive until the β-hydrogen elimination reaction occurs. However, such centers can be reactivated after chain transfer with hydrogen:

$$M-CH(CH_3)CH_2CH_2CH(CH_3)R + H_2 \rightarrow M-H + CH_3CH_2CH_2R \qquad (2)$$

This article proposes an alternative explanation of the hydrogen activation effect in propylene polymerization reactions with Ti-based Ziegler-Natta catalysts. The mechanism is based on our earlier studies of propylene polymerization kinetics with supported Ti-based catalysts [4] and the studies of ethylene polymerization kinetics with similar Ti-based catalysts [17,18].

2 Proposed Mechanism of Hydrogen Activation Effect

As discussed earlier [17,18], a large volume of kinetic data on ethylene/α-olefin copolymerization reactions can be rationalized on the basis of a single assumption, that a growing polymer chain containing one ethylene unit, the $Ti-C_2H_5$ group, inserts ethylene molecules relatively slowly. The stability of the Ti–C bond in this group can be the result of a strong β-agostic interaction between the H atom of its CH_3 group and the Ti atom. This addition to the standard reaction scheme provides plausible explanations for most peculiarities of ethylene polymerization reactions with Ti-based catalysts [17,18]. The feature of these reactions most pertinent to the present subject is the hydrogen effect on the reaction kinetics. In contrast to propylene polymerization, hydrogen always depresses ethylene polymerization reactions due to a more frequent generation of Ti–H bonds, which leads, in the first ethylene insertion step, to the formation of species with stabilized $Ti-C_2H_5$ groups.

If the same reasoning is applied to propylene polymerization, one can expect a different hydrogen effect. The primary propylene insertion into the Ti–H bond bypasses the formation of the stabilized $Ti-C_2H_5$ group:

$$Ti-H + CH_2=CHCH_3 \rightarrow Ti-CH_2CH_2CH_3 \qquad (3)$$

Indeed, addition of propylene to an ethylene polymerization reaction produces a significant increase in activity, similarly to the effects of other α-olefins [19]. According to the proposed reaction mechanism [26,28], this signifies that the Ti–C bond in the Ti$-n$-C$_3$H$_7$ group is much more reactive in ethylene insertion compared to the same bond in the Ti$-$C$_2$H$_5$ group.

However, the Ti$-$H bond in a catalytic center can be less regioselective in propylene insertion than the Ti$-$C bond. If the secondary propylene insertion into the Ti$-$H bond takes place:

$$Ti-H + CH_3CH=CH_2 \rightarrow Ti-CH(CH_3)_2 \qquad (4)$$

The resulting Ti$-iso$-C$_3$H$_7$ species should exhibit the same feature as the Ti$-$C$_2$H$_5$ species, significant stabilization due to a strong β-agostic interaction between the H atom of one of its CH$_3$ groups and the Ti atom. The fate of the Ti$-iso$-C$_3$H$_7$ species depends on the nature of the catalyst and on reaction conditions. Most Ti-based catalysts exhibit a negligible tendency for the monomer unit inversion; propylene polymers produced with such catalysts have a very low number of "head to head" units in their chains [20]. Therefore, one can consider the Ti$-iso$-C$_3$H$_7$ group as a sleeping center in a propylene polymerization reaction. It eventually either decomposes through β-hydrogen elimination with the expulsion of propylene (a slow process because the β-H atom is removed from the CH$_3$ group) or reacts with a cocatalyst. However, if hydrogen is present, the Ti$-iso$-C$_3$H$_7$ group can be hydrogenated with the formation of propane and restoration of the Ti$-$H bond. Indeed, generation of propane and significant consumption of hydrogen in propylene polymerization reactions with Ti-based catalysts [1,21] is well known.

One can also expect that ethylene insertion into the Ti$-iso$-C$_3$H$_7$ species, although relatively slow, is possible:

$$Ti-CH(CH_3)_2 + CH_2=CH_2 \rightarrow Ti-CH_2CH_2CH(CH_3)_2 \qquad (5)$$

Reaction 5 reactivates the sleeping center and makes it available for further chain growth reactions, both with ethylene and with propylene. Indeed, addition of small quantities of ethylene to an established propylene polymerization reaction results in an increase of the propylene consumption rate [1].

Below we present experimental data which confirm that Reactions 4 and 5 indeed take place in ethylene/propylene copolymerization reactions and that the reaction mechanism including Reaction 4 is a viable alternative mechanism which explains the hydrogen activation effect in propylene polymerization reactions.

3 Experimental Part

Two types of Ti-based catalysts were used, δ-TiCl$_3$ and a supported TiCl$_4$/dioctyl phthalate/MgCl$_2$ catalyst containing 2.3 wt.% of Ti. Copolymerization experiments were carried out in a 0.5-liter stainless-steel autoclave equipped with a magnet-driven propeller stirrer, a manometer, and an external electric heating jacket. In a typical experiment, solvent (*n*-heptane) was added to the clean reactor under nitrogen flow (200 or 100 ml) and the nitrogen atmosphere was replaced with propylene. AlEt$_3$ (hexane solution, 1.5 mmol) was added as a cocatalyst. In some experiments, it was modified with PhSi(OEt)$_3$ at an [Al]:[Si] ratio of 20. After the cocatalyst addition, the reactor was heated to a desired temperature, propylene and hydrogen were added to it followed by a pre-weighed quantity of a dry catalyst (ca. 0.001 g) or catalyst slurry in hexane, and, finally by ethylene which was continuously supplied to maintain a constant reaction pressure. The reactions were usually carried out to a low product yield to avoid propylene depletion in the reactor.

IR spectra were recorded with a Perkin-Elmer Paragon 1000 spectrophotometer. Copolymer compositions were calculated from IR data according to ref. 22. GC analysis of oligomers was carried out with a Hewlett-Packard 5890 gas chromatograph equipped with a 60-meter MTX-1 column, (Restek Co.) and FID.

4 Identification of Secondary Propylene Insertion

The goal of this research was identification of a particular "starting" chain end in ethylene/propylene copolymers, Pol-CH$_2$-CH$_2$-CH(CH$_3$)$_2$. This chain end is formed in a sequence of steps starting with Reactions 4 and 5. GC analysis of low molecular weight fractions of these copolymers, ethylene/propylene co-oligomers, affords identification of this chain end due to ability to differentiate various chains.

Three such differentiation options exist potentially:

1. If the growth of a co-oligomer chain starts in Reaction 4, includes several ethylene insertion steps and culminates in β-hydrogen elimination,

$$\text{Ti–CH}_2\text{CH}_2\text{-E}_n\text{-CH(CH}_3)_2 \;\rightarrow\; \text{Ti–H} + \text{CH}_2\text{=CH-E}_n\text{-CH(CH}_3)_2 \tag{6}$$

the oligomer molecules have odd carbon atom numbers and contain the vinyl group on one end and the isopropyl group on the other end.

2. If the growth of a co-oligomer chain starts in Reaction 4, includes several ethylene insertion steps, one primary propylene insertion and β-H elimination,

$$\text{Ti–CH}_2\text{CH(CH}_3)\text{-E}_n\text{-CH(CH}_3)_2 \rightarrow \text{Ti–H} + \text{CH}_2\text{=C(CH}_3)\text{-E}_n\text{-CH(CH}_3)_2 \tag{7}$$

the oligomer molecules have even carbon atom numbers and contain the vinylidene group on one end and the isopropyl group on the other end.

3. If the same sequence of reactions culminates in chain transfer with hydrogen,

$$\text{Ti–CH}_2\text{CH(CH}_3)\text{-E}_n\text{-CH(CH}_3)_2 + \text{H}_2 \rightarrow \text{Ti–H} + \text{(CH}_3)_2\text{CH-E}_n\text{CH(CH}_3)_2 \tag{8}$$

the oligomer molecules have even carbon atom numbers and contain isopropyl groups on both chain ends.

Our earlier GC work with ethylene/propylene co-oligomers produced with Ni-containing catalysts affords an unambiguous identification of these product families [23,24] and differentiation between these products and such "standard" oligomers as $(CH_3)_2CH-E_n-CH_2-CH_3$. The latter are alkanes with odd carbon atom numbers and with the isopropyl group on one of the chain ends.

4.1 Ethylene/Propylene Copolymers Produced with Ti-Based Catalysts

According to our ^{13}C NMR data, polypropylene produced with the highly isospecific $TiCl_4$/dioctyl phthalate/$MgCl_2$ catalyst has no signs of secondary propylene insertion into the Ti–C bond. We carried out several ethylene/propylene copolymerization reactions with this catalyst and with δ-$TiCl_3$. The reactions were run at relatively high temperatures and at high P_H in order to increase the yields of oligomers. The results are given in Table 1. IR analysis of the copolymers as well as ^{13}C NMR data for the copolymer produced in run 4 showed that they all have very low levels of unsaturation. The double bonds are $CH_2=C(CH_3)-$ and $CH_2=CH-$ at a ratio ranging from 4 to 6.

4.2 Standard Chain-Growth Reactions

Figure 1 shows a part of the gas chromatogram of the oligomers formed in run 4. Most peaks are assigned to standard oligomer structures:

1. n-C_n alkanes with even n numbers, n-decane and n-dodecane, are ethylene homo-oligomers H-$(E)_n$-H. Respective α-olefins $CH_2=CH-(E)_{n-1}$-H, 1-decene and 1-dodecene, are present in low quantities, in accordance with the IR data.

2. n-C_n alkanes with odd n numbers, such as n-undecane, are H-$(E)_{n-1}$-Pr-H co-oligomers. Respective $CH_2=CH-(E)_{n-2}$-Pr-H oligomers are also present.

Table 1. Reaction conditions of ethylene/α-olefin copolymerization reactions

Run	Catalyst	Temp., °C	P_{Ol}, MPa	P_E, MPa	P_H, MPa	Time, min	Yield, g/g cat	C_{Pr} mol %
	Ethylene/propylene copolymerization with a Ti-based catalyst (cocatalyst AlEt$_3$).							
1	δ-$TiCl_3$	85	0.42	0.41	0.31	12	357	19
2	" - "	105	0.66	0.28	0.34	5	864	25
3	$TiCl_4$/dioctyl phthalate/$MgCl_2$	95	0.72	0.29	0.14	60	990	37
4	" - "	100	0.59	0.24	0.28	12	4090	47
	Ethylene/1-butene copolymerization with a Ti-based catalyst (cocatalyst AlEt$_3$).							
5	$TiCl_4$/dioctyl phthalate/$MgCl_2$	95	0.34	0.21	0.28	30	3400	-

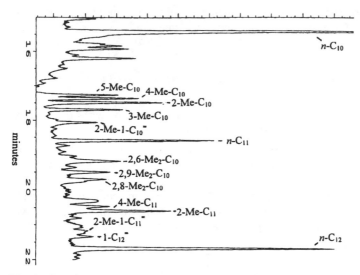

Fig. 1. Gas chromatogram (C_{11}-C_{12} range) of ethylene/propylene co-oligomers produced with $TiCl_4$/dioctyl phthalate/$MgCl_2$-$AlEt_3$ system at 100°C (Table 1, Run 4)

3. Vinylidene olefins $CH_2=C(CH_3)-(E)_{n-1}-H$ (2-methyl-1-decene) and $CH_2=C(CH_3)-(E)_{n-2}-P-H$ (2-methyl-1-undecene) are present in low quantities whereas isoalkanes $(CH_3)_2CH-(E)_{n-1}-H$ (2-methyldecane) and $(CH_3)_2CH-(E)_{n-2}-P-H$ (2-methylundecane), have the highest yields among branched alkanes, as expected.
4. Saturated oligomers with one propylene unit in the middle of the chain, $H-(E)_i-P-(E)_j-H$, were also identified. In the C_{11} range they are 3-, 4- and 5-methyl-substituted decanes. Products of this type containing two propylene units, $H-P-(E)_i-P-(E)_j-H$, are also present, 2,6- and 2,8-dimethyldecane.

4.3 Secondary Propylene Insertion into Ti—H Bond

The C_{12} range in Fig. 1, as well as other ranges, gives clear support to reaction sequences which start with the secondary insertion of a propylene molecule into the Ti–H bond and are followed by insertion of several ethylene molecules. When this reaction sequence is terminated in Reaction 8, it produces $(CH_3)_2CH-(E)_n-CH(CH_3)_2$ oligomers: 2,9-dimethyldecane (Fig. 1), 2,7-dimethyloctane, 2,11-dimethyldodecane, etc. Yields of these oligomers were used to evaluate the probability of the secondary propylene insertion into the Ti–H bond, as described below. Peaks of respective olefins, $CH_2=C(CH_3)-(E)_{n-2}-CH(CH_3)_2$, are very small and were clearly observed only in the C_{10} range, 2,7-dimethyl-1-octene.

The second GC-based proof of the possibility of the secondary insertion of an α-olefin molecule into the Ti–H bond was produced by examining oligomers formed in an ethylene/1-butene copolymerization reaction (Table 1). These oligomers contain only C_n alkanes with even n numbers. Figure 2 shows that main peaks in the C_{12} range (and in all other ranges) belong to H-$(E)_n$-H and H-$(E)_{n-1}$-Bu-H oligomers (n-dodecane) and CH_3-$C(C_2H_5)$-$(E)_{n-1}$-H oligomers (3-methyl-undecane). 3-, 4- and 5-ethyldecane are co-oligomers with one 1-butene unit in the middle of the chain. The chromatogram also contains the peak of 3,8-dimethyl-decane, the (CH_3)-$CH(C_2H_5)$-$(E)_2$-$CH(C_2H_5)$-CH_3 oligomer. The respective product in the C_{10} range is also present, 3,6-dimethyloctane. Such molecules can be formed only after the secondary insertion of a 1-butene into the Ti–H bond.

The third confirmation of the possibility of the secondary insertion of an α-olefin molecule into the Ti–H bond was found by examining oligomers formed in an ethylene/4-methyl-1-butene copolymerization reaction with a $TiCl_4$-based catalyst designed for synthesis of LLDPE resins. The structures of the same oligomers produced with a V-based catalyst were reported earlier [25]. Gas chromatograms of both the V-catalyzed oligomers (Fig. 3 in ref. 25) and the Ti-catalyzed oligomers show that they contain a peak in the C_{14} range assigned to 2,4,7,9-tetramethyldecane. This isoalkane can only be formed in the following series of reactions: the secondary insertion of a 4-methyl-1-pentene molecule into the Ti–H bond, insertion of one ethylene molecule, the primary insertion of another 4-methyl-1-pentene molecule, and chain transfer with hydrogen.

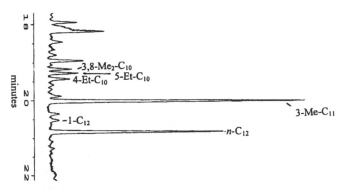

Fig. 2. Gas chromatogram (C_{12} range) of ethylene/1-butene co-oligomers produced with $TiCl_4$/dioctyl phthalate/$MgCl_2$-$AlEt_3$ system at 95°C (Table 1, run 5)

4.4 Quantitative Estimation of Secondary vs. Primary Propylene Insertion

The goal of the quantitative analysis is the estimation of the ratio of two rate constants, those for the primary propylene insertion into the Ti–H bond (Reaction 3) and the secondary insertion (Reaction 4). One oligomer pair is best suited for the purpose in terms of uniqueness of the oligomer structure: $(CH_3)_2CH-(E)_m-CH_2-CH_2-CH_3$ (the chain growth starts with Reaction 3) and $(CH_3)_2CH-(E)_m-CH(CH_3)_2$ (the chain growth starts with Reaction 4). As we discussed earlier [18], it is convenient to carry out kinetic analysis of co-oligomer yields in terms of formation probabilities (γ) of different chains. For these two products, the formation probabilities are, respectively:

$$\gamma[(CH_3)_2CH-(E)_m-CH_2-CH_2-CH_3] = P_{Pr}^{primary} \cdot P_{E/p-Pr} \cdot P_{E/E}^{m-1} \cdot P_{p-Pr/E} \cdot P_{p-Pr}^{H} \qquad (9)$$

$$\gamma[(CH_3)_2CH-(E)_m-CH(CH_3)_2] = P_{Pr}^{secondary} \cdot P_{E/s-Pr} \cdot P_{E/E}^{m-1} \cdot P_{p-Pr/E} \cdot P_{p-Pr}^{H} \qquad (10)$$

where the two probabilities of propylene insertion into the Ti–H bond are $P_{Pr}^{primary}$ and $P_{Pr}^{secondary}$, the probability of a monomer A insertion into a polymer chain with the last B monomer unit is $P_{A/B}$, and P_{p-Pr}^{H} is the probability of the chain transfer with hydrogen after the primary propylene insertion. Each probability is the linear function of the rate constant of the respective reaction and the concentration of a particular monomer. The oligomer yield ratio (from Eqs 9 and 10) is:

$$[(CH_3)_2CH-(E)_m-CH_2-CH_2-CH_3]/[(CH_3)_2CH-(E)_m-CH(CH_3)_2] = \qquad (11)$$

$$(P_{Pr}^{primary}/P_{Pr}^{secondary}) \cdot (P_{E/p-Pr}/P_{E/s-Pr})$$

where the $P_{Pr}^{primary}/P_{Pr}^{secondary}$ is the regioselectivity parameter of propylene insertion into the Ti–H bond we seek. To determine this parameter, one has to estimate the second term in Eq. 11, $P_{E/p-Pr}/P_{E/s-Pr}$.

The $P_{E/p-Pr}$ probability reflects the rate constant of the following reaction:

$$Ti-CH_2CH_2CH_3 + CH_2=CH_2 \rightarrow Ti-CH_2CH_2CH_2CH_2CH_3 \qquad (12)$$

It is approximately equal to the probability of ethylene insertion into a polymer chain with the last ethylene unit, $P_{E/E}$. The $P_{E/s-Pr}$ probability reflects the rate constant value of Reaction 5 and it is expected to be lower than the $P_{E/E}$ value.

The literature provides a means for an approximate estimation of the $P_{E/p-Pr}/P_{E/s-Pr}$ value. Natta et al. described ethylene/2-butene copolymerization in the presence of V-based catalyst systems [26] and kinetics of this reactions. 2-Butenes do not homopolymerize with these catalysts but produce copolymers with ethylene. At low [ethylene]:[2-butene] ratios, the copolymers have a nearly alternating structure and most of ethylene is consumed in the following reaction:

$$V-CH(CH_3)CH(CH_3)CH_2CH_2R + CH_2=CH_2 \rightarrow \tag{13}$$
$$V-CH_2CH_2CH(CH_3)CH(CH_3)CH_2CH_2R$$

Ethylene consumption rate in Reaction 13 is relatively low. However, as the [ethylene]:[2-butene] ratio increases, another chain growth reaction takes place:

$$V-CH_2CH_2CH(CH_3)CH(CH_3)R + CH_2=CH_2 \rightarrow \tag{14}$$
$$V-CH_2CH_2CH_2CH_2CH(CH_3)CH(CH_3)R$$

Its rate constant is higher than that for Reaction 13, therefore, the ethylene consumption rate increases. Kinetic treatment of this effect was the source of an approximate estimation of the $P_{E/p-Pr}/P_{E/s-Pr}$ value. The $k_{E/E}/k_{E/B}$ ratio in ethylene/*cis*-2-butene copolymerization reactions was estimated as ca. 1.4. Ti-based catalysts do not copolymerize ethylene and 2-butenes. Therefore, one can expect that Ti-based catalysts exert a more strict steric control in Reaction 5 (which is similar to Reaction 13) and that the $k_{E/E}/k_{E/B}$ ratio for them will be significantly higher than 1.4. Respectively, the $P_{E/p-Pr}/P_{E/s-Pr}$ value for Ti-based catalysts, which is approximated by the $k_{E/E}/k_{E/B}$ value, is also higher than 1.4.

Now, using Eq. 11 and the approximate estimation of the $P_{E/p-Pr}/P_{E/s-Pr}$ value, one can calculate regioselectivity of the Ti–H bond in the propylene insertion reaction, the $P_{Pr}^{primary}/P_{Pr}^{secondary}$ ratio. The results are given in Table 2. They show that for both Ti-based catalysts the secondary insertion of a propylene molecule into the Ti–H bond at 80-100°C is only 3-4 times less probable compared to its primary insertion.

Table 2. Estimation of $P_{Pr}^{primary}/P_{Pr}^{secondary}$ value, regioselectivity parameter of Ti–H bond in propylene insertion reactions

Run	Catalyst	Temp.,°C	$P_{Pr}^{primary}/P_{Pr}^{secondary}$
1	δ-TiCl$_3$	85	<2.6-3.2
2	" - "	105	<3.7-4.0
3	TiCl$_4$/dioctyl phthalate/MgCl$_2$	95	<3.5-4.0
4	" - "	100	<2.2-3.0

5 Conclusions

The quantitative estimation of regioselectivity of propylene insertion into the Ti–H bond adds credibility to the proposed explanation of the hydrogen activation effect in propylene polymerization reactions:

1. Two parallel reactions are possible in every chain initiation step (propylene insertion into the Ti–H bond), the primary insertion leading to the propagation center Ti–$CH_2CH_2CH_3$ (Reaction 3) and the secondary insertion leading to the Ti–$CH(CH_3)_2$ species (Reaction 4). The two species are formed in a ~3:1 ratio.

2. The Ti–$CH(CH_3)_2$ species is stable due to the β-agostic interaction of the H atom in its CH_3 group and the Ti atom, it does not insert propylene molecules.

3. Decomposition routes of the Ti–$CH(CH_3)_2$ species in the absence of hydrogen, β-hydrogen elimination and reactions with a cocatalyst, are slow, and a large fraction of active centers is dormant. Ti–$CH(CH_3)_2$ species accumulate in the catalysts until the equilibrium is reached between the rates of their formation and decomposition. Judging by the kinetic data [4], only ~10% of all potentially active species are engaged in chain growth reactions in the absence of hydrogen.

4. Addition of hydrogen, which reacts with the Ti–$CH(CH_3)_2$ species and restores the Ti–H bond, results in activation of a propylene polymerization reaction.

The proposed mechanism can also be applied for the explanation of several some other features of propylene polymerization, such as their activation in the presence of AlR_2H [27] (it leads to the Ti–H species) and the activation effect of ethylene, which is due to Reaction 5 [1].

Opposite effects of hydrogen in two polymerization reactions, depression of ethylene polymerization and activation of propylene polymerization, are caused by the same feature of Ti–$CH(CH_3)R$ groups (where R is either H or CH_3). These groups have relatively low reactivity in olefin insertion reactions due to the strong β-agostic interaction between the Ti atom and the H atom in the CH_3 group. Hydrogen depresses ethylene polymerization because it produces the Ti–H bond, the step which invariably leads to the formation of the stable Ti–CH_2CH_3 group. Hydrogen activates propylene polymerization because, when the stabilized Ti–$CH(CH_3)_2$ group is formed, hydrogen reacts with it and restores the Ti–H bond, which, in turn, preferably inserts a propylene molecule in the primary fashion.

Acknowledgments

The supported $TiCl_4/MgCl_2$ catalyst was prepared by Dr. S. A. Sergeev (Institute of Catalysis, Novosibirsk, Russia). ^{13}C NMR spectra were recorded by Dr. A. J. Brandolini (Mobil Chemical Co., Edison, NJ).

References

1   Spitz R, Masson P, Bobichon C, Guyot A (**1989**) *Makromol Chem* 190: 707, 717
2   Guyot A, Spitz R, Journaud D (**1994**) *Stud Surf Sci Catal* 89: 43
3   Bukatov GD, Goncharov VS, Zakharov VA, Dudchenko VK, Sergeev SA (**1994**) *Kinet Katal* 35: 329
4   Rishina LA, Vizen EI, Sosnovskaya LN, Dyachkovsky FS, Dubnikova IL, Ladygina TA (**1994**) *Eur Polym J* 30: 1309, 1315
5   Samson JJC, Bosman PJ, Weickert G, Westerterp KR (**1999**) *J Polym Sci, Part A: Polym Chem* 37: 219
6   Parsons IW, Al-Turki TM (**1989**) *Polym Comm* 30: 72
7   Guastalla G, Giannini U (**1983**) *Makromol Chem, Rapid Comm* 4: 519
8   Kioka M, Kashiwa N (**1991**) *J Macromol Sci-Part A (Chem)* A28: 865
9   Chadwick JC, Morini G, Albizzatti E, Balbontin G, Mingozzi AC, Sudmeijer O, van Kessel GMM (**1996**) *Macromol Chem Phys* 197: 2501
10   Chien JCW, Nozaki T (**1991**) *J Polym Sci, Part A: Polym Chem* 27: 1499
11   Bukatov GD, Goncharov VS, Zakharov VA (**1995**) *Macromol Chem Phys* 196: 1751
12   Imaoka K, Ikai S, Tamura M, Yoshikyio M, Yano T (**1993**) *J Mol Catal* 82: 37
13   Busico V, Cipullo R, Corradini P (**1992**) *Makromol Chem, Rapid Comm* 12: 15
14   Chadwick JC, Miedema A, Sudmeijer O (**1994**) *Macromol Chem Phys* 195: 167
15   Kojoh S, Kioka M, Kashiwa N, Itoh NM, Mizuno A (**1995**) *Polymer* 36: 5015
16   Mori H, Tashino K, Terano M (**1995**) *Macromol Chem Phys* 196: 651
17   Kissin YV, Mink RI, Nowlin TE, Brandolini AJ (**1999**) In Kaminsky W (ed) Metalorganic Catalysts for Synthesis Polymerization, Springer, Berlin, p 60
18   Kissin YV, Mink RI, Nowlin TE (**1999**) *J Polym Sci, Part A: Polym Chem* 37: 4255
19   Spitz R, Duranel P, Masson P, Darricades-Llauro MF, Guyot A (**1988**) In Quirk RP (ed) TransitionMetal Catalyzed Polymerizations, Cambridge University Press, New York, p 719
20   Kissin YV Isospecific Olefin Polymerization with Heterogeneous Ziegler-Natta Catalysts, Springer, New York, **1985**, Chapter 3
21   Randall JC, Ruff CJ, Vizzini JC, Speca AN, Burkhardt TJ (**1999**) In Kaminsky W (ed) Metalorganic Catalysts for Synthesis and Polymerization, Springer, Berlin, p 601
22   Gossl T (**1960**) *Makromol Chem* 42: 1
23   Kissin YV (**1986**) *J Chromat Sci* 24: 53, 278
24   Kissin YV (**1989**) *J Polym Sci, Part A: Polym Chem* 27: 605, 623
25   Kissin YV, Mink RI, Nowlin TE (**1993**) *Macromolecules* 26: 2125
26   Natta G, Dall'Asta G, Mazzanti G, Ciampelli F (**1962**) *Kolloid Z* 182: 50
27   Rishina LA, Vizen EI, Sosnovskaya LN, Dubnikova IL (**1996**) Kinet Katal 37: 421

Elementary Steps of Ziegler-Natta Catalyst Intermediates Formation. Incorporation of Magnesium Alkoxides with [Ti(dipp)$_4$], Ph$_3$SiOH, [Al(CH$_3$)$_3$], and MCl$_{(3)4}$ (M = V, Zr)

Sławomir Szafert, Józef Utko, Jolanta Ejfler, Lucjan Jerzykiewicz, Piotr Sobota[*]

Faculty of Chemistry, University of Wrocław, 14 F. Joliot-Curie, 50-383 Wrocław, Poland
E-mail: plas@wchuwr.chem.uni.wroc.pl

Abstract. Reactions of MgBu$_2$ or Mg with hydroxy-ethers or hydroxy-ketones **1a-H-1d-H** (L^2-H) give complexes of [Mg(L^2)$_2$] stoichiometry **2a-d**. Spectroscopic and crystallographic data show **2a-d** to have tetrameric [Mg$_4$ (μ_3,η^2-L^2)$_2$(μ,η-L^2)$_4$(η^1-L^2)$_2$] structure with two five- and two six-coordinate metal centers. Reactions of **2b** with [Ti(dipp)$_4$] or Ph$_3$SiOH show that substitution of one μ,η^2 coordinated alkoxo ligand occurs at five coordinate center. Reaction of **2b** with VCl$_3$ also proceeds at coordinately unsaturated center to give [V$_2$Mg$_2$(μ_3,η^2-thffo)$_2$(μ,η^2-thffo)$_4$Cl$_4$] (**7**) with V$_2$Mg$_2$ core. Instead ZrCl$_4$ and AlMe$_3$ splits dicubane structure of **2b** to give tetrameric [M$_3$Mg(μ_3-O)(μ,η^2-thffo)$_x$X$_6$] (M/x/X = 8, Zr/6/Cl; 9, Al/3/Me).

1. Introduction

Ziegler-Natta catalysis is today one of the most important and profitable processes. Despite its long history it is also the fastest growing segment of the polymer industry [1,2]. A few generations of Z-N catalysts have already been developed. Each generation contributed with higher productivity of the process and quite often with significant improvement of the stereospecifity of the α-olefin polymerization. Metallocene revolution strongly reoriented research objectives towards well defined single site catalysts [3-7]. Nonetheless, these very convenient and highly active homogenous catalysts appear to bring polymer improvement at much higher price than it was originally predicted so their market penetration is still limited although some of these catalysts are finally reaching commercialization step. This simple fact has spurred the development of non-metallocene single-site catalysts [8]. Towards these ends, a novel titanium, zirconium and vanadium catalysts have been obtained in our research group by reaction of group IV and V metal chlorides with alkoxo magnesium open dicubane complexes. Our research has been projected on determining the role of each of the catalyst component in the supported MCl$_{(3)4}$/MgX$_2$/AlR$_3$/SiO$_2$ (M = V, Ti, Zr; X = alkoxo group) Ziegler-

Natta polymerization catalysts that are extensively used in the polyolefin industry [6,9]. Also interaction of alumoxane activator and silica support with magnesium complexes were investigated.

This article reviews our recent efforts in the design and synthesis of well-defined magnesium supported heterogeneous catalysts for α-olefin polymerization.

2. Alkoxo ligands.

Most of our research was performed utilizing commercially available: 2,3-dihydro-2,2-dimetyl-7-benzofuran alcohol (dbbfo-H; **1a-H**), tetrahydrofurfuryl alcohol (thffo-H; **1b-H**), 3-hydroxy-2-methyl-4-pyrone (maltol-H; **1c-H**), and 2-methoxy-phenol (guaiacol-H; **1d-H**) (Scheme 1). Deprotonation of these hydroxides affords bidentate O,O' ligands henceforth abbreviated as L^2.

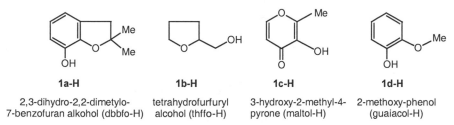

1a-H	**1b-H**	**1c-H**	**1d-H**
2,3-dihydro-2,2-dimetylo-7-benzofuran alkohol (dbbfo-H)	tetrahydrofurfuryl alcohol (thffo-H)	3-hydroxy-2-methyl-4-pyrone (maltol-H)	2-methoxy-phenol (guaiacol-H)

Scheme 1. Precursors of bidentate O,O' alkoxo ligands.

Our choice was based on the fact that heterometallic alkoxides have been postulated to act as catalysts in Ziegler-Natta polymerization or olefin metathesis reactions [10,11]. The selected ligands seem to have very attractive features. Each contains two donor atoms: one alkoxo oxygen and either ether (**1a-c**) or ketone (**1d**) oxygen atom. Both donor atoms are always separated by two-carbon spacer, what enables each ligand to act as an chelating agent or be tethered between metal atoms in μ_3,η^2, μ,η^2, μ_3,η^1, or μ,η^1 mode to create multinuclear species. Some other coordination modes as well as coordination by neutral R-OH can also be imagined broadening the spectrum of possible motifs of polymetallic species. Moreover all ligands represent a family of moderately weak Lewis bases and establish good leaving groups.

3. Reactions of L^2–H with dialkyl magnesium or magnesium turnings.

It is well documented that complexation of a Z-N catalyst component $MgCl_2$ by internal donor results in its polynuclear aggregation. Structure of $[MgCl_2(thf)_{1.5}]$ can serve as an explicit example [12]. Identical behavior was expected for **1a-H-1d-H** ligands. Reactions of **1a-H-1d-H** with di-*n*-butyl magnesium or magnesium turnings proceed easily to give analytically pure complexes of $[Mg(L^2)_2]$ (**2a-d**) stoichiometry as white powders. IR spectra of these compounds show strong bands between 300-700 cm^{-1} what suggests polynuclear character of the resulted species. Crystallization of **2a** indeed showed tetranuclear compound $[Mg_4(\mu_3,\eta^2\text{-}dbbfo)_2(\mu,\eta^2\text{-}dbbfo)_4(\eta^1\text{-}dbbfo)_2]$ with open dicubane geometry (Scheme 2 and Figure 1) [13].

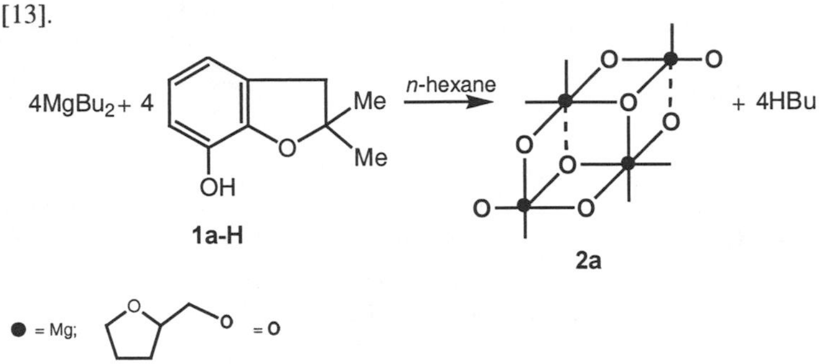

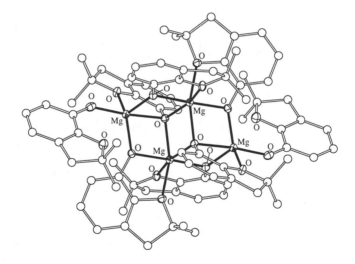

Scheme 2. Reaction of MgBu$_2$ with dbbfo-H.

Fig. 1. The perspective view of complex **2a**.

Centrosymmetric molecule of **2a** possesses two five-coordinate magnesium atoms of trigonal bypyramid geometry and two six-coordinate octahedral metal sites. This nearly regular Mg$_4$ rhombus is bridged by two μ_3- and four μ_2-oxygen atoms of six aryloxo groups to form open dicubane. The presence of coordinately unsaturated metal sites is the most interesting feature of these compounds. Two dbbfo ligands in **2a** are in μ_3,η^2 coordination mode. Four of them are in μ,η^2 and two are in η^1 fashion. As shown in Fig. 1 the equatorial plane of each bipyramid is formed by two μ_2-bridging aryloxo groups and one ether oxygen from ddbfo ligand. Axial positions are filled by μ_3-bridging aryloxo ligand and η^1 fashioned dbbfo ligand. The coordination sphere around six-coordinate magnesium is a slightly distorted octahedron formed by two μ_3- and two μ_2-bridging aryloxo group. Two remaining sites are occupied by two ether oxygens from two ddbfo ligands (in *cis* position). The average bond lengths and bond distances are of the order of the corresponding magnesium-oxygen distances observed in other magnesium compounds [14-16].

4. Reactivity of open dicubanes $[Mg_4(L^2)_8]$.

Complexes **2a-d** are sensitive towards chlorinated hydrocarbons. CH_2Cl_2 for example causes substitution of η^1-coordinated ligands by chlorine atoms to give in case of **2b** molecular complex $[Mg_4(\mu_3,\eta^2\text{-thffo})_2(\mu,\eta^2\text{-thffo})_4Cl_2]$ **(3)** shown in Scheme 3 [14].

Interestingly analogous compound was obtained when $MgCl_2$ was used instead of CH_2Cl_2 [14]. Based on reaction stoichiometry and final yield we postulate that in reaction of **2b** with $MgCl_2$ at least two products are formed. Main product **3** can immediately be isolated due to its poor solubility and accompanying ionic byproduct stays in solution. In order to be precipitated a suitable counterion is needed. Addition of $FeCl_2$ fulfills that need and polynuclear magnesium aggregate **4** can easily be isolated as shown in Scheme 3 [17].

This magnesium compound with $[Mg_7O_{24}]^{2+}$ core is the largest structurally characterized magnesium alkoxide reported to date. Similar $[Zn_7X_{24}]^{2+}$ (X = O and N) core was found in the product obtained by crystallization of ZnL_2 (L = pyridylmethanolate) from CH_2Cl_2 [18].

Also very interesting is reactivity of magnesium open dicubanes with other alcohols *e.g.* methanol. This however will not be discussed here [13].

4.1. Reaction of $[Mg_4(L^2)_8]$ with $[Ti(dipp)_4]$ and Ph_3SiOH.

As an aid to explain the formation of catalytically active center as well as to reveal the interaction of silica surface OH groups with magnesium catalyst component, we were studying the binding of $[Ti(dipp)_4]$ (dipp = 2,6-diisopropylphenoxo group) and Ph_3SiOH to the unsaturated magnesium centers in **2b** [13,19]. The reaction course in both cases is similar and gives adducts **5** and **6** that are presented in Scheme 4, Fig. 2, and Fig. 3.

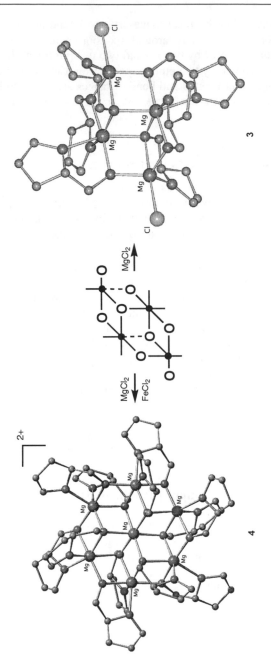

Scheme 3.   Reactivity of magnesium open dicubane with $MgCl_2$ and $FeCl_2$.

It is instructive to compare structures of **5** and **6** with the structure of **3**. In each compound the $[Mg_4(thffo)_6]^{2-}$ core is similar. The only significant difference among them is that the two terminal Cl atoms that are bonded to the five-coordinated magnesium centers in **3** are substituted by $[OTi(dipp)_3]^-$ in **5** or $[OSiPh_3]^-$ groups in **6**. Interestingly both magnesium atoms remain five coordinated.

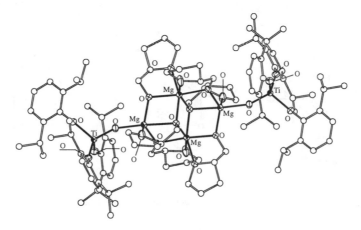

Fig. 2. The perspective view of complex **5**.

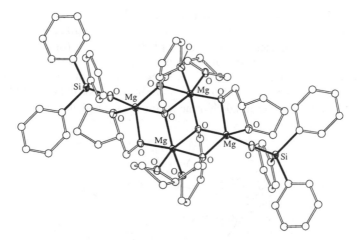

Fig. 3. The perspective view of complex **6**.

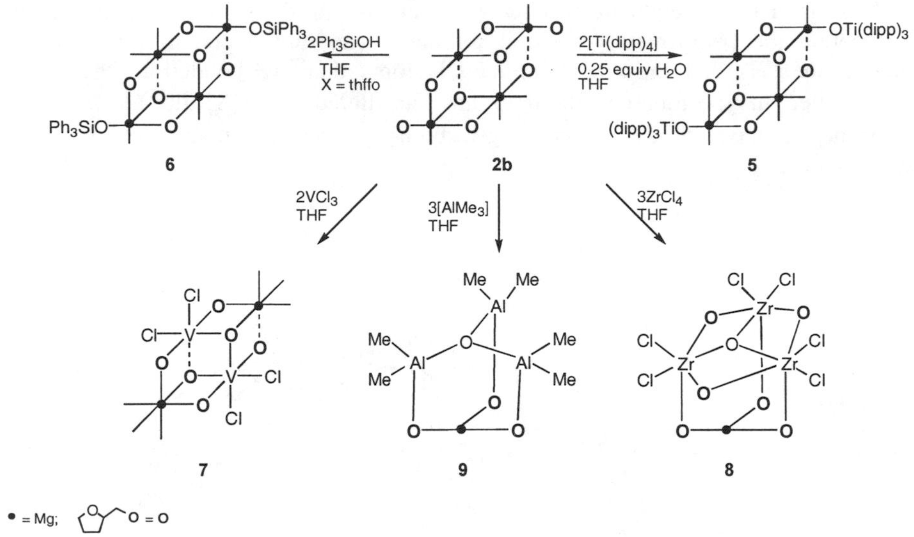

Scheme 4. Reactions of magnesium dicubane [Mg₄(thffo)₈].

4.2. Reaction of magnesium open dicubanes with MCl$_x$ (M= V, x= 3; M= Zr, x= 4).

Reactions of Lewis acids VCl$_3$ or ZrCl$_4$ with **2a-d** in thf give structurally different from **5** and **6** products (Scheme 4). Deep violet $[V_2Mg_2(\mu_3,\eta^2\text{-thffo})_2(\mu,\eta^2\text{-thffo})_4Cl_4]$ (**7**) was obtained in the reaction of **2b** with VCl$_3$ (or $[VCl_3(thf)_3]$) [20]. As shown in Fig. 4 the centrosymmetric complex molecule consists of M$_4$ rhombus in which two magnesium atoms were replaced by vanadium atoms. Each metal center is six-coordiante. Four thffo ligands are in μ,η^2 coordination mode and two are μ_3,η^2 fasioned. Four terminal chlorine atoms are located at vanadium centers.

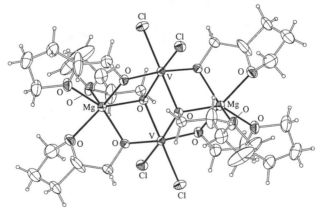

Fig. 4. The perspective view of complex **7**.

The reaction of **2b** with $ZrCl_4$ proceeds due to a different reaction course (Scheme 4). In this case tetranuclear magnesium core splits completely to give $[Zr_3Mg-(\mu_3-O)(\mu,\eta^2-thffo)_6Cl_6]$ (**8**). As shown in Fig. 5 compound **8** is a neutral tetranuclear species [19].

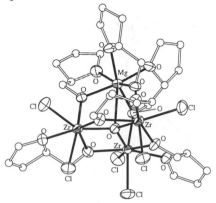

Fig. 5. The perspective view of complex **8**.

In the complex molecule three $[Zr(\mu,\eta^2-thffo)Cl_2]^+$ moieties form a threenuclear macrounit of C_3 symmetry that is held together by μ_3-oxo ligand. It is then capped with $[Mg(thffo)_3]^-$ moiety by μ-alkoxide oxygen atoms of the three thffo ligands. Each zirconium atom in **8** is seven-coordinated.

During the preparation of alkoxometal complexes formation of products containing "lone" oxo ligands, which are encapsulated in ensembles of between three or more metal atoms is not uncommon [16b].

4.3. Reaction of magnesium open dicubanes with $[Al(CH_3)_3]$.

The reaction of **2b** with $[Al(CH_3)_3]$ also results in the ultimate splitting of the tetranuclear magnesium core (Scheme 4) to give $[Al_3Mg(\mu_3-O)(\mu,\eta^2-thffo)_3Me_6]$ (**9**) [19]. As shown in Fig. 6 the tetranuclear Al_3/Mg compound forms an assembly that is similar to that of **8**.

Three $[Al(CH_3)_2]^{2+}$ moieties held together by μ_3-oxo ligand form a macrounit that is capped with $[Mg(thffo)_3]^-$ motif by μ-alkoxide oxygen atoms of the three thffo ligands. The $[Al_3(\mu_3-O)(CH_3)_6]^+$ cation could be considerate as a part of methylalumoxane (MAO) of the general formula $[CH_3AlO]_n$. Methylalumoxane is a poorly characterized species which consists of linear, cyclic and cross-linked compounds, containing predominately three and four-co-ordinate oxygen and aluminum centers, respectively [21]. Some alkylalumoxanes have been characterized crystallographically by Barron and coworkers, who obtained them by partial hydrolysis of aluminum alkyls; thus, hydrolysis of $AlBu^i_3$ gave tetranuclear $[Al_4(\mu_3-O)_2(Bu^i)_8]$ [22].

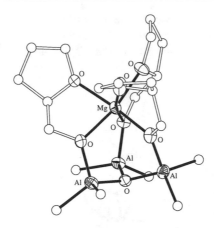

Fig. 6. The perspective view of complex **9**.

5. Summary, Conclusion, and Outlook.

A series of magnesium open dicubanes was obtained in the reactions of dialkyl magnesium or magnesium turnings with bidentate O,O' hydroxides. Crystallographic studies have shown that five and six coordinated metal sites are present in the complex molecules. All compounds show interesting reactivity towards halogenated hydrocarbons and other heterogeneous Z-N catalyst components. Reactions proceed with or without retention of dicubane structure. It has also been proven that five coordinate metal centers play a key role in a complexation of incoming reagents.

The nature of the $[Mg_4(L^2)_8]$ in solution does not seem simple. It is highly possible that in solution $[Mg_4(L^2)_8]$ undergoes ionization according to equation 1 and resulting ions undergo further reaction.

$$[Mg_4(L^2)_8] \leftrightarrow 2[Mg(L^2)_3]^- + 2[Mg(L^2)]^+ \qquad (1)$$

The equilibrium would depend upon the least soluble species precipitation of which causes the shift of the reaction equilibrium (*e.g.* zirconium trinuclear macrounit reacts with $[Mg(thffo)_3]^-$ anion). Additional conformation of this conclusion comes from a structure of $[Mg_4(\mu,\eta^2\text{-thffo})_4(\eta^1\text{-thffoH})_4Cl_4]$ (**9**) that was prepared by treatment of $[Mg_4(thffo)_8]$ with $[Zr_2(\mu,\eta^2\text{-thffo})_2(\eta^2\text{-thffoH})Cl_6]$. The crystals of **9** are composed of four $[Mg(thffo)]^+$ moieties held together by four μ_3-alkoxides to form dicubane [22].

In the course of the studies it has been shown that the existence of five coordinated magnesium centers is very important. These centers are very sensitive towards substitution. $[Ti(dipp)_4]$ and Ph_3SiOH are able to replace one of the μ,η^2 coordinated alkoxo ligands without splitting of dicubane structure. In the resulted complexes two magnesium centers remain coordinately unsaturated. We have already shown that such "invitation" is very welcomed by CH_3OH molecule which easily coordinate to this atom [13].

Atom arrangement in the presented magnesium alkoxides seems also important in view of the fact that magnesium compounds with Mg_4X_6 core are good additives boosting the activity of the catalytic mixture. Preliminary results of an ethylene polymerization test on $[Mg_4(\mu_3,\eta^2\text{-thffo})_2(\mu,\eta^2\text{-thffo})_4Cl_2]/TiCl_4/AlEt_3$ catalyst gave *ca.* 170 kg polyethylene per g Ti h^{-1} while only 11.6 kg polyethylene per g Ti h^{-1} was obtained under the same conditions when $MgCl_2/TiCl_4/AlEt_3$ catalyst was used [14].

Further experiments are being planned. In the next step we will consequently "complicate" the system and aim at the synthesis of $Si/Mg_4/Ti$ system. Also other than titanium, vanadium and zirconium transition metals will be investigated.

Recent discoveries on non-metallocene iron and cobalt olefin polymerization catalysts prove that there is no limit in terms of metal site that can give a polymerization active center. We believe that careful and rational ligand design can create enormous possibilities.

Acknowledgement.

The authors thank the State Committee for Scientific Research for financial support of this work (grant No 3 T09A13115).

References

1   A. M. Thayer, *Chem. Eng. News* **1995**, *73*, 15.

2   J. Schumacher, in *Chemical Economics Handbook*; SRI International: Menlo Park, CA, **1994**, 50.

3   M. Bochmann, *J. Chem. Soc. Dalton Trans.* **1996**, 255.

4   H. H. Britzinger, D. Fischer, R. Mülhaupt, B. Rieger, R. M. Waymouth, *Angew. Chem.* **1995**, *107*, 1255; *Angew. Chem. Int. Ed. Engl.* **1995**, *34*, 1143.

5   W. Kaminsky, *J. Chem. Soc. Dalton Trans.* **1998**, 1413.

6   K. Soga, T. Shiono, *Prog. Polym. Sci.* **1997**, *22*, 1503.

7   R. F. Jordan, *Organomet. Chem.* **1991**, *32*, 325.

8   a) L. V. Cribbs, B. P. Etherton, G. G. Hlatky, S. Wang, *Proc. Ann. Technol. Conf. – Soc. Plast. Eng.* **1998**, *56(2)*, 1871; b) G. J. P. Britovsek, V. C. Gibson, D. F. Wass, *Angew. Chem., Int. Ed. Engl.* **1999**, *38*, 428; c) G. J. P. Britovsek, M. Bruce, V. C. Gibson, B. S. Kimberley, P. J. Maddox, S. Mastroianni, S. J. McTavish, C. Redshaw, G. A. Solan, S. Strömberg, A. J. P. White, D. J. Williams, *J. Am. Chem. Soc.* **1999**, *121*, 8728; d) L. Deng, P. Margl, T. Ziegler, *J. Am. Chem. Soc.* **1999**, *121*, 6479.

9   a) A. Toyota, N. Kashiwa, *Japan Pat. kokai* **1975**, *75-30*, 983; b) P. D. Gavens, M. Botrrill, J. W. Kelland, in *Comprehensive Organometallic Chemistry I, Vol. 3* (Eds.-in-Chief: G. Wilkinson, F. G. A. Stone, E. W. Abel), Pergamon, Oxford, **1982**; c) U. Giannini, E. Albizzati, S. Parodi, F. Pirinoli, *U.S. Patents* **1978**, 4,124,532 and **1979**, 4,174,429; d) K. Yamaguchi, N. Kanoh, T. Tanaka, N. Enokido, A. Murakami, S. Yoshida, *U.S. Patent* **1976**, 3,989,881; e) G. G. Arzoumanidis, N. M. Karayannis, *ChemTech.* **1993**, *23*, 43; f) P. Sobota, *Macromol. Symp.* **1995**, *89*, 63.

10 M. Takeda, K. Imura, Y. Nozawa, M. Hisatone, N. Koide, *J. Polym. Sci.* **1968**, *25C*, 741.

11 H. Sinn, W. Kaminsky, *Adv. Organomet. Chem.*, 1980, **18**, 99.

12 J. Toney, G. D. Stucky, *J. Organomet. Chem.* **1971**, *28*, 5.

13 P. Sobota, J. Utko, K. Sztajnowska, J. Ejfler, L. B. Jerzykiewicz, *Inorg. Chem.* **2000**, *39* 235.

14 P. Sobota, J. Utko, Z. Janas, S. Szafert, *Chem. Commun.* **1996**, 1923.

15 a) P. Sobota, T. P•uzi•ski, T. Lis, *Inorg. Chem.* **1989**, *28*, 2217; b) J. Utko, P. Sobota, T. Lis, K. Majewska, *J. Organomet. Chem.* **1989**, *359*, 295.

16 a) K. G. Caulton, L. G. Hubert-Pfalzgraf, *Chem. Rev.* **1990**, *90*, 969; b) W. A. Herrman, N. W. Huber, O. Runte, *Angew. Chem. Int. Ed. Engl.* **1995**, *34*, 2187.

17 Z. Janas, L. B. Jerzykiewicz, P. Sobota, J. Utko, *New J. Chem.* **1999**, 185.

18 M. Tesmer, B. Muller, H. Vahrenkamp, *Chem. Commun.* **1997**, 721.

19 P. Sobota, J. Utko, J. Ejfler, L. B. Jerzykiewicz, in preparation.

20 Z. Janas, P. Sobota, M. Klimowicz, S. Szafert, K. Szczegot, L. B. Jerzykiewicz, *J. Chem. Soc. Dalton Trans.* **1997**, 3897.

21 a) H. Sin, W. Kaminsky, *Adv. Organomet. Chem.*, **1980**, *18*, 99; b) S. Pasynkiewicz, *Polyhedron* **1990**, *9*, 429, and references therein.

22 a) M. R. Mason, J. M. Smith, S. G. Bott, A. R. Barron, *J. Am. Chem. Soc.*, **1993**, *115*, 4971; b) Y. Koide, C. J. Harlan, M. R. Mason, A. R. Barron, *Organometallics* **1994**, *13*, 2957; c) S. G. Bott, A. R. Barron, *Organometallics* **1996**, *15*, 2213.

4. Polymerization Mechanisms, Catalyst Structure and Polymer Structure Relationships

Mechanistic Aspects of Olefin Polymerization with Metallocene Catalysts: Evidence from NMR Investigations

Incoronata Tritto*, Maria Carmela Sacchi, Paolo Locatelli, Fabrizio Forlini

Istituto di Chimica delle Macromolecole del CNR, Via E. Bassini, 15, I-20133 Milano, Italy, E-mail: Tritto@icm.mi.cnr.it

Abstract. Isotopically ^{13}C enriched $Cp_2M^{13}CH_3X$ metallocenes (M = Ti, Zr; X = $^{13}CH_3$, Cl) have been used as a probe for the reactivity of metallocene-methylaluminoxane catalysts for olefin polymerization. ^1H and ^{13}C NMR studies of the reactions between metallocenes and Lewis acids such as methylaluminoxane (MAO), $AlMe_3$, and the well-defined $B(C_6F_5)_3$ have been performed in order to clarify the role of MAO in the activation of metallocenes. The isotopic enrichment has permitted us to study these systems in conditions as close as possible to usual polymerization conditions. Evidence of the formation in solution of "cationic" metallocene species having [X⁻MAO]⁻ as counterion has been provided. In situ-polymerization of ^{13}C enriched ethylene permitted us to gain an insight into the role of the observed products in the catalytic activity and to make the first direct observation of methylenes of polymeryls of Cp_2Zr-polymeryl species. A survey of these results is presented. Recent results of the study of the reactivity of two β-alkyl substituted aluminoxancs with two metallocenes Cp_2ZrCl_2 and Cp_2*ZrCl_2 are also reported. It was found that the Cp_2*Zr-i-BuCl and Cp_2*Zr-i-OctCl stability is greater than that of Cp_2Zr-i-BuCl and Cp_2Zr-i-OctCl. The polymerization activities exhibited by different cocatalysts could be explained.

Introduction

In contrast with the traditional multi-site heterogeneous catalysts, the homogeneity and the well-defined structure of modern metallocene catalysts make them suitable for fundamental studies. Indeed, they have allowed great advances to be made in the understanding of olefin polymerization mechanisms. Especially the fundamental influence of the structure of the organic ligands on the achievement of stereo- and regiocontrol, on polymerization activity and on polymer molecular weight has been clarified [1]. Metallocenes, when combined with the traditional aluminium halide cocatalysts, exhibit low-activities for ethylene polymerization and thus their use was limited to kinetic studies until Kaminsky observed that they afford highly active catalysts for polymerizing α-olefins when combined with methylalumoxane [-Al(Me-O-]n (MAO) [2]. Because of its unique effectiveness MAO has become a very important cocatalyst for metal-catalyzed olefin polymerization. The homogeneity of these systems has also allowed us to shed light on metallocene activation, and on the understanding of the active species involved in catalytic processes, especially when NMR and X-ray analysis of catalyst precursors and of model systems are used [3]. Early studies [4], and the synthesis of stabilized group 4 metallocene ionic complexes $[Cp_2MMe]^+[X]^-$, in which X^- is a "noncoordinating" counterion, supported the hypothesis that the cationic site is the active component in these systems [3, 4].

Due to multiple equilibria present in MAO solutions, metallocene-MAO systems have been considered far too complex to allow identification of the species produced. This article will survey the results of our approach to clarifying the activation of metallocenes by MAO cocatalyst [5]. The complex equilibria that can be established between the species present in these systems, that is, metallocenes, cocatalyst, the solvent and the olefin will also be considered.

MAO is an oligomeric species, whose structure is still not entirely clear, represented for simplicity as having linear chain or cyclic ring structures. MAO contains some residual $AlMe_3$ (TMA) from the synthesis, which appears to participate in equilibria that interconvert various MAO oligomers. There are two types of TMA present in typical MAO solutions: "free" TMA and "associated" TMA (eq 1).

(1) $[-Al(Me)-O-]_n \cdot x\,AlMe_3 \rightleftharpoons [-Al(Me)-O-]_n \cdot (x-y)AlMe_3 + y\,AlMe_3$
          " associated"   "free"

More complex three-dimensional structure were recently proposed [6a,b] on the basis of structural similarities with isolable and characterizable cage stuctures of t-butylaluminoxanes [6c,d] and of DFT quantum-chemical calculations .

Our strategy to get a direct insight into the activation of metallocenes with MAO consists in an investigation of the chemistry of isotopically ^{13}C enriched $Cp_2M^{13}CH_3X$ metallocenes (M = Ti, Zr; X = $^{13}CH_3$, Cl) with MAO, $AlMe_3$, and the well-defined $B(C_6F_5)_3$, Lewis acids by NMR spectroscopy at various temperatures, Al/M ratios (from $0 \div 60$), catalyst concentrations ($0,01 \div 0.07$ [M]), solvents and $AlMe_3$ amounts, that is, by varying the parameters which influence the polymerization activity. The MAO used has been freed from residual TMA by vacuum distillation. In situ polymerization of ^{13}C enriched ethylene were performed in order to gain an insight into the role of the observed products in the catalytic activity.

NMR studies of titanocene based catalysts

Initially, our interest in clarifying the actual role of the "free" $AlMe_3$ in MAO solutions in polymerization activity in these homogeneous systems prompted us to study the reactions of $Cp_2TiMeCl$ (1) and Cp_2TiMe_2 (2) with $AlMe_3$ and with MAO [5a-d]. The comparative NMR studies of reactions between $Cp_2Ti(CH_3)Cl$ + $Al(CH_3)_3$ and $Cp_2Ti(CH_3)Cl$ + MAO showed the formation of the alkylated product Cp_2TiMe_2, and of species $Cp_2TiMeCl\cdot AlMe_3$ (3) and $Cp_2Ti^{13}CH_3^+X\cdot MAO^-$ (4).

Species **3** has downfield Ti-Cp (117.08 ppm) and Ti-Me (60.22 ppm) displacements.

(2) $Cp_2TiMeCl + AlMe_3$ $\rightleftharpoons$ $Cp_2TiMe_2 + AlMe_2Cl$

(3) $Cp_2TiMeCl + AlMe_3$ $\rightleftharpoons$ $Cp_2TiMeCl·AlMe_3$

The use of $Cp_2Ti^{13}CH_3Cl$ having 90% ^{13}C enriched methyl ligands allowed us to obtain the direct evidence for the formation of $Cp_2Ti^{13}CH_3^+X·MAO^-$ (**4**) (Fig. 1A). This species has Cp and Me resonances at 117.96 and 64.6 ppm, respectively, that is, at even lower field than those of complex **3**. Such a further downfield shift of NMR signals indicates the decrease of electron density at the metal center typical of cation-like species.

(4) $Cp_2TiMeCl + MAO$ $\rightleftharpoons$ $Cp_2TiMe_2 + MAO\ Cl$

(5) $Cp_2TiMeCl + MAO$ $\rightleftharpoons$ $Cp_2TiMe^+Cl·MAO^-$

It was found that MAO is a superior alkylating agent and has a greater capacity to produce and stabilize cation-like complexes. Unlike MAO activation, the equilibrium involved with TMA lies far to the left. MAO is a significantly stronger Lewis acid, and the resulting counterion is far less coordinating than that of TMA.

Therefore, since the MAO based system is much more active for ethylene polymerizations than the practically inactive $AlMe_3$ based system, the "free" TMA is not the actual cocatalyst in metallocene-MAO catalysts, as was proposed by an alternative hypothesis.

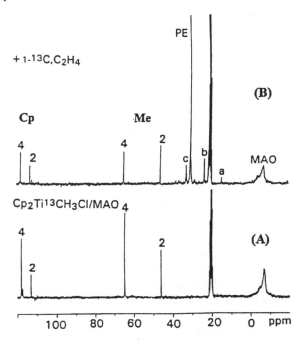

Fig. 1. ^{13}C NMR spectra in toluene-d_8 at –20 °C of $Cp_2Ti^{13}CH_3Cl/MAO$ (A) and $Cp_2Ti^{13}CH_3Cl/MAO/1$-$^{13}C, C_2H_4$ (B). [Ti] = 0.07 mol.L^{-1}, [Al]/[Ti] = 20, $^{13}C, C_2H_4$ = 0.4 mol.L^{-1}

NMR studies showed that by increasing Al/M ratios and the temperature from -78 to -20 °C the cationic species **4** increases along with polymerization activity. The in situ polymerization of ^{13}C enriched ethylene with the appearance of the polyethylene peak, as expected, at about 30 ppm, and of the relative chain end groups (a, b, c) associated with the simultaneous decrease of the Ti-$^{13}CH_3^+$ signal of **4** showed that these cation-like species are active for olefin polymerization (Fig. 1B).

The formation of Cp_2TiMe^+-MeMAO$^-$ (**5**) was evidenced by comparison of the products of Cp_2TiMe_2/MAO and Cp_2TiMe_2/B(C$_6$F$_5$)$_3$ catalytic systems. Different methyl resonances of ion pairs $Cp_2TiMe^+Cl^-MAO$ (**4**) and Cp_2TiMe^+-Me-MAO$^-$ (**5**) at 64.6 ppm and 69.0 ppm, respectively, indicate that they are not completely separated ion pairs but that they could be better described as weakly coordinated ion pairs.

4   **5**

By increasing the temperature from –20 °C to 0 °C, Tebbe-like species (**6**) and (**7**) are formed with resonances at 187.89 ppm and at 178.96 ppm, respectively, that is, in the region of methylidene carbons. In situ $^{13}C_2H_4$ polymerization and norbornene

6   **7**

oligomerization as well as batch polymerization experiments evidenced that (**6**) and (**7**) are not active for addition polymerization [5e].

NMR studies of zirconocene based catalysts

Comparative studies of the reactivity of $Cp_2Zr(^{13}CH_3)_2$ (**8**) [5f] and $Cp_2Zr^{13}CH_3Cl$ (**9**) with AlMe$_3$, B(C$_6$F$_5$)$_3$ and with MAO have been performed since zirconocenes are more stable and more easily used than Ti compounds. Apparently $Cp_2Zr(^{13}CH_3)_2$ does not react with AlMe$_3$. However, the scrambling of the ^{13}C enrichment of dimethylzirconocene with AlMe$_3$, revealed by the decrease of the relative intensity of the methyl peak, indicates a methyl exchange reaction.

The reaction equilibria between zirconocenedimethyl and B(C$_6$F$_5$)$_3$ have been studied by ^{13}C NMR at different B/Zr mole ratios between 0.5 and 2. Both dinuclear $[(Cp_2ZrMe)_2(\mu\text{-Me})]^+[MeB(C_6F_5)_3]^-$ (**10**) and mononuclear $[Cp_2ZrMe]^+[MeB(C_6F_5)_3]^-$ (**11**) ion pairs already detected by Bochmann and Brintzinger [7] were observed. Only the mononuclear ion pair is formed when B(C$_6$F$_5$)$_3$ is used in eccess (Fig. 2 A).

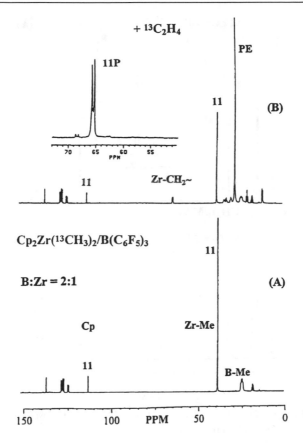

Fig. 2. ^{13}C NMR spectra in toluene-d_8 at –20 °C of Cp$_2$Zr(^{13}CH$_3$)$_2$/ B(C$_6$F$_5$)$_3$ (A) and Cp$_2$Zr(^{13}CH$_3$)$_2$/B(C$_6$F$_5$)$_3$ /^{13}C$_2$H$_4$ (B) [Zr] 0.07 mol.L^{-1}, [B]/[Zr] = 2, ^{13}C$_2$H$_4$ = 0.4 mol.L^{-1}

Our ^{13}C NMR spectroscopic studies have given direct evidence of the formation in solution of mononuclear [Cp$_2$ZrMe]$^+$[MeMAO]$^-$ (**14**) and dinuclear [(Cp$_2$ZrMe)$_2$(Me)]$^+$[MeMAO]$^-$ (**15**), as well as of the heterodinuclear [Cp$_2$Zr(Me)$_2$AlMe$_2$]$^+$[MeMAO]$^-$ (**16**) ionic species. The results obtained by changing temperatures, metal concentrations, and Al/Zr ratios have shown that the dinuclear **15** decreases on increasing the T, the Al/Zr ratio and on decreasing zirconium concentration. The two dinuclear complexes are possible dormant states for the active site olefin polymerization (Scheme 1), as proposed [7a].

In in-situ polymerization of ^{13}C-enriched ethylene in the presence of Cp$_2$Zr(^{13}CH$_3$)$_2$ with MAO and with B(C$_6$F$_5$)$_3$ we have observed the formation of polyethylene under all conditions and signals of ^{13}C-enriched multiplets of methylenes bonded to Zr of the Zr-polymeryl species [5g]. Comparative in situ

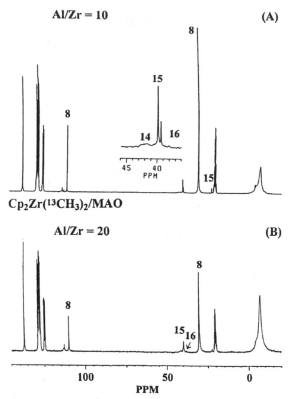

Fig. 3. ^{13}C NMR spectra in toluene-d_8 at –20 °C of Cp$_2$Zr(^{13}CH$_3$)$_2$/ MAO: [Al]/[Zr] = 10 (B) and [Al]/[Zr] = 20 (A) [Zr] 0.07 mol.L^{-1}.

polymerizations have allowed us to assign the signal at about 65 ppm to methylenes of the Zr-polymeryl species in the ionic mononuclear forms (**11P, 16P**) (Scheme 2). The signals at about 55 ppm may be assigned to the methylene of the polymeric chain bonded to the Zr in the non ionic form (**8P**) (Scheme 2). ^{13}C NMR spectra of the in situ polymerization in the presence of B(C$_6$F$_5$)$_3$ or MAO are very similar, apart from the starred signals which are due to the methylene carbons of polymeryl chain transferred to aluminum when MAO is used as cocatalyst. Chain transfers in the presence of the borane cocatalyst do not seem to occur at this temperature. This comparison allowed us to assign the Cp$_2$Zr-polymeryl species formed in the in situ polymerization in the presence of MAO. They derive from the ethylene insertion into the Zr-Me bond, and they seem to be in equilibrium with each other (Scheme 2).

Cp$_2$Zr^{13}CH$_3$Cl is more reactive than the dimethylzirconocene complex towards all three Lewis acids used. It reacts completely with AlMe$_3$ and gives a complex which shows NMR signals shifted towards lower fields. The reaction pattern with $B(C_6F_5)_3$ is rather complex: Ion pairs from both chlorine and methyl abstraction

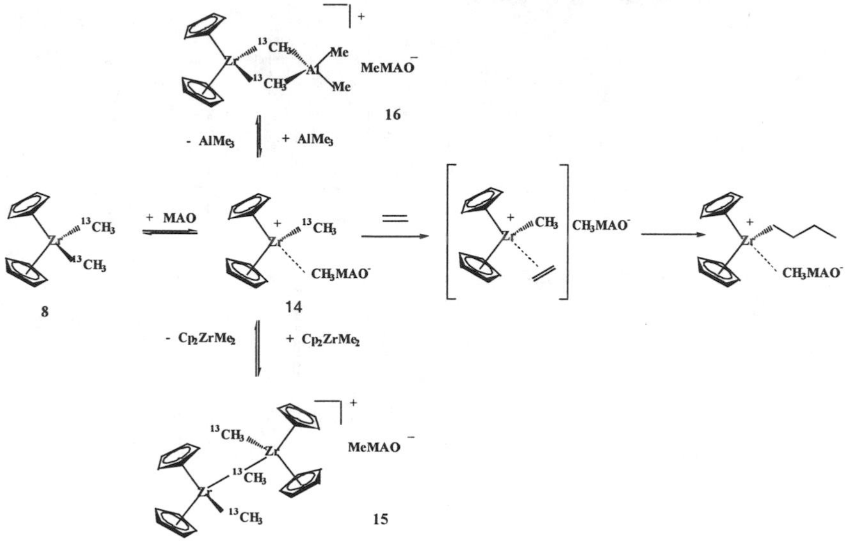

Scheme 1

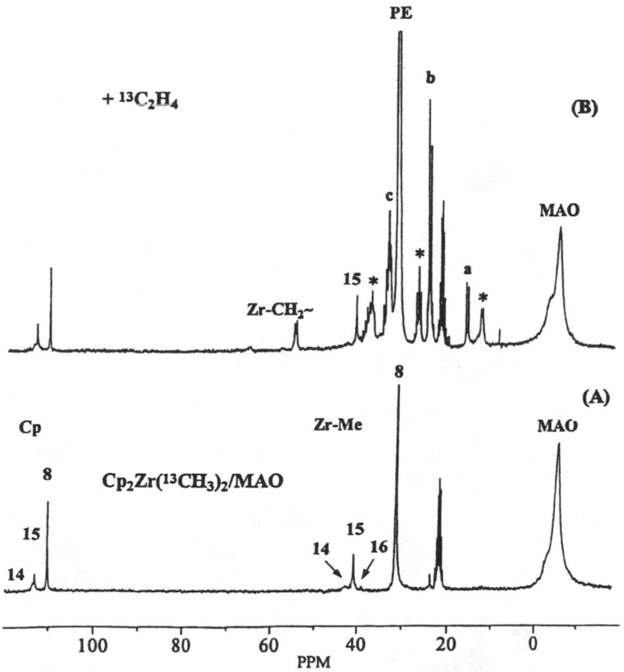

Fig. 4. ^{13}C NMR spectra in toluene-d_8 at –20 °C of Cp$_2$Zr($^{13}CH_3$)$_2$/MAO (A) and Cp$_2$Zr($^{13}CH_3$)$_2$/ MAO /$^{13}C_2H_4$ (B) [Zr] 0.07 mol.L^{-1}, [Al]/[Zr] = 20, $^{13}C_2H_4$ = 0.4 mol.L^{-1}

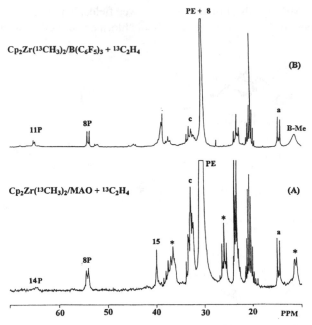

Fig. 5. ^{13}C NMR spectra in toluene-d_8 at –20 °C of $Cp_2Zr(^{13}CH_3)_2/MAO/^{13}C_2H_4$ (A) and $Cp_2Zr(^{13}CH_3)_2/B(C_6F_5)_3/^{13}C_2H_4$ (B). [Zr] 0.07 mol.L^{-1}, [B]/[Zr] = 20 $^{13}C_2H_4$ = 0.4 mol.L^{-1}.

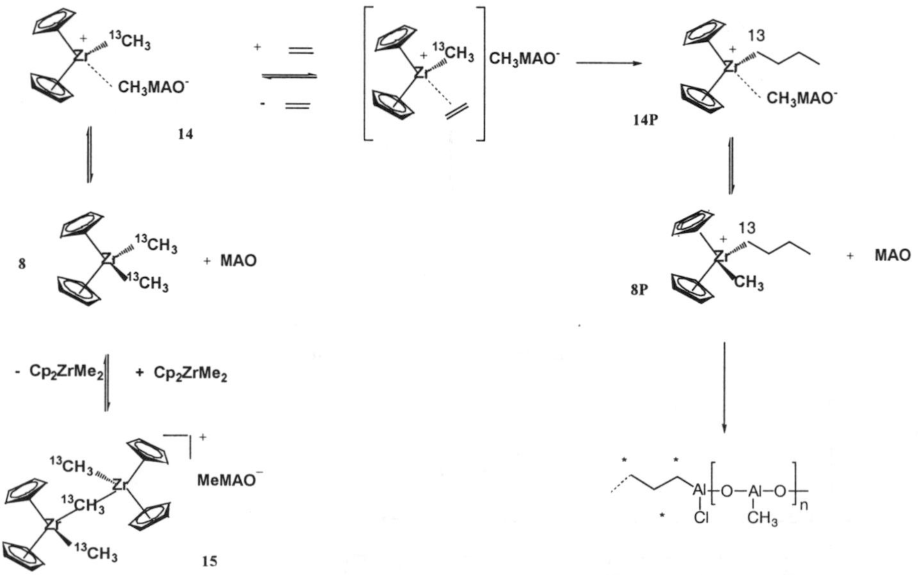

Scheme 2

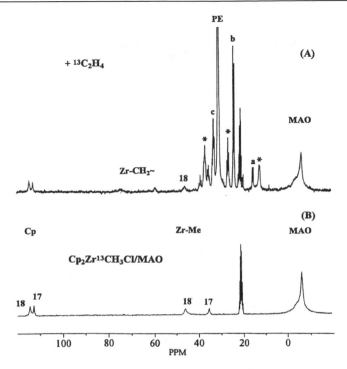

Fig. 6. ^{13}C NMR spectra in toluene-d_8 at –20 °C of Cp$_2$Zr(^{13}CH$_3$Cl/ MAO (A)   and Cp$_2$Zr(^{13}CH$_3$Cl/ MAO /^{13}C$_2$H$_4$ (B) [Zr] 0.07 mol.L^{-1}, [Al]/[Zr] = 20, ^{13}C$_2$H$_4$ = 0.4 mol.L^{-1}

and products due to counterion exchange are formed as well. Dinuclear complexes are not observed under these conditions. The reaction with MAO yields two species Cp$_2$ZrMeCl MAO (**17**) and [Cp$_2$ZrMe]$^+$[ClMAO]$^-$ (**18**) (Fig. 6A). The ion pair **18** increases along with temperatures and Al/Zr ratios. Dilution reduces the formation of all reaction products. Indeed, at Al/Zr equals 10, even at 0 °C we observe only the unreacted complex.

The in situ polymerization of enriched ethylene in the presence of the Cp$_2$ZrMeCl/MAO showed two signals due to methylene carbons bonded to two Zr-polymeryl species, apart from the features of the polyethylene related peaks and of methylene carbons of the polymer chain bonded to aluminum (starred signals). The two Zr-polymeryl species (**17P** and **18P**) observed have the structures shown in Scheme 3, one of which is ionic. They should be in equilibrium with each other, like the complexes which do not contain the polymer chain (Scheme 3).

The mechanism shown in Schemes 1-3 is similar to the intermittent chain growth mechanism proposed by Fink [4c]. This study has permitted us to provide evidence of new spectroscopically detectable complexes, among which the Zr-polymeryl ion pairs such as **16P** and **11P** which are either the propagating active species or intermediates closely related to the propagating active species.

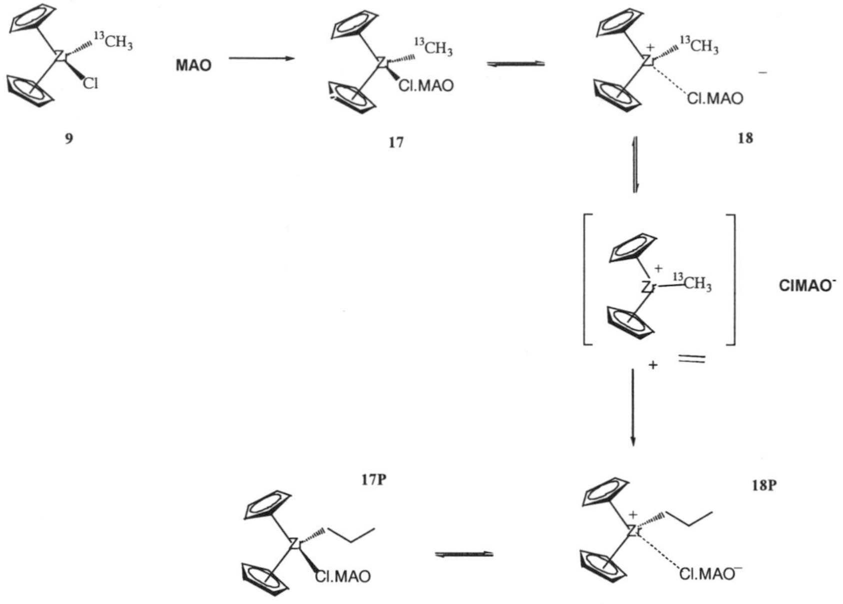

Scheme 3

The observed ion pairs **16** and **18** are weakly coordinated since their Zr-CH$_3$ resonances are different 45.6 and 41.8 ppm, respectively. Ion pairs containing polymeryl chain are weakly coordinated as well, since the signals of their Zr-CH$_2$-P methylene carbons are quite different 65.6 ppm and 74.0 ppm, in **16P** and **18P**, respectively. This is in agreement with Zr-C distances of roughly 2.4-2.5 angstrom measured from X-rays of zirconocene ion pair structures [3]. These values fall in between non bonded and polar covalent Zr-C distances. NMR line-shape analysis of substituted Cp systems obtained by Marks [3] indicate that contact ion pairs appear to dominate in these equilibria in non polar solvents such as toluene and even in polar chlorinated solvents such as dichloromethane.

All these results suggest that the olefin polymerization active site is a cationic metal alkyl and that it is not an isolated cationic site. Olefin and solvent are competitive with the counteranion. The overall scenario for the activation of a zirconocene dichloride can be represented as in Scheme 4.

The importance of ion-pairing and the presence of solvated ion pairs is often suggested by polymerization data such as activity, polymer and copolymer microstructures. Increasing polar media and higher dilution have been shown to favor the solvent-separated ion pairs over the contact ion pairs by variable dynamic NMR studies of substituted Cp systems in solvents with different polarities and coordinating abilities [3]. The ion – pair reorganization is greatly increased on going from toluene to more polar solvents. Examples of π-arene complexes of Zr and Hf have also been observed.

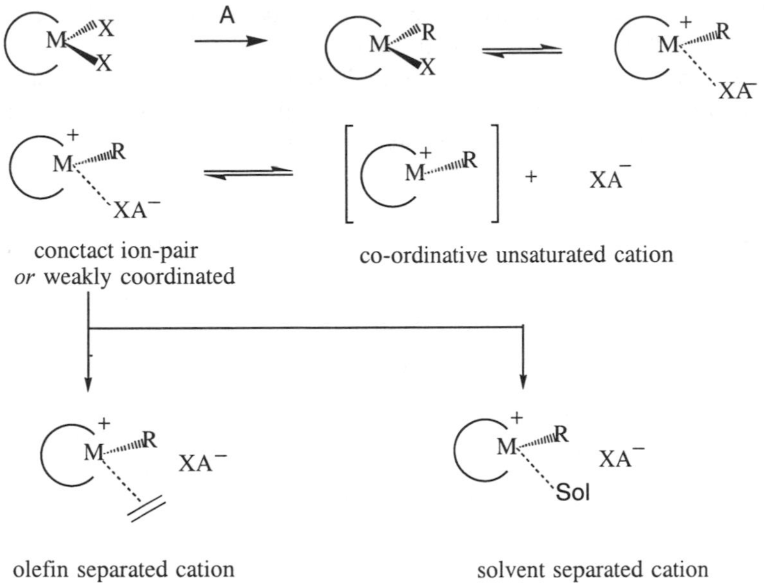

Scheme 4

Contact and solvent separated ion pairs were observed by Eisch for $Cp_2TiRCl\cdot AlCl_3$. He first suggested that solvent-separated ion pairs are more active sites than contact ion pairs. The presence of solvatated contact ion pairs has been deduced from polymerization data. We, like Brintzinger [7b], have observed different signals in NMR spectra assignable to contact and solvent separated ion pairs only for dinuclear zirconocenes which have borane as counterions, but we did not observe different signals in the presence of MAO.

There are only a few examples of olefin π-complexes [8]. Jordan's findings showed that the olefin binding affinities of coordinated vinyl alkoxides are greatly dependent on the length of the alkyl chain.

Application of NMR study to reactions between zirconocenes and new β-branched alkylalumoxanes

Studies aimed at replacing MAO as cocatalyst showed that branched β-alkyl substituted aluminoxanes, prepared *in situ* from hydrolysis of the parent aluminum alkyls, could be alternatives to MAO of variable effectiveness depending on the metallocene structure [9]. Impressive variations in the polymerization activity of a great number of unbridged and bridged metallocenes were found by using different

AlR$_3$/H$_2$O mixtures as cocatalysts [9]. They could not be explained by steric and electronic effects of metallocene substituents. We applied our NMR approach to understand the way in which the changes in the alkyl group bonded to aluminum and the changes in the ancillary π ligands of metallocenes affect the cocatalytic ability of new β-branched alkylalumoxanes [10].

TIBAO and TIOAO obtained at the mole ratio of 2 give the highest catalytic activities [9] and are represented as dimers, although they probably consist of mixtures of different oligomers associated in more complex structures.

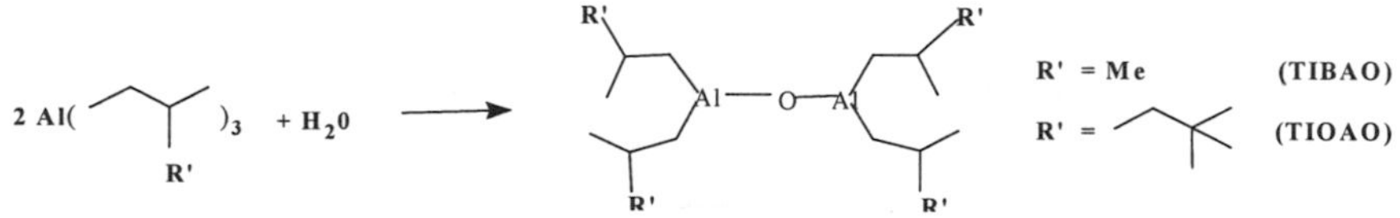

2 Al($\quad$)$_3$ + H$_2$O ⟶

R' = Me (TIBAO)

R' = (TIOAO)

Scheme 5

The reaction equilibria between two β-alkyl substituted aluminoxanes TIBAO and TIOAO - prepared *in situ* at various Al/H$_2$O ratios (R = *i*-Bu), (R = *i*-Oct) - with two simple metallocenes Cp$_2$ZrCl$_2$ (**19**) and Cp$_2$*ZrCl$_2$ (**20**) (Cp* = Me$_5$Cp) with great differences in sterics as well as in electronics, were studied. The reactions of the parent Al(*i*-Bu)$_3$ (TIBA), and Al(*i*-Oct)$_3$ (TIOA) were studied as well in a wide range of temperatures (from -78 to 40 °C) and Al/Zr ratios (from 1 to 60). Cp$_2$ZrCl$_2$ reactivity toward β-alkyl substituted aluminoxanes is represented in Scheme 6. (TIOAO behavior, besides some quantitative differences, is similar to that of TIBAO):

TIOAO ⟶   + TIAOCl ⟶ HYDRIDES +

Scheme 6

The products of the alkylation reactions between Cp$_2$ZrCl$_2$ with TIBA, TIBAO, TIOA and TIOAO are observed only at temperatures as low as -20 °C. On raising the temperature and the Al/Zr ratio in all reactions with the two β−alkyl-substituted systems we observe the formation of several types of decomposition products of the alkylated zirconocenes which tend to be irreversibly transformed into a single species (^{13}C NMR: Cp ~ 104 ppm) [11]. Decomposition processes are less favored in the presence of aluminoxanes especially in the presence of TIOAO. Thus, the incapacity of TIBAO to activate Cp$_2$ZrCl$_2$ is due to the instability of the alkylated species such as Cp$_2$Zr-i-BuCl.

With respect to the unsubstituted Cp, per-methylation in Cp$_2$*ZrCl$_2$ reduces the reactivity towards the same aluminum alkyls and the same aluminoxanes. The Cp* steric hindrance which inhibits β-hydrogen elimination during olefin

polymerization probably inhibits β-hydrogen elimination and decomposition of alkylated zirconocene ion pairs (Scheme 7).
Evidence of the greater capacity of TIOAO with respect to TIBAO for yielding the ion pairs responsible for polymerization activity were produced.

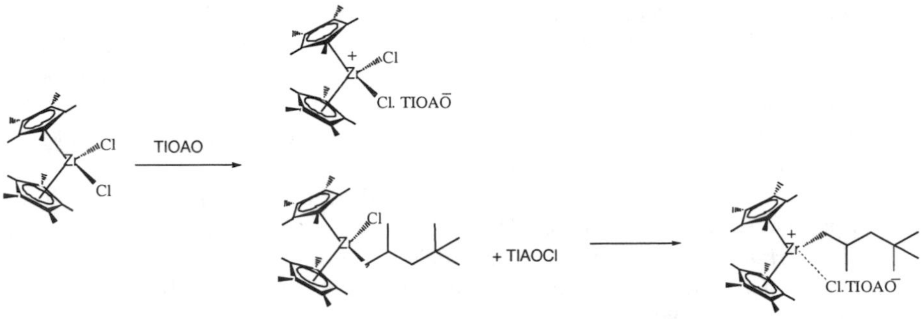

Scheme 7

In conclusion the changes in the alkyl group bonded to aluminum of aluminoxanes affect the cocatalytic ability of aluminoxanes by influencing the active species formation, that is, the alkylating power and the capability ofstabilizing the ion pairs, while changes in the metallocene ligands affect the reactivity towards the aluminoxanes and the stability of the alkylated species.

Acknowledgements

The authors would especially like to thank Giulio Zannoni for his expert NMR assistance and Montell for a partial financial support.

References

1. For recent reviews, see for example: (a) Resconi, L.; Cavallo, L.; Fait, A.; Piemontesi, F.; *Chem. Rev.* **2000,** *100,* 1253-1346. (b) Coates, G. *ibid.* 1223-1252 (c) Angermund, K. Fink, G.; ibid (d) Kaminsky, W., Ed. *Metalorganic Catalysts for Synthesis and Polymerization: Recent Results by Ziegler-Natta and Metallocene Investigations,* Springer-Verlag: Berlin, 1999. (e) Hlatky, G. *Coord. Chem. Rev.* **1999,** *181,* 243. (f) Brintzinger, H. H.; Fisher, D.; Mulhaupt, R.; Rieger, B.; Waymouth, R. M. *Angew. Chem., Int. Ed. Engl.* **1995,** *34,* 1143.
2. Sinn, H.; Kaminsky, W. *Adv. Organomet. Chem.* **1980,** *18,* 99.
3. Recent review: Chen, E. Y.; Marks, T. J. *Chem. Rev.* **2000,** *100,* 1391-1434.
4. (a) Dyachcovskii, F. S.; Shilova, A. K.; Shilov, A. E. *J. Polym. Sci., Part C* **1967,** *16,* 2333. (b) Eisch , J. J. Piotrowski, A. M.; Brownstein, S. K.; Gabe, E. J.; Lee, F. L. *J. Am. Chem. Soc.* **1985,** 107, 7219. (c) Jordan, R. F.; Bajgur, C. S.; Willett, R.; Scott, B.

J. Am. Chem. Soc. **1986**, *108,* 7410. (d) Jordan, R. F. *Adv. Organomet. Chem.* **1991**, *32,* 325 (e) Mynott, R.; Fink, G.; Fenzl, W. *Angew.Makromol. Chem.* **1987**, *154,* 1. (f) Fink, G.; Fenzl, W.; Mynott, R. Z. Z. *Naturfosch.* C **1985**, *40b,* 158. (g) Fink, G.; Rottler, R. *Angew. Makromol. Chem.* **1981**, *94,* 25.

5.   (a) Tritto, I.; Li, S.; Sacchi, M.C.; Zannoni, G. *Macromolecules* **1993**, *26,* 7111; (b) Tritto, I.; Sacchi, M.C.; Li, S. *Macromol. Rapid Commun.* **1994**, *15,* 217; (c) Tritto, I.; Sacchi, M. C.; Locatelli, P.; Li, S. X. *Macromol. Symp* **1995**, *89,* 289; (d) Tritto, I.; Li, S. X.; Sacchi, M. C.; Locatelli, P.; Zannoni, G. *Macromolecules* **1995**, *28,* 5358; (e) Tritto, I.; Li, S. X.; Boggioni, L.; Sacchi, M. C.; Locatelli, P.; O'Neill A. *Macromol. Chem. Phys.* **1997**, *198,* 1347. (f) Tritto, I.; Donetti, R.; Sacchi, M.C.; Locatelli, P.; Zannoni, G. *Macromolecules* **1997**, *30,* 1247; (g) Tritto, I.; Donetti, R.; Sacchi, M.C.; Locatelli, P.; Zannoni, G. *Macromolecules* **1999**, *32,* 264.

6.   (a) Sinn, H. *Macromol. Symp.* **1995**, *97, 27.* (b) Zakharov, I.I.; Zakharov, V.A.; Zhidomirov, G.M. in "Kaminsky W. Ed. *Metalorganic Catalysts for Synthesis and Polymerization"* Springer Verlag: Berlin 1999 p. 128. (c) Mason, M. R.; Smith, J. M.; Bott, S. G.; Barron, A. R. *J. Am. Chem. Soc.* **1993**, *115,* 4971(d) Barron, A. R. *Organometallics* **1995**, *14,* 3581.

7.   (a) Bochmann, M.; Lancaster, S. J. *Angew. Chem. Int. Ed. Engl.* **1994**, *33,* 1634. (b) Beck, S.; Prosenc, M. H.; Brintzinger, H. H.; Goretzki, R.; Herfert, N.; Fink, G. *J. Mol. Cat.A* Chem. **1996**, *111,* 67.

8.   (a) Horton, A. D.; Orpen A. J. *Organometallics* **1992**, *11,* 8. (b) Wu, Z.; Jordan, R: F: Peterson, J. *J. Am. Chem. Soc.* **1995**, *117,* 5867.

9.   (a) Resconi L.; Giannini U.; Albizzati E.; U.S. Pat 5,126,303 to Himont 1992. (b) Dall'Occo T., Galimberti M., Resconi L., Albizzati E., Pennini, G., WO 96/02580 to Montell Technology B.V. 1996. (c) Resconi L, Dall'Occo T., Giannini U. in "J. Scheirs, W. Kaminsky Eds *Preparation, Properties & Technology of Metallocenes-Based Polyolefins"* John Wiley & Sons: 2000 p. 69.

10.   Tritto, I.; Zucchi, D.; Destro, M.; Sacchi, M.C.; Dall'Occo, T.; Galimberti, M. *J. Mol. Cat.* **2000** *in press*

11.   On the basis of the ^{1}H NMR data this compound seems to be a neutral dihydride, of the type characterized by Schwartz [12], which should be inactive for polymerization.

12.   Shoer L. I.; Gell K. I.; Schwartz J.; *J. Organomet. Chem.* **1977**, *136.* C19-C22.

Syndiotactic Specific Structures, Symmetry Considerations, Mechanistic Aspects*

Abbas Razavi, Didier Baekelmans, Vincenzo Bellia, Yves De Brauwer, Kai Hortmann, Marine Lambrecht, Olivier Miserque, Liliane Peters, Martine Slawinski, Stephan Van Belle

Fina Research S.A., Centre de Recherche Du Groupe TotalFinaElf, Zone Industrielle C, B-7181 Seneffe (Feluy), Belgium

Abstract: The mechanism of syndiospecific polymerization with the catalyst systems isopropylidene(cyclopentadienyl-fluorenyl)MCl$_2$; M = Zr, Hf / MAO is discussed by taking into account the structural characteristics of the metallocene molecules and the optical particularities of their cationic species within the framework of a chain migratory insertion mechanism. A generally accepted transition state structure that respects the relative importance of different steric interactions of the active participants in the polymerization process, ligand, growing polymer chain and the coordinating monomer is discussed. It is shown that the substitution-free two top quadrants left and right to the cyclopentadienyl and the free space in the central position of the fluorenyl are, among others, essential factors to syndiospecificity of the catalyst. The model is examined on a new syndiospecific catalyst system, η1,η5-tert-butyl(3,6-bis-tert-butylfluorenyl-dimethylsilyl)amidodichloro-titanium / MAO and its validity is confirmed.

Introduction

Syndiotactic polypropylene was first isolated by Natta and co-workers as a minor by-product of an isotactic polypropylene produced with a TiCl$_3$ based catalyst [1]. The nature of the active site and the mechanism of formation of this polymer, which was discovered nearly forty years ago, are still under debate. It is believed that it is formed on the sites with low chlorine coordination via a chain end controlled mechanism. Later Zambelli and coworkers produced syndiotactic polypropylene directly using a vanadium based catalyst [2]. In this case more is known about the nature of the active site and the mechanism of the polymerization is elucidated satisfactorily. It is now certain that the polymer chains are formed at low temperature on homogeneous active sites according to a mechanism that is controlled also by the chirality of the last inserted monomer unit at the end of the polymer chain. No X-ray structure of the catalyst precursor is, however, available due to its very temperature sensitive nature. After the discovery of the first bridged cyclopentadienyl-fluorenyl metallocene based syndiospecific catalyst and the resulting syndiotactic polypropylene [3] one had for the first time the opportunity to make even more accurate statements about the nature of the active site and the mechanism of the polymerization of this fascinating yet very complex systems. By studying the available X-ray structure data of the metallocenes and their stabilized alkylmetallocenium cation (considered as the catalyst precursors) [3b] it has become possible to make reasonable deductions on the nature of the active site and its behavior during the polymerization. On the other hand facile availability of large syndiotactic polymer samples prepared at very different and precise polymerization conditions and statistical analysis of their high-resolution ^{13}C NMR spectra provided the means for accurate statements on mechanism of the polymerization. The concept of active site model and the mechanism of the polymerization that had been proposed by us initially have been refined gradually and continuously as new syndiotactic specific metallocenes were discovered and more elaborate calculation methods were applied. In this contribution we report the latest progress on catalysts development and mechanistic aspects of syndiospecific polymerization commencing with a review of the original model proposed for the isopropylidene(cyclopentadienyl-fluorenyl)MCl$_2$ systems.

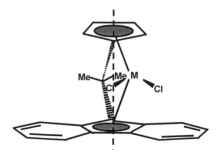

Fig. 1. Isopropylidene(cyclopentadienyl-fluorenyl)MCl$_2$ M = Zr, Hf.

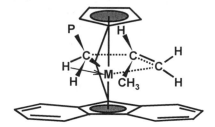

Fig. 3. Syndiospecific transition state structure

Results and discussion

Figure 1 represents the molecular structure of the metallocenes isopropylidene-(cyclopentadienyl-fluorenyl)MCl$_2$; M = Zr, Hf [3a] that were used to prepare the first ever highly crystalline syndiotactic polypropylene samples. We have been exhaustively discussing the structure, catalytic properties and the origin of the formation of syndiotactic specific chains with a ...rrrrrrrrmrrrrrrrrrrrrrrrrmmrrrrrrr... microstructure since their discoveries in numerous papers and oral presentations [3,6,9,10,11b,13]. However, for the sake of continuity of the line of evidence the highlights of these discussions will be reiterated in this paper combined with the proposals made by other authors: 1) The stereorigid metallocene procatalysts

(Figure 1) is prochiral and possesses a bilateral symmetry. 2) The chiral, cationic alkyl metallocenium species [3b] which are formed after the activation are composed of equal numbers of R and S enantiomers and are monomer π-face selective (Figure 2). The activation of metallocene dichloride proceeds, with MAO for example, via a two step dialkylation and a final alkyl abstraction and create the necessary coordination position and fragment orbital for the interaction with the incoming propylene. 3) The re- and si–face selectivity is induced by the unique steric arrangement of the chelating ligand engulfing the resident chiral transition metal center via a delicately balanced, cooperative, and non-bonded steric interaction mechanism among different parts of the "living" catalytic species, ligand, polymer chain, and coordinating monomer. The non-bonded steric interactions govern the whole scenery of syndiospecific polymerization processes. 4) Since according to these assumptions each enantiomer, independently, would produce isotactic chains (yet exclusively syndiotactic polymer is formed), it had to be concluded that the active enantiomeric species isomerize and interconvert after each monomer insertion. 5) The systematic transformation of the two antipodes into one another implies that the relative positions of at least two of the four ligands surrounding the transition metal are exchanged continuously. 6) Since the η^5 bonded aromatic ligands are tied together by a structural bridge and their rearrangement is not possible, such an isomerization can take place only when the alkyl group (polymer chain) and the coordinating monomer exchange their positions uninterruptedly (when no excessive steric restrictions are imposed). 7) The meso triad enantiomorphic site stereochemical type errors, mm, are formed whenever the said balanced non-bonded steric interaction is perturbed, and the correct alignment of the substituents of all three main participants (ligand, polymer chain and the monomer) has not been realized. In such case a monomer with "wrong face" will be inserted and a unit with inverted configuration is enchained.[‡] The ability to reverse facial selectivity emanates from inherent static structural factors and is independent of monomer concentration.

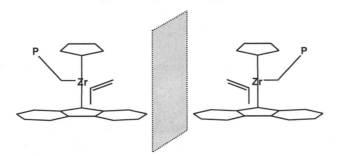

Fig. 2. Enantiotopic active species

[‡] Other origins such as chain epimerization have also been proposed for the stereo-errors, particularly for Polymers made at lower monomer concentration see V. Busico, R Cipullo, *J. Am. Chem. Soc.* **116**, 9329 (1994)

To understand how these many factors operate in a concerted manner to create the conditions for syndiospecificity we need to construct a model that explains in a plausible manner the behavior of this complex system. Figure 3 depicts the model representing the transition state structure for propylene polymerization. This model is constructed based on relative importance of the non-bonded steric interactions operating on different parts of the catalytic species and its - in the polymerization active participants - aromatic ligand, polymer chain and the coordinating monomer in the following order. The steric interaction between the flat and spatially extended fluorenyl ligand forces the growing polymer chain to be oriented towards the free space left (or right) to the unsubstituted cyclopentadienyl moiety of the ligand. The incoming monomer - to avoid excessive steric exposure - orients itself in a manner to have its methyl group trans positioned with respect to the growing polymer chain. The system reaches in this way a minimum energy state. In this orientation the coordinated monomer points with its methyl group head down into the empty space in the central region of the fluorenyl ligand. The confirmation for the head down orientation of the monomer was determined after extensive molecular mechanics and force field calculations performed by Corradini and coworkers [4]. The model underwent later additional refinement and took its current form after experiments conducted by several groups [5] supporting an α C-H agostic assistance in the transition state for the propylene polymerization.

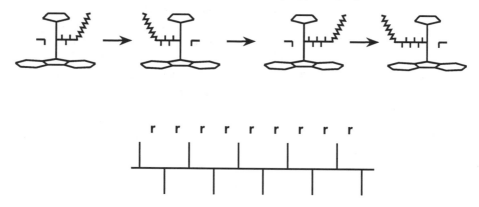

Fig. 4. Mechanism of syndiospecific polymerization (top). Fischer projection of a perfect syndiotactic chain (bottom).

This model combined with the chain migratory insertion mechanism now is perfectly fit to explain the formation of syndiotactic polymers with metallocenes of Figure 1. According to the scheme shown in Figure 4 the regularly alternating enantiofacial preference for monomer that leads to syndiotactic polymer arises from olefin insertion taking place at regularly alternating sides of the wedge of the pseudo tetrahedral structure of the active site. Despite its simplicity and convenience the model has a major handicap. It has been conceived of images coming from X-ray analysis performed on molecules contained in solid state structure. These rigid and static images reflect only a "moment-aufname" of the constantly vibrating, bending and "breathing" molecules in all directions. They should be considered only as frozen images of dynamic molecules that reflect only one aspect, the general shape and outline of the molecules. They do not reflect the whole "reality" of individual molecules freely floating in a solvent medium without the restrains of the crystal packing effects. Both IR and NMR data and chemistry of these compounds are indicative of constant bonding related movements of different part of the molecules and their fluxionallity. They are often neglected intentionally for the sake of simplicity and the fact that to prove their existence is very difficult. However, one should be aware that at least two dynamic phenomena are actively involved in one way or the other in different steps of the polymerization and influence its mechanism in one or the other direction. For example, the phenomenon of hapticity change or variation of bond order between the transition metal and the aromatic ligands should be seriously considered to be involved in certain metallocene catalysts as we have demonstrated [6] (See Figure 5). The haptotropy and ring slippage can influence the electronic properties of the active site and the steric environment surrounding it temporarily or permanently and have an impact on the molecular weight and tacticity of their polymers.

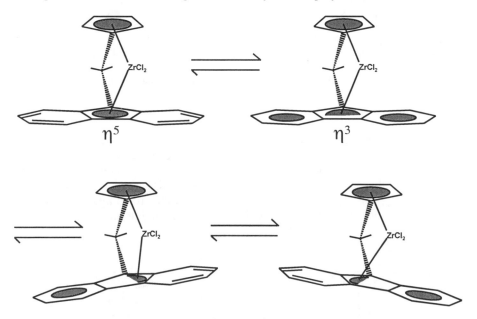

Fig. 5. Haptotropy (top) and ring slippage (bottom).

The second phenomenon is related to geometry change of the catalysts during the coordination and insertion steps. The pseudo tetrahedral geometry which is assumed for the tetracoordinated transition metal in the transition state, can not be further extended to the step just after insertion. At this stage the tetracoordinated structure collapses due to the disappearance of a ligand leaving a tricoordinated species behind in which the repulsive forces acting upon the bonding electron pairs are different and require a new geometry. The most logical structure that can be suggested for this step would be a pyramidal one (Figure 6 right). After the next monomer coordination again the structure will adopt to a tetrahedral geometry (Figure 6 left). This change in geometry, operating on all metallocenic catalyst systems, has probably more importance for the syndiospecific case where dynamic processes such as chain migration and site epimerization (vide infra) are vital for its existence.

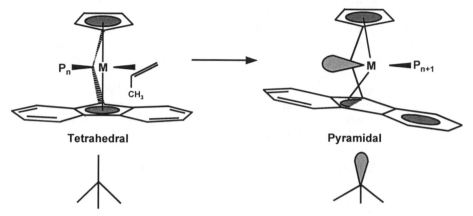

Fig. 6. Systematic structural change and geometry variation during coordination and insertion steps.

Finally, another, ligand / transition metal related, dynamic behavior that can be envisaged to be acting on the transition structure is the lateral displacement of the whole ligand system around the transition metal discovered by Petersen [7] is also noteworthy in this context. This movement that can be described as a kind of wind shield wiper type oscillation of the MCl_2 moiety within the fixed ligand system can facilitate or influence the site epimerization or chain migration mechanism. (Figure 7)

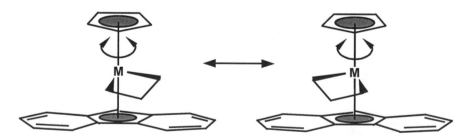

Fig. 7. Lateral displacement of ligand system around the centroids-Zr bond axis (the bridge is omitted for the sake of clarity)

The working hypothesis, active site model and the transition state structure discussed in the preceding paragraphs account perfectly for the syndiospecificity of the catalysts and formation of the syndiotactic polypropylene chain. However, for a model and a mechanism to be acceptable it is not sufficient to correctly describe the polymerization behavior and polymer's microstructure in general. It must also justify the details of the formation of microstructural stereo errors and monomer misplacements in the backbone of the polymer chain. From two types of stereo errors - the meso triads (mm) and meso dyads (m) - encountered in the back bone microstructure of the syndiotactic polymers prepared with metallocene catalysts the former – we have discussed it above briefly - is related to the enantiofacial misinsertion and is monomer concentration independent. These types of errors have been detected and explained long ago in connection with isotactic polymers prepared with the classical ZN systems. The origin of the formation of the monomer concentration dependent stereo errors the so-called meso diad (m) type errors was more complex and more difficult to discern. Their investigation, however, turn out to be more exiting and conclusive for the understanding of the functioning of the syndiospecific catalysts. Their formation can be explained according to site epimerization scheme shown in Figure 8.

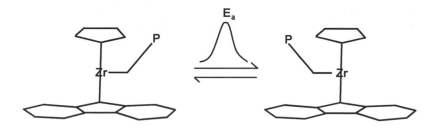

Fig. 8. Endothermic active site epimerization

The R and S configured active sites are equi-energetic and can interconvert (epimerize) during the polymerization particularly in the absence of a coordinating monomer or solvent molecules. If the rate of this interconversion is faster than the actual rate of the monomer insertion then occasionally the sites epimerize before any insertion occurs and two (in case of Hf catalyst more) consecutive insertions will take place at the same enantiomorphic coordination position enchaining two monomer units with the same prochiral face. This causes the formation of meso dyads within the polymer backbone (isotactic blocks for Hf). Even though the interconverting species are equi-energetic the epimerization requires an activation energy since the chain, while swinging back or forth, will be exposed to different parts of the ligand and has to overcome the resulting non-bonded steric repulsion. That is why at lower polymerization temperature the concentration of the meso dyad is relatively low and only with increasing temperature their number increases rapidly. The concept of site epimerization is very close to the mechanism of chain back skip proposed already by Cossee and Arlman in 1964 [8] for the non-symmetric sites in $TiCl_3$ catalyst systems. The only difference between the two mechanisms is related to the forces that drives the back skip or the chain epimerization in each system. The in the rigid ionic lattice imbedded active TiCl$_5$R sites have no real contribution to the chain back migration and only the relative ionicity / non-ionicity of the immediate environment of the chain is the driving force for its back migration (back skip). Whereas in the case of metallocene based catalysts the flexible structure of the aromatic ligand and the haptotropic nature of its bonding can and probably will contribute a whole lot to the epimerization process. We have discussed static (contact ion pairing) and steric (ligand / polymer chain substituent) interactions as the possible causes for the epimerization elsewhere [9] here we shall focus only on its impact. The existence of the site epimerization process can be easily demonstrated and / or explored by intentionally increasing the epimerization rate and consequently enlarging the isotactic sequence length to the extreme of obtaining "pure" isotactic polymers [10]. This can be done by implanting selectively a substituent at one of the distal positions of the smaller cyclopentadienyl group. Two substituents, the tertiary butyl and trimethyl silyl groups have been proven to be very expeditious in this respect. In these cases the process of epimerization can be manipulated to an extent that the tactic behavior of the catalysts is completely inverted and an originally syndiospecific catalyst produces instead isotactic or isotactic-syndiotactic block polymers. The increasing isospecificity [11a,b] of these C$_1$ catalysts with increasing polymerization temperature - in drastic contrast to the behavior of the bridged bisindenyl type catalyst systems – can be seen as a clear evidence for the operating epimerization process. Additionally, the complete absence of rmrm pentad sequences (an indication of low probability of the chain being at the more crowded coordination position) in the microstructures of isotactic polypropylene produced with the same catalysts [11b,c] provides further clue suggesting the presence of epimerization process. Which of course increases its rate with increasing polymerization temperature and provokes the formation of ever longer isotactic sequences. The mechanism of the formation of these polymers can be explained by complete disfunctioning of the chain migration replaced by rapid and continuous epimerization of the site due to steric repulsion between the chain and the substituent during the insertion process (Figure 9). From the point of view of the incoming monomer the chain seems "stationary" by attacking it always from the same position.

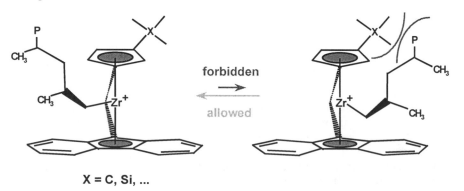

X = C, Si, ...

Fig. 9. Non-bonded steric repulsion between polymer chain and β-substituent provoking the site epimerization

It is apparent that syndiospecificity as discussed here is the result of the combined actions of several factors; the absence of each could lead to the disfunctioning of the catalyst and destruction of the syndiospecific process. The delicate steric balance between the three main participants, ligand, polymer chain and the monomer as well as proper functioning of the involved dynamic processes are all essential for its existence. Therefore, it should not be surprising if occasionally catalysts that have been prepared, that although the metallocenes are structurally "acceptable", do not work in this way. An early disappointing example of this kind was the complex dimethysilyl(cyclopentadienyl-tetramethyl-cyclopentadienyl)ZrCl$_2$. After its activation with MAO this catalyst did not behave syndiospecific at all and the product of its propylene polymerization was a purely atactic polypropylene. Here obviously the constantly rotating, voluminous methyl groups, particularly those two placed in the distal positions of the larger cyclopentadienyl interfere vigorously with proper coordination mode of the monomer and disrupt completely the stereospecificity of the system. Another example for demonstrating the delicacy and fragile nature of the steric balance for the syndiospecificity is shown below.

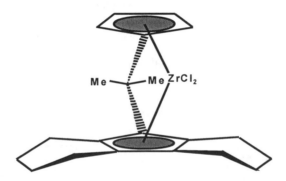

Fig. 10. Structure of diphenylmethylidene(cyclopentadienyl-octahydrofluorenyl)ZrCl$_2$

The metallocene diphenylmethylidene(cyclopentadienyl-octahydrofluorenyl)-ZrCl$_2$ whose molecular structure is presented in Figure 10 fulfils all the structural and symmetry requirements one would imposed on a would-be syndiospecific system. Having potentially all the properties required for being syndiotactic specific we were surprised that it produced perfectly atactic polypropylene. In this case, one would be led to believe that the minor change in size of benzenic CH groups to slightly larger cyclohexylic CH$_2$ groups in the fluorenyl distal positions - after hydrogenation – would be tolerated by the system and would not disturb the syndiospecific process. Nevertheless, it should be noted that in this case the substituent change is concomitant with another dynamic phenomenon that might be the real disturbing factor. The constant boat / chair conformational interconversion of the flexible cyclohexylic parts of the hydrogenated fluorenyl probably interfere more with the proper head down coordination mode of the monomer and its enantio-selective coordination than the mere slight size increase of the substituents.

This situation of course does not favor systematic and consecutive enchainment of the propylene monomers in a syndiotactic manner.

To emphasize further the vital role of the presence of the steric balance and the free central space and to avoid giving any superstitious impression that the syndiospecificity is somehow linked to the presence of the fluorenyl groups, we shall discuss briefly the syndiospecific system reported by Bercaw [12]. The double bridged zirconocene molecules shown in Figure 11 (a) fulfil all structural and symmetry prerequisites for the syndiospecificity. They resemble very much in their structures the molecules of Figure 1 in having one unsubstituted cyclopentadienyl linked to a symmetrically substituted cyclopentadienyl (replacing fluorenyl) via a structural bridge. The structures given in Figure 11 (a) are in fact syndiospecific and produce highly syndiotactic polypropylene yet they do not contain any fluorenyl group. What is essential, however, that the substituted cyclopentadienyl group which is replacing the fluorenyl maintains the free space in the central position to accommodate the methyl group of the coordinating monomer. And additionally, the smaller cyclopentadienyl group has no distal substituents prohibiting the proper orientation of the chain in the upper left and right quadrant of the catalyst (the regions above the plan encompassing, $ZrCl_2$ moiety and the bridge left and right to the σ'_v plane). The importance of the free space in central region of the larger cyclopentadienyl group for monomer's head-down coordination is reiterated by the molecule depicted on the Figure 11 (b). Because of the presence of the substituents blocking the central position this metallocene does not fulfil the imposed requirements and does not behave in its activated form syndiospecific.

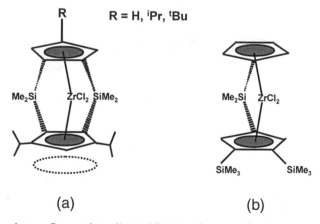

(a)   (b)

Fig. 11. Bercaw's non-fluorenyl syndiospecific (a) and non-syndiospecific (b) structures

Finally, an interesting syndiospecific case will be presented to check the validity of the working hypotheses, active site model and the transition state structure. The main ingredients of this new syndiospecific catalyst system are the complexes $\eta 1,\eta 5$-tert-butyl(2,7-bis-tert-butylfluorenyl-dimethylsilyl)amidoMCl$_2$ and $\eta 1,\eta 5$-tert-butyl(3,6-bis-tert-butylfluorenyl-dimethylsilyl)amidoMCl$_2$ (M = Ti, Zr). Figure 12 represent the single crystal X-ray structure of the $\eta 1,\eta 5$-tert-butyl(3,6-

bis-tert-butylfluorenyl-dimethylsilyl)amidodichlorozirconium. After the activation with MAO both Zr and Ti complexes polymerize propylene to syndiotactic polypropylene very efficiently, however, only Ti related catalysts produce high molecular weight polymers [6,13]. The syndiotacticity of the polymer that it produces at different polymerization temperatures measured as the concentration of racemic pentads rrrr is even higher than that reported recently by us for η1,η5-tert-butyl(2,7- bis-tert-butylfluorenyl-dimethylsilyl)amidodichlorotitanium. With the structure just mentioned above they are the only examples of a titanium based syndiospecific catalysts. It is in many other aspects different from the systems discussed above.

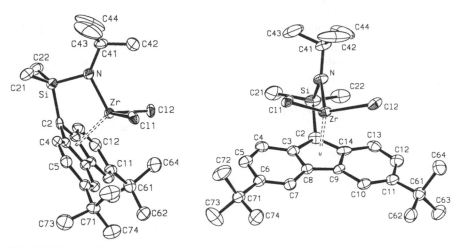

Fig. 12. Side and front view of the molecular structure of the complex η1,η5-tert-butyl(3,6-bis-tert-butylfluorenyl-dimethylsilyl)amidodichlorozirconium

It is a half sandwich molecule, contains amido type N-Ti bond and is a 12 electron system (14 electron system at the most if one considers the participation of the lone pair electrons on the nitrogen to N-Ti bond) leaving 10 (maximum 12) electrons for the active site. Nevertheless, a glance at the structure shown in Figure 12 teaches us that it has all structural and symmetry characteristics to be syndiospecific. This latest addition to the syndiospecific catalyst systems demonstrates that no matter what the make up of the metallocene precursor is, once its structure fulfils the prerequisites discussed based on the models given above, it will act syndiotactic specific.

Conclusion

It is shown that the static model that has been proposed for the active site and the transition state structure explains elegantly the syndiospecific character of the catalysts and the origin of the microstructural particularities of the syndiotactic chains. Being conceived of images taken from the solid state x-ray structure analysis, the model is too static to provide any explanation for dynamic phenomena such a chain migration and site epimerization omnipresent during all stereospecific polymerization performed with metallocene based catalysts. It is therefore, proposed that a more dynamic model should take into consideration the haptotropic nature of the transition metal / aromatic rings and ring slippage phenomena as well as transition state structural collapse leading to permanent tetrahedral / pyramidal geometry change. The origin and / or the driving force for the chain migration and site epimerization may well be lying here. The validity of the models and the working hypothesis for syndiospecific polymerization of the propylene is tested on two other syndiospecific catalyst systems, the double bridged, differently substituted biscyclopentadienyl zirconium dichloride reported by Bercaw and the new titanium based half sandwich complexes, η1,η5-tert-butyl(2,7-bis-tertbutylfluorenyl-dimethylsilyl)amidodichlorotitanium and η1,η5-tert-butyl(3,6-bis-tertbutylfluorenyl-dimethylsilyl)amidodichlorotitanium. Both systems demonstrate clearly and confirm the validity of the proposed ideas no matter the make up of the catalysts. Once the structural requirements imposed by the model proposed here is fulfilled, the catalysts behave syndiospecific independent of the nature of the transition metal and the presence of varying number of the aromatic rings in the π ligand. Figure 14 compares the structure of three systems discussed high lighting the main factors for the syndiospecificity. Of course the degree of syndiospecificity of the catalyst and stereoregularity of the resulting polymer depends on stereorigidity of the bridging ligand and the size, type and position of the substituent(s).

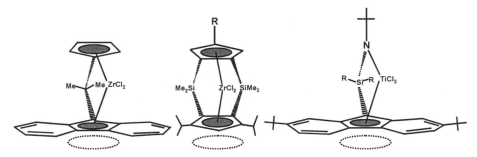

Fig. 14. Three different syndiotactic specific molecules with similar structural characteristics.

References

1   G. Natta, I. Pasquon, P. Corradini, M. Peraldo, M. Pegoraro, A. Zambelli, *Atti Accad. Nazl. Lincei Rend. Classe Sci.Fis. Mat. Nat.*, [8] **28**, 539 (1960)

2   G. Natta, I. Pasquon,A. Zambelli, *J. Am. Chem. Soc.* **84**, 1488 (1962) A. Zambelli, G. Natta, I. Pasquon; *J. Polymer Sci.*, **4**, 411 (1964)

3   (a) A. Razavi, J. L. Atwood, *J. Organomet. Chem.*, **459**, 117 (1993), (b) A. Razavi, U. Thewalt, *J. Organomet. Chem.* **445**, 111 (1993). (c) A. Razavi, J. J. Ferrara, *J. Organomet. Chem.* **435**, 299 (1992). (d) J. A. Ewen, L. R. Jones, A. Razavi, *J. Am. Chem. Soc.* **110**, 6255 (1988)

4   L. Cavallo, G. Guerra, M. Vacatello, P.Corradini, Macromolecules, **24**, 1784 (1991)

5   (a) W. E. Piers, J. E. Bercaw, *J. Am. Chem. Soc.*, **112**, 9406 (1990), (b) H. H. Brintzinger, H. Krauledat, *Angew. Chem., Int. Ed. Engl.*, **29**, 1412 (1990). (c) H. H. Brintzinger, M. L. Leclerc, *J. Am. Chem. Soc.*, **117**, 1651 (1995), (d) B. J. Burger, W. D. Cotter, E. B. Coughlin, S.T. Chascon, S. Hajela, T. A. Herzog, R. O. Koehn, J. P. Mitchell, W. E. Piers, P. J. Shapiro, J. E. Bercaw, J. E. in Ziegler Catalyst; G. Fink, R. H. Muellhaupt, H. H. Brintzinger, Eds; Springer verlag; Berlin, 1995. (e) R. H. Grubbs, G. W. Goates, *Acc. Chem. Res.*, **29**, 85 (1996).

6   A. Razavi, V. Bellia, Y. De Brauwer, K. Hortmann, M. Lambrecht, O. Miseque, L. Peters, S. Van Belle, In "Metalorganic catalysts for Synthesis and Polymerization. Ed. W. Kaminsky, Springer, Berlin (1999)

7   A. Kabi-catpathy, S. Bajgur, K. P. Reddy, J. L. Petersen, *J. Organomet. Chem.*, **364**, 105 (1989).

8   E. J. Arlman, *J. Catal.*, **3**, 89 (1964). E. J. Arlman, P. Cossee, *J. Catal.*, **3**, 99 (1964)

9   A. Razavi, L. Peters, L. Nafpliotis, D. Vereecke, K. Den Daw, *Makromol. Symp.* **89**, 345 (1995).

10 A.Razavi, J. L. Atwood, *J. Organomet. Chem.* **520**, 115 (1996).

11 a) R. Kleinschmidt, M. Reffke, G. Fink, *Macromol, Rapid Commun.* **20**, 284 (1999). b) A. Razavi, D. Vereecke, L. Peters, K. Den Daw, L. Nafpliotis, J. L Atwood, in Ziegler Catalysts; Fink, G.; Muelhaupt, R.; Brintzinger, H. H., Eds.; Springer Verlag; Berlin, 1993. c) V. Busico, R. Cipullo, G. Talarico, *Macromolecules*, **30**, 4787 (1997)

12 T. A. Herztog, L. Zubris, J. E. Bercaw, *J. Am. Chem. Soc.* **118**, 11988 (1996). Veghini, D. Bercaw, J. E. Polymer reprints vol. 39 Nr. 1, 210 (1998).

13 Razavi, A. European Patent application Nr. 96111127,5 and international patent application PCT/EP97/03649(WO 98/02469).

The Particularities of Isospecific Polypropylene Synthesis in Bulk with Ansa-metallocene Catalysts

Polina M. Nedorezova, Valentina I. Tsvetkova, Alexander M. Aladyshev, Dmitrii V. Savinov, Dmitrii A. Lemenovskii[1]

Semenov Institute of Chemical Physics RAS, Kosygin str. 4, 117977 Moscow, Russia
[1]Department of Chemistry Lomonosov Moscow State University, 119899 Moscow, Russia
E-mail: pned@center.chph.ras.ru

Abstract. hiral ansa-metallocenes with indenyl ligands are useful as catalysts for producig isotactic polypropylene (*iso*-PP). In this work the kinetics of propylene polymerization in bulk with *ansa*-zirconocenes C_2 symmetry at 30-80°C with using of different cocatalysts, mainly the polymethylaluminoxane, have been investigated. The following aspects were in focus of our interest: the nature of the bridge, the presence of substituents in the bridge and on the indenyl ligands; conditions of the pre-activation stage and the polymerization process and the study of relations between these parameters and PP properties. Using the $Me_2Si(4-Ph-2-Et-Indenyl)_2ZrCl_2$ (*rac* : *meso* = 1 : 2)/MAO the high molecular *iso*-PP was produced. The activity during *iso*-PP synthesis reaches 900 kgPP/gcat.h. Influence of hydrogen on activity and molecular weight characteristics also were studied. We compared properties of metallocene *iso*-PP with the same of PP produced with heterogeneous catalytic systems on the base of high effective Ti/Mg catalysts.

Introduction

Metallocene catalysts have created new impulses in the investigation of stereospecific polymerization and copolymerization of α-olefins [1-5]. New catalysts have allowed to control microstructure, molecular weight characteristics and the tailoring of polyolefins properties. The metallocene catalysts (MC) allow to synthesize almost all kinds of stereoregular polymers. These catalytic systems posses high activity and in contrast to the heterogeneous metallocomplex catalysts have a single active centers (AC). MC allow the rational design of multi-site catalysts by mixture of different metallocenes. This leads to production of PP with mono-, bi- and polymodal molecular mass distributions and this broadens the PP applications areas on a base of homo and propylene copolymers [6].

Activity, molecular weight and stereoregularity of polymers produced on metallocene catalysts, in contrast to the heterogeneous catalysts, are very sensitive to the polymerization conditions: polymerization temperature, monomer and catalyst concentrations, nature of solvent, ratio of Al/Zr etc. PP with the higher molecular weight and higher yield one can produce at polymerization with high concentration of propylene, namely in a medium of a liquid monomer. High concentration of propylene at polymerization in bulk promotes increasing of the degree of PP regularity also [7-9].

In the present work during study of *iso*-PP synthesis in medium liquid propylene with *ansa*-metallocene catalyst the following aspects were in focus of our interest: the nature of the bridge, the presence of substituents in the bridge and in the indenyl ligands; the influence of conditions of the pre-activation stage and the polymerization process (temperature, ratio Al to Zr, concentration of metallocene and hydrogen) on the kinetic characteristics, catalytic systems activity and PP properties.

We studied propylene polymerization in bulk with the next *ansa*-metallocenes: *rac*-(CH$_2$)$_2$(Ind)$_2$ZrCl$_2$ (**MC-1**), *rac*-Me$_2$C(Ind)$_2$ZrCl$_2$ (**MC-2**), Me(*c*-C$_3$H$_5$)C(Ind)$_2$-ZrCl$_2$ (**MC-3**), *rac*-(CH$_2$-C$_6$H$_4$-CH$_2$)(Ind)$_2$ZrCl$_2$ (**MC-4**) *rac*-Me$_2$Si(Ind)$_2$ZrCl$_2$ (**MC-5**) and Me$_2$Si(4-Ph-2-Et-Ind)$_2$ZrCl$_2$ *(rac: meso*= 1:2) - (**MC-6**).

A comparison with some property of *iso*-PP produced with heterogeneous Ti/Mg catalyst also was done.

Experimental

Propylene of polymerization-grade without additional purification was used. Content of admixtures - CO, CO$_2$, H$_2$O was less than 5 ppm.

Polymethylaluminoxane (MAO) from Witco (10 % w solution in toluene) was used without further treatment.

Rac-(CH$_2$)$_2$Ind$_2$ZrCl$_2$ - ethylenebis(1-indenyl)zirconium dichloride; *rac*-Me$_2$CInd$_2$ZrCl$_2$, - dimethylcarboniumbis(1-indenyl)zirconium dichloride; Me(*cyclo*-C$_3$H$_5$)CInd$_2$ZrCl$_2$ - methylcyclopropylcarboniumbis(1-indenyl)zirconium dichloride; *rac*-(CH$_2$-C$_6$H$_4$-CH$_2$)Ind$_2$ZrCl$_2$ – *o*-xylilidenbisindenylzirconim dichloride; *rac*-Me$_2$SiInd$_2$ZrCl$_2$ - dimethylsilyl-bis(indenyl) zirconium dichloride and Me$_2$Si(4-Ph-2-Et-Ind)$_2$ZrCl$_2$ *(rac: meso*=1:2) - dimethylsilylbis(4-phenyl-2-ethyl-indenyl) zirconium dichloride were prepared as described in [10-12].

Modified Ti/Mg catalyst containing 2,3% Ti in combination with AlEt$_3$ was used. Dibutylphtalate and propyltrimethoxisilane (PrTMS) were used as internal and external donors [13].

Propylene was polymerized in a liquid monomer medium at a pressure exceeding that of the propylene saturated vapor pressure by original method described in [11, 14]. PP characterization was described in [11, 13].

Results and discussion

All complexes studied for *iso*-PP synthesis were chiral *ansa*-metallocenes with C$_2$ symmetry having indenyl ligands. This type of symmetry is favorable for the isospecific polymerization of propylene. However exists the high difference in propylene polymerization behavior between metallocenes with unsubstituted (MC-1-MC-5) and substituted ligands (MC-6).

PP produced with using MC-1 - MC-5 catalysts activated by MAO has low Mw (Table 1) because in the case of catalytic systems with unsubstituted bisindenyl zirconocenes the ratios of the constant rates of the reaction chain propagation to the constant rates of the reactions chain transfer on monomer and on zirconium are low values. PP with maximal Mw about 60000 was obtained with MC-5 containing the Si-bridge. In the case of MC with C-bridge the PP Mw is smaller (15000). The content of isotactic pentad "mmmm" in PP obtained varies from 81% (MC-3/MAO) to 90% (MC-5/MAO). The difference in behavior MC with Si-bridge and C-bridge was caused by more opened Zr atom due to the lowest angle Cp-Zr-Cp in MC with C-bridge (MC-2, MC-3). As result one can see the increasing of the termination reaction rate due to β-hydride elimination on monomer and the increasing of the regiomistakes number in polymer chains due to 2-1 interaction [15]. The low regioregularity of PP produced with MC with C-bridge leads to decreasing of the PP melting points to 132 °C.

Table 1. Propylene polymerization with $XInd_2ZrCl_2/MAO$ in liquid monomer at 50°C. [MAO]- 0,6-1 g/l.

X- bridge	[MC]10^4, mol/l	Activity, kgPP/mmolZr h	Mw 10^3	Tm, °C	α(Cp-Zr-Cp), [16]
Me_2C	1,4	132	15	134	118
Me(*cyclo*-C_3H_5)C	1,6	110	16	132	118
$(CH_2)_2$	1,0	115	37	141	125
Me_2Si	2,0	125	58	144	128

The important parameter is the distance between Zr and Cp rings. The increase of these distances leads to decreasing of catalytic systems activity or one can see the complete stop of propylene polymerization as in case of MC-4/MAO catalytic system. At the same time we observed the high activity in ethylene polymerization process using this MC activated by MAO.

Investigation of influence of catalytic systems forming ways on the propylene polymerization process and on the properties of the PP produced is important for determination the optimal conditions the propylene polymerization with MC-catalyst.

With using of MC-5 we have analyzed next ways of the complex activation:
1 - the process was conducted by injecting the toluene solution of metallocene directly inside the reactor with a liquid propylene and MAO;
2 - preliminary dissolving the MC in toluene solution of MAO and injecting the resulting mixture into the reactor with a liquid propylene and MAO;
3 - preliminary dissolving the MC in toluene solution of MAO and injecting the resulting mixture into the reactor with liquid propylene and (i-C_4H_9)$_3$AlCl – TIBA.

The stable polymerization rate in the case 1 and 2 indicates that MAO was needed for reactivation of inactive complexes formed by the hydrogen transfer reactions and to prevent bimolecular process. In the presence of TIBA the bimolecular deactivation is evident. The respective data on catalytic activity and properties of *iso*-PP are presented in Table 2.

Table 2. Influence of the method of forming catalytic system on the base of $Me_2SiInd_2ZrCl_2$ on the activity and PP properties.[*]

Method of forming	$\frac{Al(MAO)}{Zr}$	$\frac{Al(TIBA)}{Zr}$	Activity, kgPP/mmolZrh	Mw 10^{-3}	$\frac{Mw}{Mn}$	Tm, °C
MC/tol+MAO	12600	-	38	90,5	2,0	144
MC/MAO+MAO	12000	-	130	58,4	1,8	145
MC/MAO+TIBA [17]	250	1050	100	60,0	2,1	143

[*] Propylene polymerization in bulk at temperature 50°C, the reactor volume 0,25 l., $Zr-6\cdot10^{-7}$ mol.

One can see, that PP have narrow MWD, close melting point in all cases, but we observed that the preliminary dissolution MC-5 in toluene MAO solution led to considerable increase in activity, and this resulted to PP with lower Mw (58400 against 90500). Polymerization at TIBA presence did not influence on Mw of PP.

Analysis of literature data, first of all of Spaleck works [12, 18], showed that the maximal catalytic activity, molecular weight and isospecificity between the known zirconocene systems have the systems with indenyl ligands substituted in position 2 and 4. The main chain termination reaction is β-hydrogen transfer with the monomer. This reaction is very effectively suppressed by substituents in position 2 of the indenyl ring. The use of aromatic substituents in position 4 results to the additional electronic effects. We synthesized the bisindenyl Zr with Si bridge with ethyl and phenyl groups in positions 2 and 4 of indenyl ligands, respectively. Our catalyst (MC-6) is a mixture *rac* and *meso* isomers in ratio 1 to 2. This catalyst was used in propylene polymerization in bulk without stage of isomers separation.

The polymerization process was conducted by preliminary dissolving the MC in toluene solution of MAO and injecting the resulting mixture into the reactor with liquid propylene and MAO. We studied the influence of different ratio Al/Zr in the pre-activation stage on activity and properties of PP.

Table 3. Influence of the preactivation conditions on the activity of catalytic system$(CH_3)_2Si(4-Ph-2-Et-Indenyl)_2ZrCl_2$ (*rac:meso*=1:2)/ MAO and PP properties.[*]

Conditions of preactivat.	Activity, $\frac{kg\ PP}{mmolZrh}$	iso-PP, %w	atact-PP, % w.	M_w 10^{-3}	$\frac{M_w}{M_n}$	Activity, (*rac*) $\frac{kg\ PP}{mmolZrh}$	Activity, (*meso*) $\frac{kg\ PP}{mmolZrh}$	$\frac{Rate(rac)}{Rate(meso)}$
MC/MAO Al:Zr=200	117	89	3,9	770	3,5	309	6,9	45
MC/MAO Al:Zr=2000	422	97	1,8	690	2,9	1240	11,5	108

[*] Propylene polymerization in bulk. Al/Zr=20000, [MAO] = 0,6-1 g/l

Data for activity and properties of PP produced at different conditions of pre-activation are shown in Table 3. As seen the increasing of Al/Zr ratio on the stage of pre-activation for MC-6 leads to increasing of catalytic system activity and PP isotacticity index from 89 to 97%. We believe that *rac*-isomer produces the *iso*-PP and achiral *meso*-isomer produces the atactic PP. As this MC represents itself mixture of rac and *meso* isomers, we supposed that increasing of ratio Al to Zr leads to activation mainly of *rac*-form of this metallocene. As one may see from Table 3 the ratio of the polymerization rate with using of *rac*–form to the polymerization rate with using of *meso*-form increases more than 2 times.

This effect can be caused by two reasons.

1. The first is the influence on the equilibrium between possible monomeric and dimeric forms of a metallocene catalyst. Spectroscopic investigations of various MC/MAO systems allowed to identify the nature of the active species. According to the work [19] in MC/MAO solution equilibrium between associated (dimeric) and dissociated (monomeric) cationic metallocene species is observed. The increasing proportion of the latter is accompanied by increasing of the catalytic activity. The higher activity under increasing of ratio Al to Zr on the pre-activation stage may be due to the more complete alkylation of the MC by MAO and increasing of the concentration of monomeric cationic MC species.

2. The second (as we suppose) can arise from *rac-meso* isomerization of MC structure. The mechanism of such isomerization may include MAO stimulated reversible dissociations of Cp-ring – and as a result the increase of AC number.

So the use of MC with mixture of *rac* and *meso* isomers we can produce *iso*-PP with high yield and degree of isotacticity. Maximal catalytic activity that we observed was 900 kg/gcat.h and content of isotactic pentad "mmmm" 96%.

Molecular weights and stereoregularities of metallocene *iso*-PP are very sensitive to the polymerization temperature in contrast to the Ti/Mg *iso*-PP. It is explained by the flexibility of ligands and by the increasing of the rate of β-hydride elimination at the higher temperature. In the Table 4 the data about influence of polymerization temperature on the activity and molecular mass characteristics of PP for the MAO activated system on a base MC with substituted ligands are shown. As one can see decreasing of polymerization temperature leads to increasing the molecular weight of PP, what is typical for MC-system. The maximum molecular weight was equal to 2000000 g/mol for PP, produced at 30°C.

Table 4. Effect of polymerization temperature on the activity and molecular weight characteristics of *iso*-PP with Me$_2$Si(4-Ph-2-Et-Ind)$_2$ZrCl$_2$ (*rac:meso*= 1:2)/MAO.

Temp. polym., °C	Activity, $\frac{kgPP}{g.MC \cdot h}$	M_w 10^{-3}	$\frac{M_w}{M_n}$
50	325	1140	2,7
60	650	720	2,9
70	900	570	2,9
80	670	430	3,0

From M_w data we found that at polymerization with catalytic system on the base of MC-6 the difference between the activation energy of chain propagation and termination -8,5 kcal/mol. This value is greater than those for catalytic system on the base of MC-5/MAO (-3,1 kcal/mol). These data are in agreement with the higher Mw of PP produced with catalytic system on a base of MC-6.

Propylene polymerization in bulk at the presence of hydrogen has been studied also when using MC-6/MAO catalytic system. PP with Mw from $7 \cdot 10^4$ to 10^6 was produced. All samples of PP had high Tm (161-164°C). Decreasing of Mw at addition of hydrogen did not influence on Tm. This result shows the high stereo- and regioselectivity of this catalytic system and confirms that the decreasing of Tm for MC with unsubstituted ligands is connected first of all with the presence of the irregular sequences in polymer chains rather than with the value of Mw. X-ray diffraction picture of *iso*-PP produced with MC-6/MAO indicates that the PP crystallizes under a α-morphological form and one can see absence of γ -form with a peak at $2\theta = 20°C$. This corresponds to high regioselectivity of studied MC-catalytic system.

At the presence of hydrogen the activity of MC-6/MAO catalytic system is increased in two times. A comparison between the data about influence hydrogen on activity and *iso*-PP molecular weight characteristics with using Ti/Mg catalyst was done also. It was found that sensitivity of MC catalytic system to the hydrogen introducing differ from the similar for Ti/Mg catalyst.

As can see from Table 5, at the same H_2 concentration in the reaction medium the molecular weight decreasing is more sharp in the case of metallocene catalysts then for Ti/Mg catalyst.

From the dependence of Mn on $[H_2]$ we can obtain ratios values of propagation rate constants and polymer chain transfer rate constants for the monomer and for the hydrogen.

$$1/P_n = k_t^m / k_p + k_t^{Zr} / k_p C_m + k_t^H C_H / k_p C_m,$$

where Pn – number average polymerization degree, Cm- monomer concentration, k_t^m, k_t^{Zr}, k_t^H – constants rates of the reactions of the polymer chain transfer on a monomer, metal and hydrogen respectively.

Table 5. Effect of hydrogen on the activity and molecular weight characteristics of *iso*-PP. [*]

Catalytic system	$[H_2] \cdot 10^2$, mol/l	Activity, $\frac{kg\ PP}{gcat.\ h}$	M_w 10^3	$\frac{M_w}{M_n}$
Me$_2$Si(4-Ph-2-EtInd)$_2$ZrCl$_2$ (*rac:meso*=1:2)/ MAO	-	200	510	2,9
	1,8	340	142	2,4
	3,6	440	71	2,1
Ti/Mg-AlEt$_3$/PrTMS	-	10,3	1580	4,7
	0,5	13,5	1050	3,9
	1,8	18,2	670	3,5
	5,0	25,7	356	3,1

[*] Propylene polymerization in bulk at 70°C, the reactor volume 0,4 l.

In the Table 6, the kinetic data are presented for a number of studied stereospecific catalytic systems. There are high difference for MC and Ti/Mg systems. As one can see for MC-catalyst the ratio of the rate propagation polymer chain constants to the rate transfer on H_2 constants is low (3) and for TMC this ratio is high (21).

Table 6. Kinetic data for propylene polymerization in bulk with different catalytic systems.

Catalyst	k_p/k_t^m	k_p/k_t^H	k_t^H/k_t^m	E_{eff}, kcal/mol	$(E_p-E_t^m)$, kcal/mol
Me$_2$Si(4-Ph-2-EtInd)$_2$ZrCl$_2$ (*rac:meso*=1:2) – MAO	4350	3	1450	10,0	-8,5
Ti/Mg-AlEt$_3$/PrTMS	8000	21	385	13,0	-

From Arrenius plot we estimated the activation energies for studied catalytic systems. These values are about 8-10 kkal/mol for propagation stage.

If to suppose that in all studied systems each zirconium atoms forms an active complex [20, 21] and taking into account the data about catalytic systems activity and Mw characteristics of polymers produced one may evaluate the minimal polymers' chains number per one zirconium molecule during polymer chain growth and the maximal time of insertion for one propylene unit. For comparison the data for syndiospecific catalytic system Ph$_2$CCpFluZrCl$_2$/MAO are presented [22]. As seen from Table 7, for studied catalytic systems one zirconium molecule produces 1000-10000 mol PP per hour. The insertion time of one propylene unit is equal to 10^{-3}–10^{-4} s. The time of growth of one polymer molecule in Zr-cene systems has very small value (1-6 s).

Table 7. The some characteristics of MC-catalytic systems.[*]

Catalytic system	Me$_2$SiInd$_2$ZrCl$_2$ – MAO	Me$_2$Si(4-Ph-2-EtInd)$_2$ZrCl$_2$ (*rac:meso*=1:2) -MAO	Ph$_2$C•pFluZrCl$_2$ - MAO
Activity, kgPP/mmolZr h	130-190	200-600	80-150
PP molecular weight, $M_n \cdot 10^{-3}$	20-30	190-420	150-160
mol PP/ mol Zr	4500-10500	500-3000	530-930
Time of polymer chain growth, *sec.*	0,3-0,8	1,0-7,0	4,0-6,5
Time of C$_3$H$_6$ interaction, *sec*	$8 \cdot 10^{-4}$-10^{-3}	$2,5 \cdot 10^{-4}$–$7 \cdot 10^{-4}$	$5,5 \cdot 10^{-4}$–$2 \cdot 10^{-3}$

[*] Propylene polymerization in bulk at 50-70°C.

The metallocene catalysts can be used for the production of new PP materials [22-24]. The particularities of metallocene based isotactic PP are the high modulus (up to 1900-2100 MPa), the high hardness and stifness. For comparison, the modulus of typical industrial PP is equal to 1200-1500 MPa only. High modulus is a result of the narrow MWD, high PP homogeneity and high PP crystallinity.

The following development of such research will be connected with supported metallocene catalytic systems [25], which allow production of *iso*-PP with high activity, properties typical for m-*iso*PP and will allow to develop the technology of "drop-in" present industrial processes.

Acknowledgement

This work was financially supported by the Russian Foundation for Basic Research (Project N 99-03-32948a).

References

1   Ewen J.A. (1984) *J. Am. Chem. Soc.***106** (21): 6355.
2   Kaminsky W., Kulper K., Brintzinger H.H., Wild F.R.W.P. (1985) *Angew.Chem., Int.Ed.Engl.,*.**24**: 507.
3   Brintzinger H.H., Fischer D., Mulhaupt R., Rieger B., Waymouth R. (1995) *Angew.Chem. Int.Ed.Engl.*, **34**: 1143.
4   Ewen J.A.,Jones R.L.,Razavi A., Ferrara J. (1988) *J.Am.Chem.Soc.,*.**110**: 6255.
5   Ziegler Catalysts (Eds.: G.Fink, R.Mulhaupt, H.H.Brintzinger), Springer-Verlag, Berlin, (1995).
6   Fischer D. Met'Con 98 "Polymers in Transition", (1998), Houston, TX, USA.
7   Resconi L., Fait A., Piemontesi F., Colonnesi M., Rychlicki M., Ziegler R. (1995) *Macromolecules*, **28**: 6667.
8   Busico V., Caporaso L., Cipullo R. and Landriani L. (1996) *J.Am.Chem.Soc.*, **118**: 2105.
9   Leclerc M., Brintzinger H.H. (1996) *J.Am.Chem.Soc.*, **118**: 9024.
10   Nifant'ev I.E., Ivchenko P.V.(1997) *Organometallics*, **16**: 713.
11   Nedorezova P.M., Tsvetkova V.I., Bravaya N.M., Savinov D.V., Dubnikova I.L., Borzov M.V., Krut'ko D.P. (1997) *Polimery*, **42** (10): 599.
12   Spaleck W., Kuber F., Winter A., Rohrmann J., Bachmann B., Antberg M., Dolle V., Paulus E.F.(1994) *Organometallics*, **13**: 954.
13   Aladyshev A.M., Isichenko O.P., Nedorezova P.M., Tsvetkova V.I., Gavrilov Yu.A., Dyachkovskii F.S. (1991) *Polymer Science*, **33** (8): 1595.
14   Aladyshev A.M., Tsvetkova V.I., Nedorezova P.M., Optov V.A., Ladygina T.A., Savinov D.V., Borzov M.V., Krut'ko D.P., Lemenovskii D.A. (1997) *Polimery*, **42** (10): 595.
15   Resconi L., Cavallo L., Fait A., Piemontesi F. (2000) *Chem. Rev.*, **100**: 1253.
16   Resconi L., Piemontesi F., Camurati I., Sudmeijer O., Nifant'ev I.E., Ivchenko P.V., Kuz'mina L.G. (1998) *J.Am.Chem.Soc.*, **120** (10): 2308.
17   Khukanova O.M., Babkina O.N., Rishina L.A., Nedorezova P.M., Bravaya N.M. (2000) *Polimery*, 5: 328.

18   Kuber F., Aulbach M., Bachmann B., Klein R., Spaleck W., (1995) Met'Con 95
      "Polymers in Transition", Houston, TX, USA.
19   Coevoet D., Cramail H., Deffieux A. (1998) *Macromol.Chem.Phys.*, **199**: 1451.
20   Chien J.C.W., Wang B.P. (1989) *J.Polymer Sci. Part A*, **27**: 1539.
21   Tait P. (1988) "Transition Metals and Organometallics as Catalyst for Olefin
      Polymerization", Ed.W.Kaminsky, Springer Press, Berlin, 309.
22   Nedorezova P.M., Tsvetkova V.I., Aladyshev A.M., Savinov D.V.,Dubnikova I.L.,
      Optov V.A., Lemenovskii D.A. (2000) *Polimery*, 5: 333.
23   Gownder M., (1997) SPO'97, Seventh International Business Forum on Specialty PO,
      Houston, USA, 289.
24   Weikert T.A., (1997) SPO'97, Seventh International Business Forum on Specialty PO,
      Houston, USA, 237.
25   Chien J.C.W. (1999) *Topics in Catalysis*, **7**: 23.

Ethene Polymerization Catalyzed by Monoalkyl and Methyl Substituted Zirconocenes. Possible Effects of Reaction Barriers and Ligand-Metal Agostic Interactions

Knut Thorshaug[*a1], Jon Andreas Støvneng[a], Hanne Wigum[a], Erling Rytter[ab]

[a] Department of Chemical Engineering, Norwegian University of Science and Technology (NTNU), N-7491 Trondheim, Norway
[b] Statoil Research Centre, N-7005 Trondheim, Norway
E-mails: knut@icm.mi.cnr.it; konjoas@statoil.com; hanne.wigum@chembio.ntnu.no; erling.rytter@statoil.com

Abstract. Ethene polymerization with the monoalkyl substituted zirconocenes $(\eta^5\text{-}C_5H_4R)_2ZrCl_2$ ($R=C_nH_{2n+1}$, $n=0\text{-}5,8,12$) and the complete set of methyl substituted complexes $(\eta^5\text{-}C_5H_{5-m}Me_m)_2ZrCl_2$ ($m=0\text{-}5$) has been performed together with polymer characterization. Monoalkylation of the cyclopentadienyl (Cp) ligand shifts the distribution from vinyl towards *trans*-vinylene. For the methyl substituted complexes, we find a particularly high *trans*-vinylene content for pentamethyl cyclopentadienyl (Cp*). A kinetically controlled competition between termination and isomerization is used to rationalize the findings. Density-functional theory calculations show that an agostic interaction between the metal and the alkyl substituent on the Cp ligand may be important when the alkyl is ethyl or longer.

Introduction

The electronic and steric surroundings of the catalytic active site, obtained when a metallocene and methylaluminoxane (MAO) are mixed, can be manipulated by changes in the ligands on the metallocene [1,2]. As a consequence, the various reaction rates may be altered in a controlled fashion and thereby both the catalytic activity and the polymer properties can be controlled. An increase in the electron withdrawing ability of the substituent has been reported to lower the propagation rate [3,4] and the molecular weight [3], whereas the activity is inversely proportional to the steric demands of the ring system [4]. For electron donating alkyl substituents, it is known that both their positions and sizes influence the catalytic properties [5].

---

[1] Current address: ICM-CNR, via E. Bassini 15, 20133 Milano, Italy.

We have polymerized ethene catalyzed by different unbridged alkyl substituted zirconocene complexes activated by MAO. The specific complexes used were monosubstituted of the general formula $(\eta^5\text{-}C_5H_4R)_2ZrCl_2$ $(R=C_nH_{2n+1}, n=0\text{-}5,8,12)$ and the complete set of methyl substituted complexes described by the general formula $(\eta^5\text{-}C_5H_{5-m}Me_m)_2ZrCl_2$ $(m=0\text{-}5)$. For the latter, the different isomers for the *di-* and *tri-*substituted complexes, i.e. 1,2-Me$_2$, 1,3-Me$_2$, 1,2,3-Me$_3$ and 1,2,4-Me$_3$, were all investigated. Data on the polymer unsaturations and catalytic activity are presented together with data obtained by density-functional theory (DFT) calculations.

Experimental and Computational Details

Experimental part

General: Toluene was refluxed over sodium/benzophenone and distilled under nitrogen atmosphere. Metallocenes (Boulder Scientific Company), MAO (Albemarle) and ethene (Borealis) were all used as received.

Polymerization: The reactor was repeatedly evacuated and purged with nitrogen before monomer and toluene were introduced. Temperature, stirring rate and pressure were set, and equilibration was allowed before MAO and catalyst were injected. The product was precipitated in a mixture of methanol and hydrochloric acid, stirred overnight, filtered, washed with methanol and dried.

Characterization: Polymer characterization is described elsewhere [6-8].

Computational part

Kinetic modeling: A kinetic model that includes activation, propagation, formation and reactivation of latent sites, and permanent deactivation was used [7,9,10]. The term *corrected activity* corresponds to the theoretical propagation rate obtained from this model.

Quantum Chemical modeling: All geometries and energies discussed in the present paper are evaluated with DFT methods [11]. We used a double-. STO basis set for C and H and a triple-. STO basis set for Zr. The 1s to 3d orbitals on Zr and the 1s orbital on C were treated within the frozen-core approximation. Geometries were optimized within the local density approximation [12]. Gradient corrections to the energy were based on the functionals proposed by Becke [13] and Perdew and Wang [14]. Further details can be found elsewhere [7,8,11].

Results and Discussion

Characterization of the polymer unsaturation provides important information about the termination reactions. During ethene polymerization, β-H transfer to either the metal center or to a coordinated monomer are possible termination paths which both give vinyl unsaturation in the polymer. To experimentally distinguish the two reactions, kinetic considerations can be made [6,7]. Chain transfer to aluminum will give a saturated end-group after work-up. The presence of other unsaturations than vinyl requires further explanation, and we have earlier suggested an isomerization mechanism (Scheme 1) in order to account for the presence of *trans*-vinylene [6,7]. The step where the polymer chain is separated from the metal center may or may not be monomer assisted, see table 1, although in Scheme 1 we have drawn it as a monomer assisted reaction.

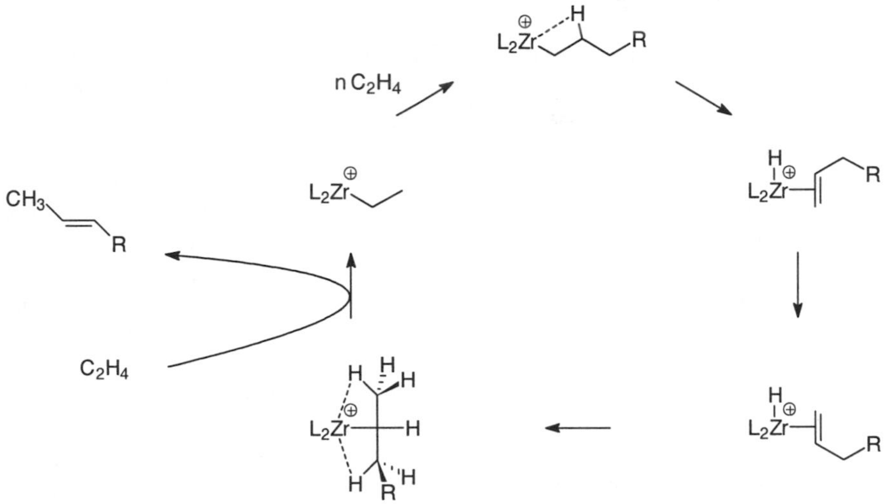

Scheme 1. Proposed isomerization path for the formation of *trans*-vinylene during ethene polymerization catalyzed by metallocene catalysis [6,7].

Large differences in the values for the polymer unsaturations were found when polyethene synthesized with Cp_2ZrCl_2 and $Cp*_2ZrCl_2$ precursors were compared. In the polymers obtained with Cp_2ZrCl_2, vinyl was found to be the most abundant unsaturation, whereas for Cp_2*ZrCl_2, the *trans*-vinylene and vinyl contents were approximately equal. This may seem surprising at first, since it could be expected that the Cp* ligands for steric reasons would hinder the isomerization to a much larger extent than the Cp ligands. A reasonable explanation can be provided by the energy barriers calculated by DFT and listed in table 1. For the calculations, we assumed the active site to be described by L_2ZrR^+ cations (L=Cp, Cp*, R $\geq C_3H_7$).

From table 1 we see that the barriers against isomerization are almost the same for the Cp and Cp* ligand, with a calculated difference of 8kJ/mol. Thus, DFT shows that the more hindered Cp* ligand does not hinder the rotation in the π-complex to a much larger extent than the Cp ligand.

Table 1. Activation energies derived by DFT calculations for reactions relevant to the formation of unaturations in polyethene [7].

Reaction	Activation energy (kJ/mol)	
	Cp_2ZrR^+	$Cp^*_2ZrR^+$
Insertion	8	24
β-H transfer to monomer	41	68
β-H transfer to Zr	115	64
Isomerization	42	50
β-H transfer to monomer after isomerization	40	75
β-H transfer to Zr after isomerization	90 – 100	50

Linear-transit calculations show that the energy barriers against monomer coordination are influenced by pentamethyl substitution on the Cp ligand. We find essentially no barrier against coordination with Cp_2ZrR^+ but a barrier of 24kJ/mol for $Cp^*_2ZrR^+$. Insertion from the π-complex occurs with a barrier of about 8kJ/mol for Cp_2ZrR^+ and with approximately no barrier for $Cp^*_2ZrR^+$. Further from table 1 we see that for Cp_2ZrR^+, the activation energy for β–H transfer to monomer before isomerization is about the same as the barrier against isomerization, whereas for $Cp^*_2ZrR^+$, the isomerization barrier is lower than any of the termination barriers. We have proposed a multi-step isomerization reaction (Scheme 1), and it therefore seems reasonable that β-H transfer to monomer is more probable than isomerization in the case of Cp_2ZrR^+ compared to $Cp^*_2ZrR^+$ where the termination and isomerization barriers are almost equal. The high *trans*-vinylene content found in the polymers synthesized by $Cp^*_2ZrR^+$ can be rationalized by the fact that the barrier against isomerization is lower than any of the termination barriers so that isomerization becomes more important for this catalyst. This also shows that the formation and distribution of the unsaturations is under kinetic control.

A qualitatively good agreement between the experiments and DFT calculations is also found for the termination mechanisms. Experimentally, we find the molecular weights to be pressure independent for Cp_2ZrR^+ [7], thus termination occurs by β-H transfer to a monomer. This agrees with the energy barriers listed in table 1 where we see that for Cp_2ZrR^+, termination by β-H transfer to the monomer has the lowest activation energy, thus it is the most likely termination path also according to the DFT calculations. For $Cp^*_2ZrR^+$ we experimentally find the molecular weight to increase with pressure at low pressures and to become pressure independent at higher pressures [7], thus termination may occur by both β-H transfer to the metal or to a monomer, where the latter dominates as the pressure is increased. Table 1 shows similar barriers for β-H transfer to monomer and to the metal for $Cp^*_2ZrR^+$, so according to DFT they are both possible termination paths, as found experimentally. Chain transfer to aluminum is experimentally found to occur for $Cp^*_2ZrCl_2$ / MAO but not for Cp_2ZrCl_2 / MAO [7]. DFT calculations were not performed for this termination path.

A systematic variation in the number and position of the methyl substituents on the Cp ligand was performed and its effect on the polymer properties was investigated. By FTIR, we could only detect vinyl and *trans*-vinylene unsaturations in the polymer. As shown in figure 1, the concentration of *trans*-vinylene, and thus also the tendency for isomerization, is low and essentially constant when *tetra-* or less substituted Cp ligands are used. When Cp* ligands were used, the *trans*-vinylene content, and the degree of isomerization, is higher as outlined in details above. For the polymer vinyl content a slight decrease is observed when the unsubstituted Cp ligands and the monomethylated MeCp ligands are compared. Further increase in the number of methyl substituents on the Cp ligands, is accompanied by a noticeable drop in the polymer vinyl content. No significant differences could be observed between the *di-*, *tri-* or *tetra*-methyl substitued Cp-rings, with small but insignificant differences in the polymer vinyl concentrations obtained with the different *di-* (i.e. 1,2-Me$_2$Cp vs. 1,3 Me$_2$Cp) and *tri*-substituted (1,2,3-Me$_3$Cp vs. 1,2,4-Me$_3$Cp) zirconocenes. Compared to the *di-*, *tri-*, and *tetra*-substituted Cp-ligands an increase in the polymer vinyl concentration was observed for Cp*.

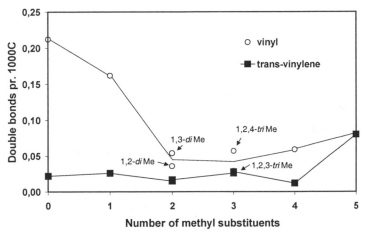

Fig. 1. Concentration of the unsaturations in polyethene synthesized with metallocene catalysts of the general form $(\eta^5\text{-}C_5H_{5-m}Me_m)ZrCl_2$ / MAO (m=0-5). By coincidence, the *trans*-vinylene contents overlap for the 1,2- and 1,3-*di*Me as well as for the 1,2,3- and 1,2,4-*tri*Me catalysts. For the 1,2,3-*tri*Me, the vinyl and *trans*-vinylene concentrations also overlap.

The total number of unsaturations in the polymer can be taken as a measure of the ease of termination. Under the given conditions, we find the following order for the $(\eta^5\text{-}C_5H_{5-m}Me_m)_2ZrCl_2$ (m=0-5) / MAO catalysts (Easiest termination: Cp). For simplicity, the zirconocenes are symbolized by their ligands:

$$Cp > MeCp > Cp^* > Me_2Cp \approx Me_3Cp \approx Me_4Cp$$

This order is confirmed by the molecular weights as measured by GPC [8].

A study of monoalkylated Cp ligands was also performed. Based on our earlier work on quantitative structure-activity relationships (QSAR) [10], we chose $(\eta^5\text{-}C_5H_4R)_2ZrCl_2$ (R=C_nH_{2n+1}, n=0-5,8,12) as catalyst precursor. Again, by FTIR we only detected vinyl and *trans*-vinylene unsaturations in the polymers with vinyl as the most abundant one. In figure 2 we have plotted the absolute values of *trans*-vinylene as a function of the length of the alkyl substituent (R) positioned on the Cp ligand. The concentration of *trans*-vinylene shows a slight increase with an increase in R, with a particularly high value for R=*n*-propyl. From a similar plot, we see that there is no systematic variation in the vinyl concentration with R, although a slightly higher value is observed for R=H than for the others. As expected, the amounts of unsaturations increase with temperature due to shorter chains.

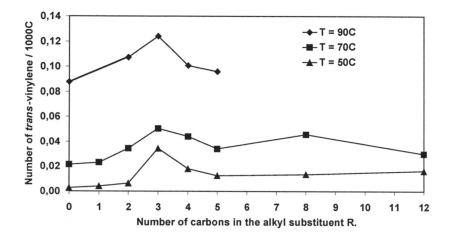

Fig. 2. *trans*-vinylene concentration as a function of the length of the monoalkyl substituent positioned on the Cp ligand

To get a better impression of the distribution between the unsaturations, it is helpful to define their relative values, or fractions, as the concentration of one type of unsaturations divided by the sum of all unsaturations. In figure 3 we have plotted these relative values as a function of the alkyl substituent (R), and it is clearly seen that both R and the reaction temperature influence the distribution of unsaturations. An increase in the alkyl length from R=H to R=*n*-propyl is accompanied by a shift in the distribution of unsaturations from vinyl towards *trans*-vinylene. Substituents longer than *n*-propyl do not shift the distribution of unsaturations further. Increasing the temperature shifts the distribution towards *trans*-vinylene, and yields lower molecular weight polymers.

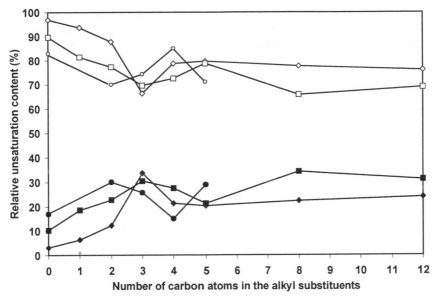

Fig. 3. Relative distribution of the different unsaturations found in polyethene synthesized by $(\eta\text{-}^5C_5H_4R)_2ZrCl_2$ ($R=C_nH_{2n+1}$, $n=0\text{-}5, 8, 12$) / MAO. Open symbols: vinyl, closed symbols: *trans*-vinylene. O/● 90°C, /■ 70°C, ◊/♦ 50°C.

An exception from the above statement is R=*n*-propyl, where essentially no shift in the distribution of unsaturations could be detected when the temperature was raised from 50°C to 90°C. Due to the high activity and strongly exothermic nature of the reaction, the experiments plotted for R=*n*-propyl at 50°C were actually run at an average temperature of 56°C. Since there are large differences in the absolute values of unsaturation in the polymers obtained at different temperatures, the temperatures at the active sites have also been different. Therefore a temperature gradient around the active site seems unlikely in order to explain the anomalous behavior of R=*n*-propyl, and we interpret the observation as a true alkyl substituent effect.

We rationalize the influence of the alkyl substituent to be related to the barriers against isomerization and termination [7]. We assumed the number of double bonds to correspond to the kinetics of the termination and isomerization reactions on the one hand, and propagation kinetics on the other. From an Arrhenius relationship, differences in activation energy between termination/isomerization and propagation were found, and they are listed in table 2.

Table 2. Termination and propagation activation energy differences for ethene polymerization catalyzed by $(\eta^5\text{-}C_5H_4R)_2ZrCl_2$/MAO ($R=C_nH_{2n+1}$, $n=0, 2\text{-}5$).

Unsaturation	$\Delta E_a = E_a^{term} - E_a^{prop}$, in kJ/mol				
	R=H	R=Ethyl	R=*n*-propyl	R=*n*-butyl	R=*n*-pentyl
Vinyl	40	41	40	50	38
trans-vinylene	85	69	36	41	49

The catalytic activity is also influenced by the choice of alkyl substituent. In figure 4, we have plotted both the average activities over one hour and the corrected activities when deactivation reactions are taken into account [7,9]. The term corrected activity is the intrinsic activity at $t=0$, assuming all metal centers to be active catalysts. n-propyl substitution on the Cp ligand increases the average activity by a factor 4 and the corrected activity by a factor 3 compared to unsubstituted Cp. From the kinetic modeling, we also find the lowest deactivation rate when n-propyl substituted Cp ligands are used. None of the substituted Cp ligands lower the average or corrected activity compared to unsubstituted Cp ligands.

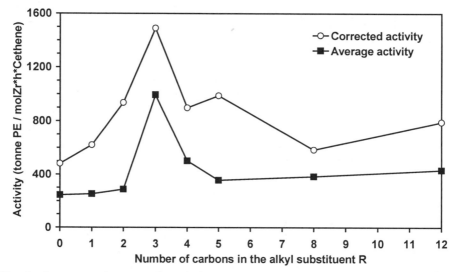

Fig. 4. Average and corrected catalytic activities for ethene polymerization at 50°C catalyzed by $(\eta^5\text{-}C_5H_4R)_2ZrCl_2$ / MAO ($R=C_nH_{2n+1}$, $n=0\text{-}5,8,12$)

During polymerization, different agostic interactions between the metal and the growing chain may be of importance [16-18] and their strengths may influence on the rates of the different reactions. If an additional ligand-metal interaction is present (Scheme 2) it is reasonable to expect this to influence on the metal-growing chain agostic interactions and thereby offer a possibility to manipulate the catalytic activity and polymer properties such as molecular weight and end-groups.

Assume the cation to be in a conformation where a β– or γ–agostic interaction is present between the metal and the growing chain. DFT calculations show an additional agostic interaction, as shown in Scheme 2, to indeed be possible for substituents such as ethyl or longer. This additional interaction is at its strongest when n-propyl is used as alkyl substituent with the cation in a β–agostic metal-growing chain conformation. The second alkyl substituent positioned on the other Cp ligand, does not interact with the metal center and it points away from the growing chain [11].

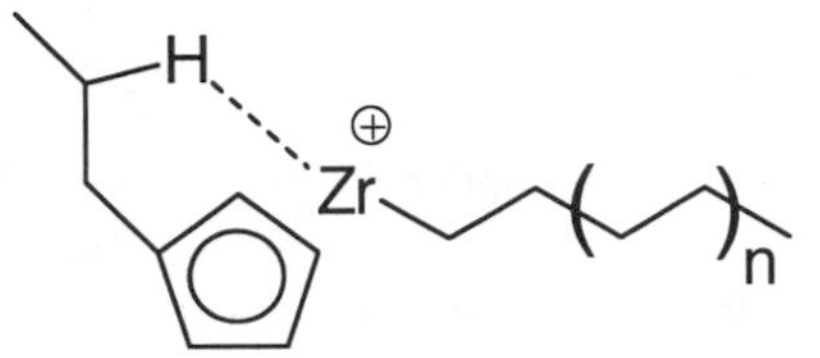

Scheme 2. Sketch of a possible metal-ligand agostic interaction (Zr$\cdots$H$_{Alkyl}$ interaction). The interaction comes in addition to a metal-growing chain α(TS)-, β-, or γ-agostic interaction (omitted for clarity). The alkyl substituent on the other ligand (omitted for clarity) points away from the growing chain and does not interact with the metal center.

The experimentally found activities and concentrations of *trans*-vinylene both increase whereas the concentration of vinyl unsaturations decrease slightly when R=H is substituted with R=*n*-propyl. At the same time the Zr$\cdots$H$_{Alkyl}$ interaction is theoretically found to be at its strongest for R=*n*-propyl when the complex is in a conformation where a β–agostic interaction between Zr and the growing polymer chain is present. It seems justified to postulate that the additional metal-ligand interaction (Scheme 2) may partly explain the experimental observations.

To the best of our knowledge, a ligand-metal interaction such as the Zr$\cdots$H$_{Alkyl}$ interaction outlined above has not been considered for olefin polymerization by transition metal catalysis prior to this study.

Conclusions

We have shown that the polymer unsaturations, molecular weight, and catalytic activity can be manipulated to different extents depending on either the number or positions of the methyl substituents or the length of a monoalkyl substituent on the Cp ligand. The observed differences in the vinyl and *trans*-vinylene unsaturations are explained in terms of energy barriers against termination and isomerization, and an isomerization mechanism has been formulated. In order to account for the high *trans*-vinylene content and activity observed with the *n*-propyl substituted metallocenes, we propose an agostic interaction to be present between the metal center and the alkyl substituent placed on the Cp ligand. All of the proposed mechanistic explanations are supported by density-functional theory calculations.

Acknowledgements

The Norwegian Research Council (NFR) under the programs for Polymer Science, Reactor Technology in the Petrochemistry and Polymer Industry (REPP), and Supercomputing, together with Borealis, financed this work.

References

1. Brintzinger, H.-H.; Fischer, D.; Mülhaupt, R.; Rieger, B.; Waymouth, R. M. *Angew. Chem. Int. Ed. Engl.* **1995**, *34*, 1143-1170.
2. Möhring, P. C.; Coville, N. J. *J. Organomet. Chem.* **1994**, *479*, 1-29.
3. Piccolrovazzi, N.; Pino, P.; Consiglio, G.; Sironi, A.; Moret, M. *Organometallics* **1990**, *9*, 3098-3105.
4. Janiak, C.; Versteeg, U.; Lange, K. C. H.; Weimann, R.; Hahn, E. *J. Organomet. Chem.* **1995**, *501*, 219-234.
5. Spaleck, W.; Antberg, M.; Aulbach, M.; Bachmann, B.; Dolle, V.; Haftka, S.; Küber, F.; Rohrmann, J.; Winter, A. In: *Ziegler Catalysts*, Fink, G.; Mülhaupt, R.; Brintzinger, H. -H. Ed.; Springer-Verlag, Berlin Heidelberg, 1995, pp. 83-98.
6. Thorshaug, K.; Rytter, E.; Ystenes, M. *Macromol. Rapid Commun.* **1997**, *18*, 715-722.
7. Thorshaug, K.; Støvneng, J. A.; Rytter, E.; Ystenes, M. *Macromolecules* **1998**, *31*, 7149-7165.
8. Wigum, H.; Tangen, L.; Støvneng, J. A.; Rytter, E. *J. Polym. Sci.: Part A: Chem. Ed.* **2000**, *38*, 3161-3172.
9. Wester, T. S.; Johnsen, H.; Kittilsen, P.; Rytter, E. *Macromol. Chem. Phys* **1998**, *199*, 1989-2004.
10. Støvneng, J. A.; Stokvold, A.; Thorshaug, K.; Rytter, E. In: *Metalorganic Catalysts for Synthesis and Polymerization*, Kaminsky, W. Ed.; Springer-Verlag, Berlin Heidelberg, 2000, pp. 274-282.
11. Thorshaug, K.; Støvneng, J. A.; Rytter, E. *Macromolecules,* in press.
12. *ADF 2.3*, Scientific Computing & Modelling, Chemistry Department, Vrije Universiteit, De Boelelaan 1083, 1081 HV Amsterdam, The Netherlands.
13. Vosko, S. H.; Wilk, L.; Nusair, M. *Can. J. Phys.* **1980**, *58*, 1200-1211.
14. Becke, A. D. *Phys. Rev. A* **1988**, *38*, 3090-3100.
15. Perdew, J. P.; Wang, Y. *Phys. Rev. B* **1992**, *45*, 13244-13249.
16. Grubbs, R. H.; Coates, G. W. *Acc. Chem. Res.* **1996**, *29*, 85-93.
17. Lohrenz, J. C. W.; Woo, T. K.; Fan, L.; Ziegler, T. *J. Organomet. Chem.* **1995**, *497*, 91-104.
18. Støvneng, J. A.; Rytter, E. *J. Organomet. Chem.* **1996**, *519*, 277-280.

A Density Functional Theory Study of the Syndiotactic-Specific Polymerization of Styrene

Luigi Cavallo,[*†] Gianluca Minieri,[†] Paolo Corradini,[†] Adolfo Zambelli,[‡]
Gaetano Guerra[‡]

[†] Dipartimento di Chimica, Università di Napoli, Via Mezzocannone 4,
I-80134, Napoli, Italy
[‡] Dipartimento di Chimica, Università di Salerno, Baronissi, I-84081 Salerno, Italy
E-mail: cavallo@chemistry.unina.it

Abstract. Preliminary theoretical results relative to a possible mechanism of syndiospecific polymerization of styrene with models based on the $CpTiP^+$ (P = Polymeryl) species are presented. The styrene-free $CpTiCH_2Ph^+$ species is characterized by a η^7 coordination of the benzyl-type chain end, which is preserved also in the most stable species upon coordination of a styrene molecule. The transition state for the insertion step is characterized by the classical four centers geometry, usually proposed for the polymerization of olefins. The insertion can occur with a energy barrier of 47 kJ/mol. As for the mechanism of stereocontrol, the transition state which leads to formation of a syndiotactic diad is favored over the transition state which leads to formation of an isotactic diad by 5 kJ/mol. The former is favored because in the transition state leading to formation of an isotactic diad the growing chain is placed within the coordination sphere of the titanium atom, and interacts repulsively with the other ligands.

1. Introduction.

Syndiotactic polystyrene is a new polymeric material[1-3] of industrial relevance, since it shows a high melting point, 270 °C, and high crystallization rates.[4] Syndiotactic polystyrene is a highly stereoregular polymer which can be obtained by using several soluble titanium, and to a less extent zirconium, compounds.

Among titanium based precursors, monocyclopentadienyl compunds of the type $CpTiCl_3$ or Cp^*TiCl_3 ($Cp = \eta^5\text{-}C_5H_5$, $Cp^* = \eta^5\text{-}C_5Me_5$) activated by methylalumoxane (MAO) or $B(C_6F_5)_3$ showed the best performances, although several substituted mono-Cp or indenyl derivative, and Cp-free compounds as $Ti(CH_2Ph)_4$, $Ti(OR)_4$ (R = alkyl, aryl) are quite active as well. In short, practically any soluble titanium compound can be used as pre-catalyst.[5-9]

Although there is some debate about the exact nature of the species active in polymerization,[10-14] many characteristics of the polymerization mechanism have been addressed beyond any doubt. NMR experiments have clearly indicated that polymerization of styrene to syndiotactic polymer occurs through a Ziegler-

Natta type polyinsertion mechanism.[15, 16] In particular, these experiments have shown that: i) the insertion mechanism occurs through the *cis* opening of the monomer double bond;[17] ii) the regiochemistry of styrene insertion is secondary;[18, 19] iii) the stereoselectivity of the insertion step is controlled by the chirality of the growing-chain end. This is clearly indicated by the analysis of the stereochemical composition of the syndiotactic polymer, which shows the presence of the *rmr* tetrad, which is consistent with chain-end stereocontrol, and the substantial absence of the *rmm* tetrad, which is consistent with site-stereocontrol.[20-22]

Despite the many experimental studies which have contributed to clarify even details of the polymerization mechanism, almost nothing has been done from a theoretical point of view. This is in sharp contrast to the considerable amount of high level computational studies which have contributed to the comprehension of fine details of 1-olefins (ethene and propene) polymerizations with both early and late transition metals. For these reasons, we decided to make a combined quantum mechanics/molecular mechanics, QM/MM, study of the mechanisms of polymerization and of stereocontrol in the syndiospecific polymerization of styrene. In particular, we have investigated insertion of styrene on the CpTi^{III}CH$_2$Ph$^+$ active species, in which the achiral –CH$_2$Ph group has been used to simulate the growing chain. This choice allowed us to explore details of the insertion reaction delaying the complicacies due to the presence of a chiral growing chain. After the basic features of the insertion step have been clarified, we afforded the issue of stereoselectivity. Of course, models with a chiral growing chain have been considered. Thus, in the second part of the paper we propose a possible mechanism of growing-chain stereocontrol, by investigating insertion of styrene on the chiral CpTi^{III}CH(CH$_3$)Ph$^+$ active species.

2. Models and Methods.

The models used to investigate the propagation mechanism are composed by monometallic cationic species. These models comprise a Ti atom in the oxidation state III, to which a spectator Cp ligand will be coordinated. Furthermore, a growing chain, simulated by the achiral –CH$_2$Ph benzyl group, will be coordinated to the metal, and the approach and insertion of a styrene molecule will be considered. In all the studies regarding the elementary steps of the propagation mechanism, for the sake of simplicity we modeled the growing chain with the achiral benzyl group. This choice allowed us to focus on the basic features of the propagation step only.

As we investigated the mechanism of stereoselectivity, however, we had to consider elements of chirality, and those which are relevant for the the present study are briefly recalled here. First of all, upon coordination the prochiral styrene molecule gives rise to non-superimposable *re* and *si* coordinations.[23] A second element of chirality is the configuration of the tertiary carbon atom of the growing chain nearest to the metal atom, and the Cahn-Ingold-Prelog *R,S* nomenclature will be used. In this study, the growing chain will be modeled with the chiral –

CH(CH₃)Ph group of both *R* and *S* chiralities, which are the chiralities which stem from insertion of a *re-* or *si*-coordinated styrene, respectively. For this reason, we will refer to the –CH(CH₃)Ph group of *R* and *S* chirality as *re-* and *si*-ending chain, respectively.

Stationary points on the potential energy surface were calculated with the Amsterdam Density Functional (ADF) program system,[24] developed by Baerends *et al*.[25, 26] The electronic configuration of the molecular systems were described by a triple-ζ basis set on titanium for $3s$, $3p$, $3d$, $4s$, and $4p$. Double-ζ STO basis sets were used for carbon $(2s,2p)$ and hydrogen $(1s)$, augmented with a single $3d$ and $2p$ function, respectively.[24] The inner shells on titanium (including $2p$) and carbon $(1s)$ were treated within the frozen core approximation. Energetics and geometries were evaluated by using the local exchange-correlation potential by Vosko *et al*,[27] augmented in a self-consistent manner with Becke's[28] exchange gradient correction and Perdew's[29, 30] correlation gradient correction. An unrestricted formalism was used for all species with an unpaired number of electrons.

The quantum mechanics (QM) and molecular mechanics (MM) calculations were performed with a modified version of ADF,[31] The partitioning of the systems into QM and MM parts, only involve the added methyl group needed to transform the achiral –CH₂Ph group used to model the achiral growing chain, into the chiral –CH(CH₃)Ph group used to model a chiral growing chain. The only MM atoms, hence, are those of the methyl group of the growing chain. In the optimization of the MM part, the ratio between the sp^2-sp^3 C–C bond crossing the QM/MM border, and the corresponding optimized C–H distance, was fixed equal to 1.35. Further details on the methodology can be found in previous papers.[31-33]

The MM potentials developed by Bosnich for bent metallocenes have been adopted.[34] This approach substantially corresponds to an extension of the Karplus's CHARMM force field[35] to include group 4 metallocenes. All the structures which will follow are stationary points on the combined QM/MM potential surface. Both geometry optimizations and transition state searches were terminated if the largest component of the Cartesian gradient was smaller than 0.002 au.

3. Results and Discussion.

The most stable structure of the CpTiCH₂Ph⁺ system is **1** of Fig. 1. In this structure the benzyl group coordinates to the metal with all the C atoms of the aromatic ring, as suggested by the short (< 2.5 Å) Ti-C(phenyl) distances. The distance between the C atom of the CH₂ group and the Ti atom, 2.60 Å, is considerably longer than standard Ti-C σ-bond distances, which are close to 2.15-2.20 Å, usually.[36] Moreover, the C(phenyl)-CH₂ distance is considerably shorter than a sp^2-sp^3 C–C bond distance, which suggests a strong double bond character for this bond. Finally, the phenyl ring is distorted towards a boat-like conformation. The overall geometry of **1** reminds of a metallocene structure in which one of the Cp rings is replaced by the phenyl ring of the benzyl group, which is η^7 coordinated to the metal. The geometry of coordination of the benzyl group in **1** is very similar to the geometry of coordination of one of the benzyl groups (the one which is η^7 coordinated) in the X-ray structure of the $[Cp^*Ti(CH_2Ph)_2]^+[B(CH_2Ph)C_6F_5)_3]^-$ compound.[37]

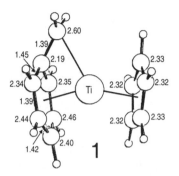

Fig. 1. Minimized energy geometries of the most stable styrene-free $CpTiCH_2Ph^+$ species. The numbers close to the C atoms represent the distance of these atoms from the metal. All distances are reported in Å.

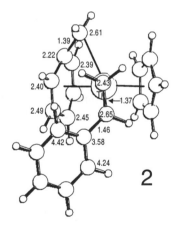

Fig. 2. Minimized energy geometries of the most stable coordination intermediate obtained by styrene coordination to the $CpTiCH_2Ph^+$ species. The numbers close to the C atoms represent the distance of these atoms from the metal. All distances are reported in Å.

As for coordination intermediates, that is species in which a styrene molecule coordinates to the $CpTi(CH_2Ph)^+$ system, we localized several structures of low energy. The most stable between them, structure **2** of Fig. 2, preserves a η^7 coordination of the benzyl-type chain end, while the monomer molecule is η^2

coordinated to the Ti atom. In fact, in **2** the C atoms of the styrene double bond are at distances of coordination from the Ti atom (smaller than 2.7 Å), whereas the shortest Ti-C distance involving a C atom of the phenyl group of styrene is longer than 3.5 Å. Moreover, in structure **2** the aromatic group of styrene and the Cp ring are on opposite sides (i.e. *anti*) of the plane defined by the Ti atom and by the two C atoms of the styrene double bond. The styrene uptake energy with respect to structure **1** and a free styrene molecule amounts to 88 kJ/mol. This uptake energy is comparable to the propene uptake energy in cationic $Cp_2MtCH_3^+$ group 4 metallocenes.[38]

The most stable transition state for the insertion reaction we found, structure **3** of Fig. 3, shows the classical four centers geometry characterizing olefins polymerization reactions. The Ti–CH$_2$(benzyl), Ti–CH(styrene) and CH$_2$(styrene)–CH$_2$(benzyl) distances are very close to the analogous distances in the transition states for olefin polymerization with group 4 metallocenes.

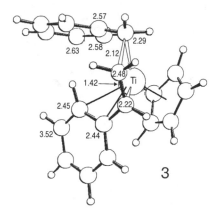

Fig. 3. Geometries of the transition state for insertion of styrene on the $CpTiCH_2Ph^+$ species. The numbers close to the C atoms represent the distance of these atoms from the metal. All distances are reported in Å.

The torsional angle Ti–CH(styrene)–CH$_2$(styrene)–CH$_2$(benzyl) assumes a value close to -8°, which indicates an almost planar transition state geometry. Moreover, an adjuvant α-agostic interaction between the Ti atom and one of the H atoms of the benzyl CH$_2$ group is present. The distance between the Ti atom and the C atom of the CH$_2$ group of the benzyl-type chain end is considerably shorter in structure **3** than in structure **2**, while the interaction scheme of the monomer molecule with the metal atom is closer to a *cis*-η^4-coordination. Finally, the aromatic rings of the benzyl group and of the styrene molecule are slightly farther and closer to the Ti atom in **3** relative to **2**, respectively. It is worthy to note that approach of the styrene aromatic ring to the Ti atom occurs in a relatively uncrowded sector, due to the *anti* disposition of this group and of the Cp ring. Energetically, **3** lies 47 kJ/mol above **2**. This barrier is slightly higher than the insertion barrier for olefin polymerization with group 4 metallocenes, which usually are in the range 0-20 kJ/mol.[39]

In the next paragraphs, we discuss on a possible mechanism of stereoselectivity. In this case, we have to introduce elements of chirality, whose combination originates diastereomeric situations of, possibly, different energy. The starting point for the following discussion will be the achiral transition state **3**. The achiral CH_2Ph benzyl group is made chiral by substituting the H atoms with methyl groups. The R or S chirality of the resulting $-CH(CH_3)Ph$ group is depending on which particular H atom has been substituted by a methyl group. However, for clarity of presentation we considered the mirror image of the structure obtained by substitution of the H atom which is involved in the α-agostic interaction in **3**. This way, the $-CH(CH_3)Ph$ group is of the same chirality, R, in both structures. However, as explained in the **Models and Methods** section, we will refer to it as *re*-ending growing chain.

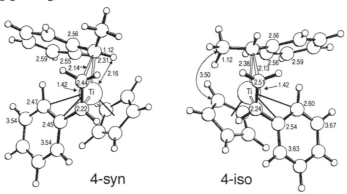

Fig. 4. Geometries of the transition states for insertion of styrene on the chiral $CpTiCH(Ph)CH_3^+$ species. The numbers close to the C atoms represent the distance of these atoms from the metal. All distances are reported in Å.

After these substitutions, we localized the two transition states **4-syn** and **4-iso** of Fig. 4, which correspond to insertion of a *si* and of a *re* coordinated styrene molecule on a *re*-ending chain, respectively. Thus, **4-syn** and **4-iso** will lead to formation of a syndiotactic *r*, and of an isotactic *m* diad, respectively.

Both transition states are still characterized by the classical four centers geometry, and present small deformations relative to the parent achiral structure **3**. However, in **4-syn** the added methyl group points out of the coordination sphere of the metal atom, and steric repulsive interactions with other ligands are minimized. In **4-iso**, instead, the added methyl group is located within the coordination sphere of the metal atom, and steric repulsive interactions with other ligands, in particular with the Cp ring, are present. This is also suggested by the short distances between the CH_3 group and the Cp ring.

Energetically, **4-syn** is favored relative to **4-iso** by 5 kJ/mol, indicating that formation of a *r* diad is favored relative to formation of a *m* diad. The $\Delta E^{\ddagger}$ in favor of the *r* diad we calculated is in good agreement with the experimental difference between the apparent energies of activation leading to *r* and *m* diads, ΔE_a, which has been determined to be 7-8 kJ/mol.[21]

Conclusions.

In this paper we have presented preliminary results of a theoretical study which is concerned with a possible mechanism of propagation and of stereoselectivity in the syndiospecific polymerization of styrene. The main conclusions can be summarized as follows:

The most stable styrene-free structure is characterized by a coordination scheme in which the benzyl-type chain end is η^7 coordinated to the metal atom. The most stable coordination intermediate preserves the η^7 coordination of the benzyl-type chain end, and it is characterized by a η^2 coordination of the styrene molecule. The monomer uptake energy amounts to 88 kJ/mol, and is comparable to the uptake energy calculated for ethene coordination to group 4 metallocenes. The insertion reaction can proceed through a transition state which is characterized by the calssical four centers transition state geometry proposed for olefins polymerization with Ziegler-Natta catalysts. A rearrangment of the benzyl-type chain end from the η^7 coordination to an almost σ-bonding scheme between the metal atom and the C atom of the CH_2 group of the chain-end is required to facilitate the insertion reaction. Moreover, the styrene molecule is almost cis-η^4 coordinated at the transition state. The energy barrier with respect to the most stable coordination intermediate is 47 kJ/mol.

The stereoselectivity is connected with different repulsive interactions between the growing chain, and the other ligands, favoring insertion of one styrene enantioface. In particular, in the transition state leading to the syndiotactic diad the polymeric growing chain develops out of the coordination sphere of the titanium, thereby minimizing steric repulsions with the other ligands, whereas in the transition state leading to the isotactic diad the remaning of the growing chain develops within the coordination sphere of the metal, giving rise to repulsive interactions with other ligands, the Cp ring in particular. Energetically, the transition state leading to the syndiotactic diad is favored by 5 kJ/mol with respect to the one leading to the isotactic diad.

Acknowledgements.

The financial support of the MURST of Italy, grant PRIN-1998, and by Montell Polyolefins, is gratefully acknowledged.

References.

1   N. Ishihara, T. Seimiya, M. Kuramoto, M. Uoi, *Macromolecules* **1986**, *19*, 2464.
2   N. Ishihara, M. Kuramoto, M. Uoi, *Macromolecules* **1988**, *21*, 3356.
3   P. Ammendola, C. Pellecchia, P. Longo, A. Zambelli, *Gazz. Chim. Ital.* **1987**, *117*, 65.
4   G. Guerra, V. M. Vitagliano, C. De Rosa, V. Petraccone, P. Corradini, *Macromolecules* **1990**, *23*, 1539.

5   S. W. Ewart, M. C. Baird, in *Metallocene Based Polyolefins, Preparation, Properties and Technology, Vol. 1*, Eds.: J. Scheirs, W. Kaminsky, John Wiley & Sons, New York, **1999**, p 119.

6   S. W. Ewart, M. C. Baird, *Top. Catal.* **1999**, *7*, 1.

7   C. Pellecchia, A. Grassi, *Top. Catal.* **1999**, *7*, 125.

8   R. Po, N. Cardi, *Prog. Polym. Sci.* **1996**, *21*, 47.

9   N. Tomotsu, N. Ishihara, T. H. Newman, M. T. Malanga, *J. Mol. Catal. A* **1998**, *128*, 167.

10   J. C. W. Chien, Z. Salajka, S. Dong, *Macromolecules* **1992**, *25*, 3199.

11   A. Grassi, C. Pellecchia, L. Oliva, F. Laschi, *Macromol. Chem. Phys.* **1995**, *196*, 1093.

12   A. Grassi, A. Zambelli, F. Laschi, *Organometallics* **1996**, *15*, 480.

13   A. Grassi, S. Saccheo, A. Zambelli, F. Laschi, *Macromolecules* **1998**, *31*, 5588.

14   E. F. Williams, M. C. Murray, M. C. Baird, *Macromolecules* **2000**, *33*, 261.

15   C. Pellecchia, D. Pappalardo, L. Oliva, A. Zambelli, *J. Am. Chem. Soc.* **1995**, *117*, 6593.

16   Q. Wang, R. Quyoum, D. J. Gillis, M.-J. Tudoret, D. Jeremic, B. K. Hunter, M. C. Baird, *Organometallics* **1996**, *15*, 693.

17   P. Longo, A. Grassi, P. A., P. Ammendola, *Macromolecules* **1988**, *21*, 24.

18   C. Pellecchia, P. Longo, A. Grassi, P. Ammendola, A. Zambelli, *Makromol. Chem., Rapid. Commun.* **1987**, *8*, 277.

19   A. Zambelli, P. Longo, C. Pellecchia, A. Grassi, *Macromolecules* **1987**, *20*, 2035.

20   A. Grassi, C. Pellecchia, P. Longo, A. Zambelli, *Gazz. Chim. Ital.* **1987**, *117*, 249.

21   P. Longo, A. Proto, A. Zambelli, *Macromol. Chem. Phys.* **1995**, *196*, 3015.

22   A. Zambelli, C. Pellecchia, L. Oliva, S. Han, *J. Polym. Sci.* **1988**, *26*, 365.

23   P. Corradini, G. Paiaro, A. Panunzi, *J. Polym. Sci., Part C* **1967**, *16*, 2906.

24   ADF 2.3.0 Users's Guide, Vrije Universiteit Amsterdam, Amsterdam, The Netherlands 1996.

25   E. J. Baerends, D. E. Ellis, P. Ros, *Chem. Phys.* **1973**, *2*, 41.

26   B. te Velde, E. J. Baerends, *J. Comp. Phys.* **1992**, *99*, 84.

27   S. H. Vosko, L. Wilk, M. Nusair, *Can. J. Phys.* **1980**, *58*, 1200.

28   A. Becke, *Phys. Rev. A* **1988**, *38*, 3098.

29   J. P. Perdew, *Phys. Rev. B* **1986**, *33*, 8822.

30   J. P. Perdew, *Phys. Rev. B* **1986**, *34*, 7406.

31   T. K. Woo, L. Cavallo, T. Ziegler, *Theor. Chem. Acc.* **1998**, *100*, 307.

32   L. Deng, T. K. Woo, L. Cavallo, P. M. Margl, T. Ziegler, *J. Am. Chem. Soc.* **1997**, *119*, 6177.

33   L. Cavallo, T. K. Woo, T. Ziegler, *Can. J. Chem.* **1998**, *76*, 1457.

34   T. N. Doman, T. K. Hollis, B. Bosnich, *J. Am. Chem. Soc* **1995**, *117*, 1352.

35   B. R. Brooks, R. E. Bruccoleri, B. D. Olafson, D. J. States, S. Swaminathan, M. Karplus, *J. Comput. Chem.* **1983**, *4*, 187.

36   A. Guy Orpen, L. Brammer, F. H. Allen, O. Kennard, D. G. Watson, R. Taylor, *J. Chem. Soc., Dalton Trans.* **1989**, S1.

37   C. Pellecchia, A. Immirzi, D. Pappalardo, A. Peluso, *Organometallics* **1994**, *13*, 3773.

38   P. M. Margl, L. Deng, T. Ziegler, *Organometallics* **1998**, *17*, 933.

39   P. M. Margl, L. Deng, T. Ziegler, *J. Am. Chem. Soc.* **1998**, *120*, 5517.

Influence of Ligands on the Deactivation of Group IV Metallocene Catalysts in the High-Temperature Polymerization of Ethene

Alexander Rau*, Thomas Wieczorek, Gerhard Luft

Darmstadt University of Technology, Department of Chemical Engineering, Petersenstr. 20, 64287 Darmstadt, Germany
E-mail: alex@bodo.ct.chemie.tu-darmstadt.de

Abstract. Ethene polymerizations were performed using catalyst systems based on different types of zirconocene dichlorides and $Al(^iBu)_3/$ $[PhNHMe_2]^+[B(C_6F_5)_4]^-$ at temperatures of 100 to 140 °C and pressures of 20 to 70 bar. The starting polymerization rate Rp_0 of the $(^nBu-Cp)_2ZrCl_2$-based catalyst increases linearly with increasing catalyst concentration. The same catalyst showed a second-order dependency of Rp_0 on the ethene concentration, $Rp_0 \sim [ethene]^{2.0}$. The rate time profiles showed a rapid decay in activity, which could be described with first-order deactivation reactions regarding the concentration of active sites. The rate constants of deactivation k_d were determined for a series of seven different metallocene catalysts. The half-lives resulting from k_d were used to compare the stability of the metallocene catalysts with different ligand systems.

Introduction

Industrial-scale ethene polymerization processes are usually performed at high temperatures with a minimum pressure of 5 bar. In gas phase- and suspension-processes, the temperatures are kept in the range from 80 to 100 °C, whereas the solution process requires temperatures above 100 °C, which are increased to up to 300 °C in some polymerization plants [1]. The kinetics and the deactivation of metallocene-based catalysts at increased temperatures and pressures are therefore of great interest for industrial-scale applications. In fact, investigations at high temperatures are rare and are usually performed at pressures below 5 bar, with metallocene dichloride/methylalumoxane catalyst systems and propene as monomer. The studies show that increasing the temperature above approximately 90 °C usually causes a decrease in monomer consumption over time, which can be explained by decomposition of the catalyst [2,3]. At lower temperatures, the decline of activity is less rapid and sometimes leads to nearly steady-state values [2,4]. In the temperature range of 20 to 60 °C, a rapid initial decay followed by a second slow deactivation was observed with the Cp_2ZrCl_2/MAO catalyst [5]. In addition to the temperature, variations of the metallocene ligand influence the run of the rate time profiles, as is shown by the comparison of $Me_2Si(2\text{-}Me\text{-}Ind)_2ZrCl_2$ and $Me_2Si(Ind)_2ZrCl_2$ under identical conditions in the high-temperature polymerization of propene [2].

Similar results with regard to the influence of temperature were observed recently in polymerizations of ethene catalyzed by $Et(Ind)_2ZrCl_2/Al(^iBu)_3/$ $[PhNHMe_2]^+[B(C_6F_5)_4]^-$ and $Ph_2C(Cp)(Flu)ZrCl_2/Al(^iBu)_3/[PhNHMe_2]^+[B(C_6F_5)_4]^-$ at 6 and 20 bar, respectively [6,7]. With both catalysts, an increase in temperature from 150 to 200 °C causes a strong decrease in catalyst productivity, in the first case from 31 to 10 kg/mmol Zr and in the latter from 172 to less than 40 kg/mmol Zr. Here we present kinetic studies of ethene polymerizations in the solution process, catalyzed by systems based on different types of zirconocene dichlorides combined with $Al(^iBu)_3/[PhNHMe_2]^+[B(C_6F_5)_4]^-$ at temperatures of 100 to 140 °C and pressures of 20 to 70 bar. The reaction orders concerning the metallocene and the monomer were determined, and comparisons were made of the thermal stability of seven different metallocene catalysts based on the rate-time profiles observed.

Experimental section

Materials

All operations were carried out in an argon atmosphere. Toluene was distilled over sodium. Ethene (99.8 %) supplied by Linde was purified by passage through columns containing molecular sieves and a copper catalyst. $Al(^iBu)_3$ was purchased from Witco. $Ph_2C(Cp)(Flu)ZrCl_2$, $rac\text{-}Me_2Si(Ind)_2ZrCl_2$, $(^nBu\text{-}Cp)_2ZrCl_2$, Cp_2ZrCl_2 and $[PhNHMe_2]^+$ $[B(C_6F_5)_4]^-$ were donated by BASF AG. $Me_2Si(Cp)_2ZrCl_2$ was prepared according to the literature [8]. A similar method was used to prepare $[(2\text{-}MeO\text{-}3\text{-}Me)Bz]_2Si(Cp)_2ZrCl_2$ and $Me[(2\text{-}MeO\text{-}3\text{-}Me)Bz]Si(Cp)_2ZrCl_2$. The purity of the metallocenes was >99 % checked by ^1H NMR.

Catalysts

The metallocenes used were activated in a two-step procedure. First, a solution of zirconocene dichloride in toluene was set to react with 200 mol equivalents of $Al(^iBu)_3$. The solution was stirred at room temperature for 30 min and, in a second step, added to a solution of $[PhNMe_2H]^+[B(C_6F_5)_4]^-$ in toluene, whereby the B/Zr ratio was adjusted to 1.2. After a preactivation time of 90 min, the resulting catalyst solutions were injected into the reactor.

Polymerization

The polymerization unit consists of a 500-ml steel reactor equipped with a stirrer and a storage autoclave for the ethene supply. The maximum working pressure of the storage autoclave was 250 bar. Before polymerization, it was pressurized to 220 bar with ethene using a compressor. The maximum working pressure and temperature of the reactor were 100 bar and 200 °C, respectively. Prior to use, the reactor was heated to 140 °C and evacuated for one hour. After cooling, it was first rinsed with a dilute solution of $Al(^iBu)_3$ in toluene and then filled with desired amounts of toluene and $Al(^iBu)_3$. After the solution was heated to the desired temperature, the ethene pressure was adjusted. The polymerization was started by injection of the preactivated catalyst solution (5 ml) into the reactor with argon overpressure. During polymerization, the pressure in the reactor was kept constant, as ethene was supplied from the storage vessel. The ethene consumption was determined by recording temperature and pressure in the storage vessel.

After a certain polymerization time, the pressure was released and the polymer precipitated by subsequently draining off the reaction mixture into methanol.

The monomer concentrations were calculated from liquid-vapor equilibria of the toluene/ethene- system with the Peng-Robinson equation of state combined with mixture rules of the Van der Waals type and confirmed by measurements. The results showed that, in the temperature- and pressure- range used, a liquid phase and a gas phase are present in the reactor.

The solubility of metallocene- polyethylene in the liquid phase at 70 bar was examined in an autoclave equipped with saphire windows [9]. A sample poly-ethylene was used, which was produced with the catalyst based on $(^nBu-Cp)_2ZrCl_2$. The Mn of the polymer was 240000 g/mol and the Mw/Mn was 2. The cloud- point temperature of the liquid phase was determined to 83 °C using a system containing 5 weight percent of the polymer. This result ensured that the polymerizations were performed in a homogeneous liquid phase.

Results and discussion

Influence of zirconium and ethene concentration

The influence of the zirconium concentration on the starting polymerization rate Rp_0 at 100 and 140 °C using $(^nBu-Cp)_2ZrCl_2$ as catalyst precursor is shown in Fig. 1. In order to keep the temperature constant and to avoid transport limitations, the catalyst concentration used had to be below 0.3 and 1.1 μmol/l, respectively. In these concentration areas, a change of the stirring speed from 1000 to 2000 rpm has no influence on the polymerization rate, yields or molecular weight distribution.

As shown in Fig. 1, Rp_0 increases linearly with increasing Zr concentration, resulting in a first-order reaction with regard to the metallocene catalyst. The same linear relationship is found, if the yields are plotted against the Zr concentration. At both temperatures, the regression lines in Fig. 1 intercept the x-axis at certain values. According to the intercept values, a certain minimum concentration of metallocene is required to start a polymerization, which can be explained by deactivation reactions before the start of polymerization, or a non quantitative activation of the metallocene compound. Considering the intercepts of the regression lines with the x-axis, the productivities of polymerizations depend on the metallocene concentration used, although Rp_0 and yields increase linearly with the metallocene concentration.

Fink et al. made similar observations in ethene polymerizations catalyzed with $Me_2Si(Ind)_2ZrMe_2/[(C_6F_5)_3C]^+[B(C_6F_5)_4]^-$, whereby the intercept values are around 1000 times higher [10]. The lower minimum concentrations in our case may be due to stabilization - and scavenger effects of the aluminum alkyl.

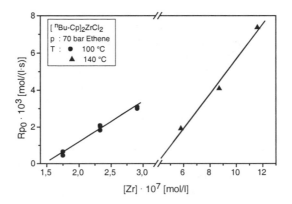

Fig. 1. Influence of the Zr concentration on the starting polymerization rate

Remarkable are the different minimum concentrations at 100 and 140 °C, which are 0.15 and 0.34 µmol metallocene/l respectively, as can be seen in Fig. 1. A possible explanation for the strong temperature dependency of the intercept values could be catalyst deactivation after injection into the reactor and before the start of propagation.

If the intercept values represent the amount of inactive metallocene $[Zr]_{inact}$ before the propagation start, the maximum value of active metallocene concentration $[Zr^*]$ can be described by subtraction of the intercept values from the total amount of metallocene $[Zr]_{tot.}$, see Equation 1.

$$[Zr^*] = [Zr]_{tot.} - [Zr]_{inact.} \tag{1}$$

Considering this, Rp_0 can be described with Equation 2.

$$Rp_0 = k_p' \cdot ([Zr]_{tot.} - [Zr]_{inact.})^x \tag{2}$$

$$\text{with } k_p' = k_p \cdot [ethene]^y$$

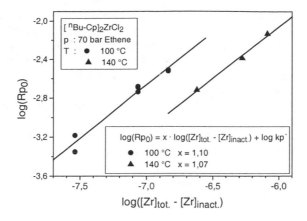

Fig. 2. Reaction order x as found from $Rp_0 = k_p \cdot [\text{ethene}]^y \cdot ([Zr]_{tot.} - [Zr]_{inact.})^x$

Fig. 2 shows that this assumption leads at both temperatures to nearly exact first-order dependence on $[Zr^*]$, $Rp_0 \sim [Zr^*]^{1.10}$ and $Rp_0 \sim [Zr^*]^{1.07}$.

In order to determine the influence of the monomer concentration on the polymerization rate, the ethene pressure was varied from 20 to 70 bar. Depending on the pressure used, the ethene concentration in the liquid phase changed from 1.4 to 5 mol/l. The results depicted in Fig. 3 show a second-order dependency of the monomer concentration on Rp_0, $Rp_0 \sim [\text{ethene}]^{2.0}$.

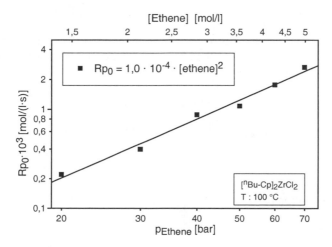

Fig. 3. Reaction order y as found from $Rp_0 = k_p \cdot [Zr^*]^x \cdot [\text{ethene}]^y$

Similar results concerning the relationship between the polymerization rate and the monomer concentration have been observed in several metallocene/MAO catalyst systems, where reaction orders between 1.2 and 2.4 were found, at much lower pressures [2,11,12,13]. As our experimental series shows, the rate approaches no saturation up to ethene concentrations of 5 mol/l, and the second-order reaction is observed over a broad concentration range of ethene. At a pressure of 70 bar, these results lead to a polymerization rate around 10 times higher than the one at 20 bar.

Within the range of 20 to 70 bar, the pressure had no influence on the deactivation of the catalyst. Consequently, the yields and the productivities show the same second-order dependence on ethene concentration as Rp_0. Thus, using the same catalyst system in technical solution polymerization of ethene, a pressure increase leads to much higher space-time yields.

Catalyst deactivation during polymerization and half-lives

According to the observed relationship between the reaction rate and zirconium- and ethene concentrations, the polymerization rate can be described by Equation 3.

$$Rp(t) = k_p \cdot [Zr^*(t)] \cdot [ethene]^2 \qquad (3)$$

As shown in Fig. 4, the polymerization rate decreases with time. At 140 °C, the decline of the polymerization rate is substantially faster than at 100 °C, so that the polymerization rate at 140 °C goes down to zero within 700 s. Since temperature, and pressure are kept constant and no transport or diffusion limitations are present, the observed decrease in the polymerization rate is due to the decreasing number of active catalyst sites. For the seven metallocene catalysts used, the experimental data obtained could be well described by a first order deactivation reaction with regard to [Zr*]; Equation 4:

$$Rp(t) = k_p \cdot [ethene]^2 \cdot [Zr^*] \cdot e^{-k_d \cdot t} \qquad (4)$$

From this Equation 5 follows:

$$Rp(t) = Rp_0 \cdot e^{-k_d \cdot t} \qquad (5)$$

According to Equation 6,

$$\ln(Rp(t)) = \ln(Rp_0) - k_d \cdot t \qquad (6)$$

the rate constant of deactivation was determined by a logarithmic plot of the polymerization rate against time. Fig. 5 shows the determination of k_d for three different zirconium concentrations, using the $(^nBu-Cp)_2ZrCl_2$-based catalyst system. Within this range, k_d was not affected by the initially applied zirconium concentration, as can be seen by the parallel formation of the slopes. This result clearly indicates first-order deactivation with regard to the catalyst.

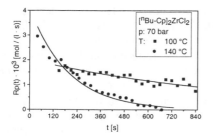

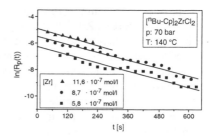

Fig. 4. Rate-time profiles

Fig. 5. Plot of ln(Rp(t)) vs. t for three different [Zr]

The deactivation constant k_d is a measure for the thermal stability of the catalysts. More descriptive than k_d are the half-lives, which were calculated according to $t_{1/2} = \ln 2/ k_d$ and depicted together with productivities in Fig. 6. Half-lives of 9 min were found with Cp_2ZrCl_2 and $(^nBu\text{-}Cp)_2ZrCl_2$, indicating that the ^{n}butyl-group attached to the cyclopentadien ring has no influence on thermal stability of the catalyst. The catalyst based on $Me_2Si(Cp)_2ZrCl_2$ showed nearly the same half-life. From this result, it can be concluded that *ansa*-metallocene catalysts are not generally more stable than those based on non-bridged metallocenes. The most stable catalysts in our experimental series were based on $Me_2Si(Ind)_2ZrCl_2$ and

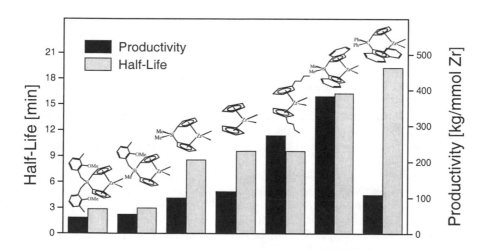

Fig. 6. Half-lives and productivities

$Ph_2C(Flu)(Cp)ZrCl_2$, which are both *ansa*-metallocenes with fused-ring-systems. Half-lives between 15 and 20 min were found with these catalysts. Short half-lives of around 3 min were found with the two non-standard metallocenes [(2-MeO-3-Me)Bz]$_2$Si(Cp)$_2$ZrCl$_2$ and Me[(2-MeO-3-Me)Bz]Si(Cp)$_2$ZrCl$_2$, which are both bearing methoxy-groups tethered by a three carbon chain to the silicon atom of the bridge-linking group. An explanation for the reduced stability may be the involvement of the methoxy-groups in deactivation reactions, probably by forming zirconium oxygen bonds leading to inactive metallocene species.

As seen in Fig. 6, the productivities do not correlate with the half-lives. This could be due to the different rate constants k$_p$ and to different zirconium concentrations used in the experiments. It is important to note that the productivties depend on the used metallocene concentration, as shown in the chapter 3.1, whereas the half-lives are independent of the metallocene concentration.

Acknowledgements

The authors wish to thank the Bundesministerium für Bildung und Forschung and BASF AG for financial support.

References

1   K. S. Whiteley, *Ullmann's Encyclopedia of Industrial Chemistry*, H.-J. Arpe (Ed.), VCH Weinheim **1992**, Volume A21, p. 487 ff.

2   T. S. Wester, H. Johnsen, P. Kittilsen, E. Rytter, *Macromol. Chem. Phys.* **1998**, *199*, 1989.

3   B. Rieger, G. Jany, R. Fawzi, M. Steimann, *Organometallics* **1994**, *13*, 647.

4   J. Huang, G. L. Rempel, *Ind. Eng. Chem. Res.* **1997**, *36*, 1151.

5   D. Fischer, R. Mülhaupt, *J. Organomet. Chem.* **1991**, *417*, C7.

6   A. Yano, M. Sone, S. Yamada, S. Hasegawa, A. Akimoto, *Macromol. Chem. Phys.* **1999**, *200*, 917.

7   A. Yano, S. Hasegawa, S. Yamada, A. Akimoto, *J. Molecular Catalysis A: Chemical* **1999**, *148*, 77.

8   H. Yasuda, K. Nagasuna, M. Akita, K. Lee, A. Nakamura, *Organometallics* **1984**, *3*, 1470.

9   The experimental procedure for determination of cloud-points is described in detail in: G. Luft, N. S. Subramanian, *Ind. Eng. Chem. Res.* **1987**, *26*, 750.

10 S. Hahn, G. Fink, *Macromol. Rapid. Commun.* **1997**, *18*, 117.

11 N. Herfert, G. Fink, *Macromol. Chem. Macromol. Symp.* **1993**, *66*, 157.

12 S. Jüngling, R. Mülhaupt, U. Stehling, H. H. Brintzinger, D. Fischer, F. Langhauser, *J. Polym. Sci. A. Polym. Chem.* **1995**, *33*, 1305.

13 J. C. W. Chien, Z. Yu, M. M. Marques, J. C. Flores, M. D. Rausch, *J. Polym. Sci. A. Polym. Chem.* **1998**, *36*, 319.

5. Polymerization, Co-Polymerization, Non-Traditional Monomers and Polymer Characterization

Correlations Between Chain Branching, Morphology Development and Polymer Properties of Polyethenes

Philipp Walter, Johannes Heinemann, Henner Ebeling, Dietmar Mäder, Stefan Trinkle, Rolf Mülhaupt

Freiburger Materialforschungszentrum und Institut für Makromolekulare Chemie of the Albert-Ludwigs University, Stefan-Meier Strasse 31, 79104 Freiburg, Germany
E-mail: mulhaupt@uni-freiburg.de

Abstract. Molecular architectures of polyethenes, in particular short and long chain branching, were varied over a very wide range by means of either metallocene-catalyzed ethene/1-olefin copolymerization or Ni- and Pd-catalyzed migration/insertion-type ethene homopolymerization. While short chain branches affected melting, glass transition, and blend compatibility, long chain branching represented the key to improved melt processability. Both the number of short and long chain branches depended upon the ligand substitution pattern of dimethylsilylene-bridged bisindenyl complexes. The degree of branching increased with variation of the substitution in 4-position, i.e., 4-napthyl > 4-phenyl > benzannelation. Variation of the 1-butene content of ethene/1-butene (EB) copolymers gave control of morphology development and properties of isotactic polypropene (iPP) blends with EB. Highly flexible, single-phase as well as stiff and tough two-phase iPP/EB (70 wt.-%/30 wt.-%) blends were obtained. Rheological studies on ethene/1-eicosene model polyethene revealed the presence of a positive comonomer effect with respect to molar catalyst activity, molecular weight, and long-chain branching. A new family of thermoplastic elastomers based upon highly branched polyethene was prepared via Pd-catalyzed ethene copolymerization with 2,2,6,6-tetramethyl-piperidineoxy (TEMPO)-functionalized 1-olefin as macromonomers to produce macroinitiators for the initiation of the controlled free radical graft copolymerization of styrene onto highly branched polyethenes. The variation of the polystyrene block length gave control on nanophase separation of the resulting branched polyethene – graft – polystyrene.

Introduction

The discovery of novel single-site catalysts based upon metallocene [1] and late transition metal complexes [2] is stimulating the development of novel linear and branched polyethene materials with tailor-made property profiles. While many conventional Ziegler-Natta catalysts gave rather complex reaction mixtures, modern single-site catalyst technology affords very uniform polymers, thus providing the base for better understanding of basic correlations between catalyst and polymer architectures. Several catalytic processes have been established to introduce branches into the polyethene backbone. Traditional approach (pathway IV) represents copolymerization of ethene with 1-olefins. Modern metallocene catalysts are offering new opportunities in ethene copolymerization such as incorporation of less reactive long-chain 1-olefins, isobutene, and even styrene [3].

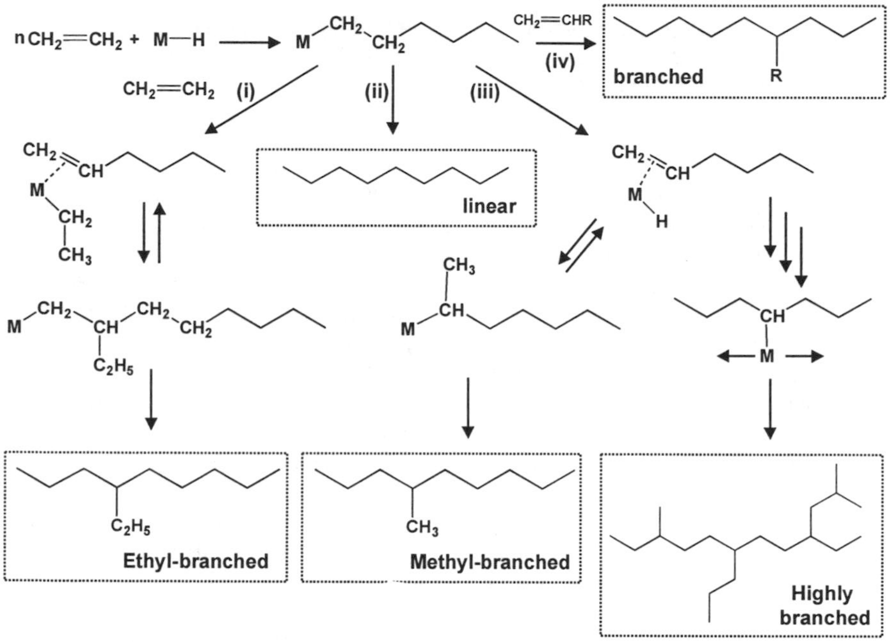

Fig. 1. Branching during ethene homopolymerization without comonomer addition

As illustrated in Fig. 1, branching has been achieved in ethene homopolymerization without requiring addition of comonomer. For instance, metallocene catalysts based upon meso-[Et(Ind)$_2$]ZrCl$_2$ were reported to produce polyethene containing isolated ethyl-branches due to β-hydride transfer to ethene, followed by reinsertion of the vinyl-terminated polyethene (pathway I) [4]. In a versatile process, known as 2,ω-polymerization or "chain walking", the transition metal alkyl migrates up and down the polyethene chain. Chain propagation occurring exclusively after migration to the less sterically hindered chain end affords methyl–branched polyethene, whereas insertion during migrations yields highly branched polyethene with linear and branched side chains (pathway III) [2,5]. End of last century highly active Ni- and Pd-diazadiene catalysts were discovered to produce branched homopolyethene via migration/insertion polymerization [6]. For Ni catalysts, the degree of branching increased with decreasing ethene pressure, whereas Pd-diazadiene catalysts gave much higher, pressure-independent degree of branching [6,7]. In contrast to group 4 catalyst, late transition metal catalysts are well known to tolerate polar comonomers. Brookhart used his Pd-diazadiene complexes to copolymerize ethene with methylacrylate [6]. This overview highlights how single-site catalyst technology is exploited to establish correlations between catalyst type, polyethene branching, polymer morphology, and mechanical properties in order to tailor polyethene properties and to improve melt processing.

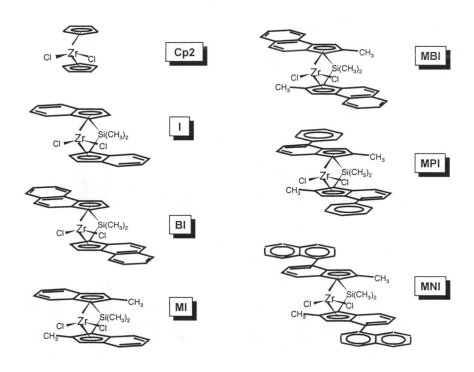

Fig. 2. The zirconocene catalyst family based upon the substituted dimethylsilylenebisindenyl ligand framework

Short chain branched polyethene

Short chain branches were introduced using metallocene and late transition metal catalysts according to the mechanistic scheme in Fig. 1. The metallocene family depicted in Fig. 2 comprised dimethylsilylene-bridiged bisindenyl zirconium dichloride catalysts activated with methylaluminoxane (MAO). Today it is well established that 2-methyl substitution, referred to as M, of the dimethylsilylene-bridged bisindenyl ligand system (I), accounts for increased molecular weight, whereas benzannelation (B) and especially 4-phenyl (P) and 4-naphtyl (N) substitution promote catalyst activity and incorporation of 1-olefin comonomers. With respect to short chain branching via 1-olefin incoporation, the following ranking was observed: 4-naphthyl > 4-phenyl > benzannelation [8-11].

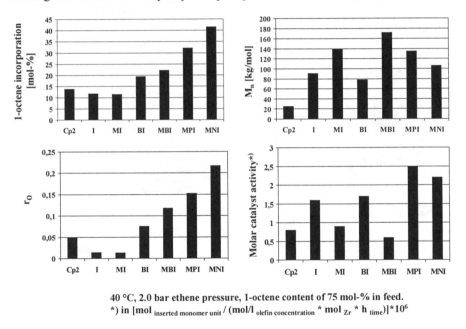

40 °C, 2.0 bar ethene pressure, 1-octene content of 75 mol-% in feed.
*) in [mol $_{inserted\ monomer\ unit}$ / (mol/l $_{olefin\ concentration}$ * mol $_{Zr}$ * h $_{time}$)]*10^6

Fig. 3. Comparison of metallocene substitution patterns in ethene/1-octene copolymerization (data from Suhm [8]).

Metallocene-catalyzed ethene/1-olefin copolymerization using the metallocenes shown in Fig. 2, hybrid catalysts with in-situ formation of 1-olefin comonomers, and migratory/insertion-type homopolymerization using Brookhart-type Ni- and Pd-catalysts produced a large variety of ethene copolymers with random distribution of short chain branches [8]. Such model systems were used to establish correlations between degree of branching and polyethene phase transition temperatures. In accord with observations by other groups [12], melting temperature depended primarily on the segment length of linear polyethene between two branches [8,13]. For ethene-rich copolymers a correlation between glass transition temperature and degree of branching was established taking into account both the number of branched C and the number of C atoms of the n-alkyl side chains [14]. The influence of short chain branching on polymer compatibility was examined using ethene/1-butene (EB) copolymers with random incorporation of 0-100 wt.-% 1-butene. Blend formation with isotactic polypropene (iPP) afforded a remarkably wide range of materials [15]. At 1-butene content exceeding 90 wt.-%, EB was miscible with iPP, thus producing highly flexible single-phase iPP/EP (70 wt.-% / 30 wt.-%) blends. While EB with very low 1-butene content gave rather poor interfacial adhesion, very effective EB rubber dispersion and interfacial adhesion of EB was observed for 1-butene content around 50 wt.-%. The resulting two-phase blends exhibited attractive combination of stiffness and low temperature impact resistance [15].

Long chain branched polyethenes

Although uniform polymers with narrow molecular weight distributions are highly desirable with respect to their low organoleptics, polymers with narrow molecular weight distribution are difficult to process. Perferably the melt viscosity should decrease drastically with increasing shear rates typical for polymer processing using extruders and injection molding. In order to improve melt processing, it is important to identify how catalyst type, catalyst support, comonomer addition, and process condition affect the rheological behavior. It should be noted, however, that frequently spectroscopic methods are not accurate enough to permit precise characterization of the degree of long chain branching. Special rheological tools are being developed to evaluate long chain branching. Several groups have addressed the influence of metallocene catalysts on long chain branching [16-18]. Multiple-single site catalysts were designed to improve melt processing [20].

When ethene was copolymerized with 1-olefins using various metallocene catalysts, it was observed that long chain branching was clearly related to the 1-olefin content and the copolymerization parameters. At high 1-olefin content linear polymers were formed, whereas with decreasing propene content, long chain branching increased. This is associated with the type of end group. At high propene content mainly vinylidene end groups are obtained. In contrast to vinyl end groups vinylidene groups are not copolymerized by metallocenes displayed in Fig. 2. Copolymerization of the vinyl end group of polyethene also depends upon the catalyst type. For example, catalysts such as $Cp^*_2ZrCl_2/MAO$ ($Cp^* = Me_5Cp$), which give very poor 1-olefin incorporation, produce mainly linear polymers, whereas catalysts based upon MNI and MPI give much higher incorporation of vinyl-terminated polyethene. Moreover, as apparent from Fig. 4, the incorporation of long chain branches follows the same substituent influence ranking as reported for the short chain branching, i.e., the degree of long chain branches increased with 4-naphthyl > 4 phenyl > benzannelation of the dimethylsilylene-bridged bisindenyl ligand framework.

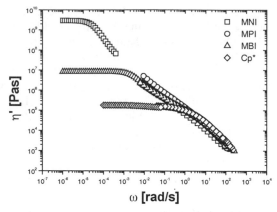

Fig. 4. Influence of metallocene type on polyethene processability (ethene was polymerized at 40 °C, ethene pressure of 2.0 bar, Al/Zr= 10000/1, c_{cat} = 2.0 μmol/l)

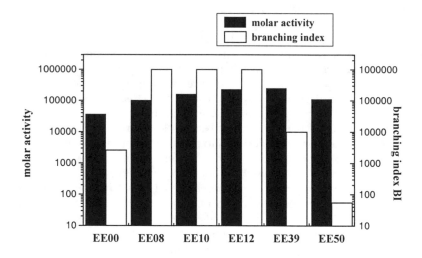

Fig. 5. The influence of 1-eicosene content on molar catalyst activities and long chain branching index in ethene/eicosene copolymerization (40 °C, 2.0 bar, in toluene, eicosene content ranging between 8 and 38 mol-%, Al/Ti = 5000/1, [CBT] = 8 – 20 μmol/l) [22]

A detailed study on ethene/1-eicosene copolymerization revealed that a positive comonomer effect existed for long chain branching [22]. At eicosene content lower than 25 wt.-%, molar catalyst activities, molecular weight and also long chain branching increased simultaneously with increasing eicosene content. However, at higher eicsoene content, due to the lower reactivity and steric hindrance of eicosene with respect to ethene, catalyst activities, molecular weight and long chain branching decreased. Obviously, comonomer addition promoted catalyst activities as well as incorporation of vinyl-terminated polyethene to produce long chain branches. Small amount of eicosene may promote formation of active sites at the expense of dormant sites and also promote diffusion and end group mobility, thus enhancing vinyl end group copolymerization to form long chain branches. At high eicosene content, predominantly vinylidene end groups were formed, thus preventing copolymerization of vinylidene-terminated poly(ethene-co-eicosene) and producing linear copolymers.

Polar branched polyethenes

Most conventional group 4 Ziegler-Natta catalysts and aluminum alkyl activators are strong Lewis acids which form strong complexes with polar comonomers. Since polar comonomers compete favorably with ethene for vacant coordination sites, such catalysts are severely poisoned by polar comonomer addition. Metallocene catalysts incoporate polar monomers such as undecenol when MAO concentration is high enough to scavenge the polar component by complexation [23]. To avoid chelate complex formation, the polar group is usually separated from the vinyl group by methylene spacers. In contrast, it is well established that the much less oxophilic late transition metals such as Ni and Pd are much less sensitive and can copolymerize catalyst poisons such as carbon monoxide. Ni- and Pd-diazadiene catalysts produced ethene/methylacrylate copolymers containing both n-alkyl and esteralkyl side chains [7,24]. While most catalysts require activators, Grubbs' salicylaldiminato nickel catalysts are activator free, produce linear and branched polyethenes, and copolymerize ethene with polar norbornenes such as norbornenol [25,26]. Mecking polymerized ethene in water using Pd catalysts without sacrificing high catalyst activity and high molecular weight [27]. In comparison to solution polymerization in methylenechloride, lower degree of branching was obtained. The groups of Mecking [27] and Spitz [28] reported the first successful attempts toward catalytic ethene emulsion polymerization.

Another synthetic route to long-chain-branched polyethene employs the macroinitiator concept to combine controlled free radical and insertion polymerization. Polypropene macroinitiators to initiate controlled free radical polymerization was first reported by Waymouth and Hawker [29] for the preparation of polypropene graft copolymers. Baumert and Heinemann [30] prepared a 2,2,6,6-tetramethylpiperdinoxy (TEMPO)-functionalized 1-olefin (cf. Fig. 6) which was used as comonomer in Pd-diazadiene-catalyzed ethene copolymerization to produce highly branched polyethene with low glass temperatures of –60 °C and 1 to 3 TEMPO-functional side chains. Such TEMPO-functional polyethenes are macroinitiators for the controlled free radical polymerization. At temperatures exceeding 100 °C, polyethene radicals are formed which initiate the polymerization of styrene or styrene containing other polar comonomers such as acrylonitrile. Due to the presence of TEMPO radicals which reversible terminate and reinitate the polystyrene graft copolymerization, no other chain transfer or chain termination reactions of the polymer radicals take place. As a consequence very uniform polystyrene branches are polymerized onto polyethene. The control of the molecular weight of grafted polystyrene with narrow molecular weight distribution represents the key to achieving control of phase separation. As a function of the graft chain length it is possible to produce nanophase-separated polyethene (cf. Fig. 7). Such branched polyethenes grafted with polystyrene or styrene copolymers are attractive thermoplastic elastomers which are also of interest in blend formation.

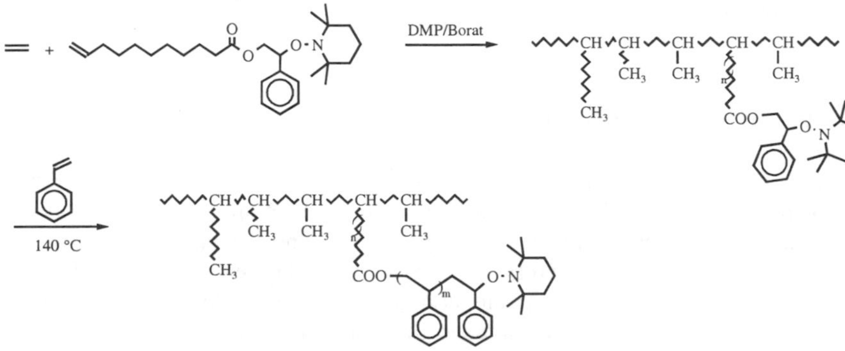

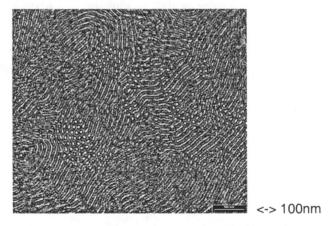

Fig. 6. Macrinitiator route to branched polyethene-graft-polysterene

<-> 100nm

Fig. 7 TEM of PE-g-PS (M_n (PE) = 14.200 g/mol; M_n (total) = 39.700 g/mol)

Conclusions

Remarkable advances in the development of single-site catalyst technology based upon metallocenes as well as late transition metal complexes offer attractive opportunities for designing of polymer molecular architectures. In contrast to many earlier catalyst generations the resulting polymers are very uniform with respect to their molecular weight and molecular-weight-independent comonomer incorporation. Today one of the major challenges in polyolefin technology is to exploit the potential of these new catalyst generations to design novel polyolefin materials exhibiting both improved property profiles and better processing. Tailoring of molecular weight distributions and long chain branching is of particular interest because some of the very uniform polymers are rather difficult to process by conventional melt extrusion or injection molding. In future the copolymerization of polar monomers and graft as well as block copolymerization by combined insertion/free rdaical processes are expected to broaden the range of polyolefin materials which will continue to compete very successfully with other polymeric materials.

Acknowledgement

The authors gratefully acknowledge support by BASF AG and by the Bundesminister für Bildung and Forschung as part of the research grants no. 03M40719 and no. 03N1028 0.

References

1   Scheirs, J, Kaminsky W (**1999**) *Metallocene-Based Polyolefins – Preparation, Properties and Technology*, John Wiley & Sons, Chichester
2   Ittel SD, Johnson LK, Brookhart M (**2000**) *Chem Rev* 100:1169
3   Suhm J, Schneider MJ, Mülhaupt R (**1999**) in Karger-Kocsis (ed.) *Polypropylene – An A-Z reference*, Kluwer Academic Publishers, Dordrecht 104
4   Izzo L, Caporaso L, Senatore G, Olivia L (**1999**) *Macromolecules* 32:6913
5   Schubbe R, Angermund K, Fink G, Goddard R (**1995**) *Macromol Chem Phys* 196:478
6   a) Johnson LK, Killian CM, Brookhart M (**1995**) *J Am Chem Soc* 117:6414, b) Gates DP, Svejda SA, Onate E, Kilian CM, Johnson LK, White VS, Brookhart M (**2000**) *Macromolecules* 33:2320
7   Mecking S, Johnson LK, Wang L, Brookhart M (**1998**) *J Am Chem Soc* 120:888
8   Heinemann J, Walter P, Mäder D, Schnell R, Suhm J, Mülhaupt R (**1999**) in Kaminsky W *Metalorganic Catalysts for Synthesis and Polymerization*, Springer Publ., Berlin, p.473
9   Schneider MJ, Mülhaupt R (**1997**) *Macromol Chem Phys* 198:1121
10   Schneider MJ, Suhm J, Mülhaupt R, Prosenc MH, Brintzinger HH (**1997**) *Macromolecules* 30:3164
11   Spaleck W, Küber F, Winter A, Rohrmann J, Bachmann B, Aulberg M, Dolle V, Paulus EF (**1994**) *Organometallics 13:954*
12   Minick L, Moet A, Hiltner A, Baer E, Chum SP (**1995**) *J Appl Polym Sci* 58:1371
13   Suhm J, Heinemann J, Thomann Y. Thomann R, Maier R.-D., Schleis T, Okuda J. Kressler J., Mülhaupt R (**1998**) *J. Mater Chem* 8:553
14   Mäder D, Heinemann J, Walter P, Mülhaupt R (**2000**) *Macromolecules* 33:1254
15   Mäder D, Thomann Y, Suhm J, Mülhaupt R (**1999**) *J. Appl. Poly. Sci.* 74:838
16   Wang WJ, Yan DJ, Zhu SP, Hamielec AE (**1998**) *Macromolecules* 31:8677
17   Wang WJ, Yan DJ, Charpentier PA, Zhu SP, Hamiliec AE, Sayer BG (**1998**) *Macromol Chem Phys* 199:2409
18   Malmberg A, Kokko E, Lehmus P, Lofgren B, Seppala JV (**1998**) *Macromolecules 31:8448*

19 Malmberg A, Liimatta J, Lehtinen A, Lofgren B (**1999**) *Macromolecules* 32:6687

20 Yan D, Wang WJ, Zhu S (**1999**) *Polymer* 40:1737

21 Beigzadeh D, Soares JBP, Duever TA (**1999**) *Macromol Rapid Commun* 20:541

22 Walter P. Trinkle S, Suhm J, Mäder D, Friedrich C. Mülhaupt R (**2000**), *Macromol Chem Phys* 201:604

23 Aaltonen P, Fink G, Lofrgren B, Seppala JV (**1996**) *Macromolecules* , 29: 5255

24 Heinemann J. Mülhaupt R. Brinkmann P, Luinstra G (**1999**) *Macromol Chem Phys* 200:384

25 Wang C, Friedrich SK, Younkin TR, Li RT, Grubbs RH, Bansleben DA, Day MW (**1998**) *Organometallics* 17:3149

26 Younkin TR, Conner EF, Hederson JI, Friedrich SK, Grubbs RH (**2000**) *Science* 287:460

27 Held A, Bauers FM, Mecking S (**2000**) *Chem Commun* (4):301

28 Tomov, Broyer JP, Spitz R (**2000**) *Macromol Symp* 150:53

29 Stehling UM, Malmstrom EE, Waymouth RM, Hawker CJ (**1998**) *Macromolecules* 31:4396

30 Baumert M, Heinemann J, Thomann R, Mülhaupt (**2000**) *Macromol Rapid Commun* 21:271

Heterogenized Bifunctional Catalysts in Olefin Polymerization

Evgeniya Mushina, Yury Podolsky, Vadim Frolov

Topchiev Institute of Petrochemical Synthesis Russian Academy of Sciences,
29 Leninsky pr., Moscow, 117912, Russia
E-mail: mushina@ips.ac.ru

Abstract. Investigation of the earlier developed bifunctional catalysts (BFC), which consist of the chromium containing compounds or titanium-magnesium catalyst, modified by nickel- (or zirconium-) aluminum oligodienyl complexes, revealed that in the gas phase polymerization of ethylene two processes take place simultaneously - oligomerization of ethylene and its co-polymerization with the oligomers formed *in situ*. The properties of the polyethylene (PE) obtained are close to the properties of linear medium density PE [1]. To continue these studies the bifunctional catalysts of new generation - bisupported catalytic systems - have been created and investigated. Similarly to BFC the bisupported catalytic systems contain two components (Cr- or Ti-containing compounds and complexes of Ni or Zr), but these components are deposited on different carriers. Two cases are possible: these carriers may be of different nature (SiO_2 and $MgCl_2$, for example) or two components of the catalyst may be deposited on two different portions of one and the same carrier followed by their mixing. The presence in one catalytic system of two active centers (Ti+Ni, Ti+Zr, Cr+Ni or Cr+Zr), deposited on different carriers, opens fundamental new and practically unbounded possibilities for controlling the micro- and macrostructure of the polymers during the polymerization process by changing nature of the components and their relative concentrations.

Introduction

Several approaches are used to construct heterogenized metal complex catalysts for polymerization of unsaturated hydrocarbons. The preparation of these catalysts is based on the process of immobilization of metal organic compounds at the surface of organic or inorganic carriers. As a rule this leads to the hybridization of properties of the homogenous and heterogeneous catalysts and to increasing their stability, activity and stereospecificity of action.

The main feature of the bifunctional catalysts in question is that they contain two different metal complex compounds deposited on the surface of inorganic carrier. The titanium-magnesium or chromium-containing catalyst is a first component of the catalyst, whereas oligodienyl complex of nickel or zirconium is used as a second one.

Experimental Methods

The procedure of preparation of the catalyst involves the synthesis of the components and their co-deposition on the surface of dehydrated microspheric silica gel.

The titanium-magnesium catalyst had been prepared by interaction of titanium chloride with freshly prepared magnesium chloride.

As chromium-containing compound the triphenylsilyl chromate on the silica gel treated with diethylaluminum ethoxide had been used.

The second component of the catalytic system – an oligodienyl complex of nickel or zirconium - was synthesized by exchange reaction between the corresponding metal chloride and $Al(i\text{-}C_4H_9)_3$ in the presence of 1,2- or 1,3-dienes in aliphatic or aromatic hydrocarbons used as solvents. The components ratio was close to the stoichiometric one.

The formation of the complex is a result of the insertion of a diene across the metal-hydrogen bond of the hydride complex of the metal. The latter one is the product of decomposition of a metal organic compound according to the mechanism of β-elimination of hydrogen.

The structure and composition of the complexes obtained have been studied by IR-spectroscopy and chromato-mass spectrometry of their decomposition products.

It has been shown that in the presence of a diene the exchange reaction between the metal chloride and triisobutyl aluminum results in the formation of π-alkenyl structure M-C and a bridged structure formed from a metal chloride and diisobutylaluminum chloride [1].

According to the results of chromato-mass analysis the oligodienyl complexes are the blends of metal complexes, containing the oligomer ligands with different molecular masses, the main part of oligomers being di-, tri- and tetramers of isoprene.

Results and Discussion

The bifunctional catalysts have been tested in polymerization of various monomers – ethylene, dienes, higher olefins (Table 1)

The presence of two active centers on the surface of silica gel makes it possible to combine different catalytic functions when using such catalytic systems in polymerization processes.

Table 1. Bifunctional Catalysts in Polymerization of Various Monomers.[*]

Catalyst	Monomer	Polymer
TiCl$_4$/MgCl$_2$/SiO$_2$ + MOC	C$_2$H$_4$	Linear low density polyethylene
The same	C$_2$H$_4$ + C$_4$H$_8$	Medium density polyethylene
The same	C$_4$H$_6$	Blend of 1,4-*trans*- and 1,4-*cis*-polybutediene
The same	C$_5$H$_8$	Blend of 1,4-*trans*- and 1,4-*cis*-polyisoprene
TiCl$_4$/MgCl$_2$/SiO$_2$ + ZrOC	4-MP-1	Poly--4-MP-1 with *iso*-specificity – 85%
TiCl$_4$/MgCl$_2$/SiO$_2$	C$_4$H$_6$	1,4-*trans*- polybutediene
The same	C$_5$H$_8$	1,4-*trans*- polyisoprene

[*] *MOC*, oligodienyl complex of Zr or Ni; *ZrOC*, oligodienyl complex of Zr; *4-MP-1*, 4-methylpentene-1.

Particularly, in the case of ethylene it had appeared that bifunctional catalytic systems could carry out simultaneously oligomerization of ethylene and its co-polymerization with oligomers formed *in situ*. Due to these processes some compositions of the catalyst produce, in contrast to usual practice, the linear medium density polyethylene without using co-monomers.

In case of dienes the bifunctional catalysts display the ability to combine such catalytic functions as *cis*- and *trans*- polymerization.

When these catalysts have been studied in polymerization of higher π-olefins the increase of stereo specificity of action was observed.

An increase of the metal ratio in the catalyst (zirconium to titanium or nickel to titanium) enhances the branching of polyethylene macromolecules (Table 2) with the corresponding decrease of the density.

Table 2. The Structure of Polyethylene

Catalyst	Metal rratio (mol/mol)	Branching and nonsaturation per 1000 atoms C		
		-CH$_3$	-C$_2$H$_5$	-CH=CH$_2$
TiCl$_4$/MgCl$_2$/SO$_2$ + ZrOC	Zr/Ti=0.5	1.1	0.18	0.08
	Zr/Ti=1.0	1.7	0.14	0.06
	Zr/Ti=2.0	5.9	1.30	0.01
TiCl$_4$/MgCl$_2$/SO$_2$ + NiOC	Ni/Ti=0.5	1.4	0.80	0.02
	Ni/Ti=2.5	2.7	1.20	0.08
Commercial catalyst S2 + NiOC	Ni/Cr=0.2	0.5	0.23	0.27
	Ni/Cr=1.0	1.7	0.33	0.28
Commercial catalyst S2 +ZrOC	Zr/Cr=0.8	1.8	0.36	0.52
	Zr/Cr=1.6	3.8	0.58	0.65

The elasticity of the polymer, estimated by the value of relative elongation at rupture, remains on the level of seven to eight hundred (Table 3).

The polyethylene obtained using titanium-magnesium containing catalyst is characterized by relatively narrow molecular-mass distribution (M_w/M_n=4-7), while usage of chromium-containing catalyst leads to formation of polymer with broad distribution (M_w/M_n=18-25) (Table 4).

Table 3. Activity of Catalysts and Mechanical Properties of Polyethylene.[*]

Catalyst	Activity[a]	Mechanical Properties			
		Melt index	Elongation, (%)	E (MPa)	Tensile Strength, (MPa)
S2	8	0.6	750	420	30
S2+MOC	60	0.6	890	560	30
$TiCl_4/MgCl_2$ + + MOC+AlR$_3$	130	0.4	740	490	36

[*] S2, Chromium-containing industrial catalyst; MOC, oligodienyl complex of Zr or Ni.
[a] Activity of catalysts is expressed in kg of polymer per g of catalyst per hour.

Table 4. Molecular-Weight Characteristics of PE.[*]

Catalytic System	Molecular Weights		M_w/M_n
	M_n	M_w	
Ti/Mg+Ni	61 000	435 000	7.1
Ti/Mg+Zr	159 000	663 000	4.1
Cr + Ni	18 800	380 000	20,4
Cr + Ni	15 900	382 000	24,0
Cr + Zr	17 100	422 000	24,7

[*] Ni and Zr, oligodienyl complexes of Ni and Zr; Cr, triphenylsilylchromate

It is well known that when the immobilized catalysts are prepared, the amount of metal complexes, which can be chemisorbed on the silica gel surface, depends on concentration of hydroxyl groups on the surface.

In the case of titanium-magnesium supported catalyst the oligodienyl metal complexes can interact either with free hydroxyl groups of silica gel, or with the titanium-magnesium component, already anchored on silica gel (Fig. 1, scheme I and II). This is why the sequence of deposition of the catalyst's components is of fundamental importance and changing of this sequence leads to products with different properties.

The catalysts with different properties could be prepared also if two components of the catalyst are deposited on different carriers or on two different portions of the same carrier (Fig. 1, scheme I and III).

The presence in one catalytic system of two active centers, deposited on different carriers, opens new possibilities for controlling micro- and macrostructure of the polymers.

To illustrate this statement, the comparison of two catalysts in ethylene polymerization has been performed. One of the catalysts was titanium-magnesium catalyst, deposited on the surface of silica gel and then treated with nickel oligodienyl complex (Fig. 1, scheme II). It appeared that the maximum activity of this catalyst lies in a very narrow range of nickel to titanium molar ratio – between 0.5 and 1 (Fig. 2).

The attempts to regulate the molecular weight of the polymer by changing hydrogen concentration proved to be unsuccessful: an increase of hydrogen content in gas phase leads to a drastically decrease of catalyst's activity.

Quite different was the behavior of another catalyst – the blend of two portions of silica gel, one of which was treated with nickel oligodienyl complex, and the other was used as a carrier for titanium-magnesium catalyst (Fig. 1, scheme III). In this case, bi-supported catalysts have been obtained, in which the molar ratio Ni/Ti may vary in the range from 1 to 6 mol/mol practically without decrease of the catalyst's activity (Fig. 3). Moreover, it was shown that in the case of bi-supported

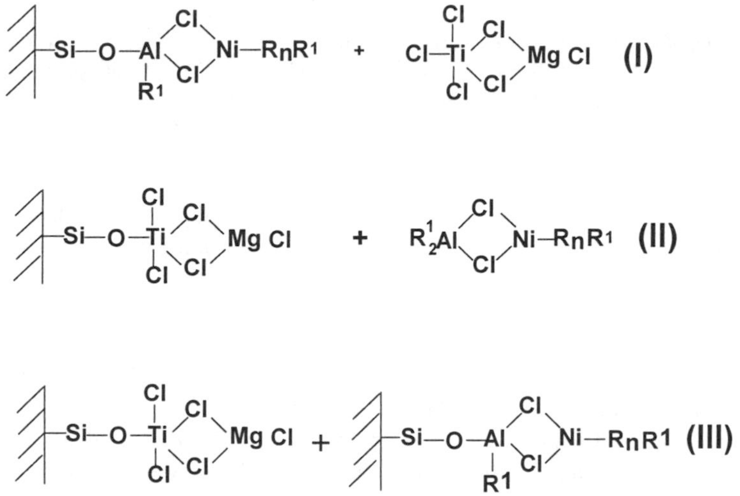

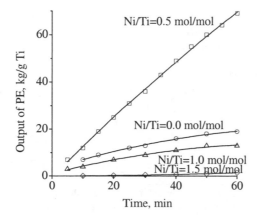

Fig. 1. Various methods for preparation of bifunctional catalysts

Fig. 2. Kinetics of ethylene polymerization (concentration of H_2 – 30%, Al/Ti=100 mol mol^{-1}; catalyst – $TiCl_4/MgCl_2/SiO_2$ + NiOC + AlR_2Cl)

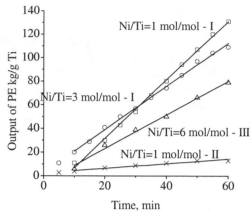

Fig. 3. Kinetics of ethylene polymerization (concentration of $H_2 - 30\%$, Al/Ti = 100 mol mol^{-1}, catalysts: I – NiOC/SiO$_2$ + TiCl$_4$/MgCl$_2$ + AlR$_2$Cl; II - TiCl$_4$/MgCl$_2$/SiO$_2$ + NiOC + AlR$_2$Cl; III - NiOC/SiO$_2$ + TiCl$_4$/MgCl$_2$/SiO$_2$ + AlR$_2$Cl.

catalysts the rate of polymerization does not decrease with increasing of hydrogen concentration (from 10 to 30 vol.%), meanwhile the melt index changed from 0.45 to 5.80 g/10 min.

That makes it possible to vary in wide range the melting index and molecular weight of the polymer obtained.

Some interesting results have been obtained when using the bifunctional catalysts in the polymerization of higher α-olefins.

Using the 4-methyl-pentene-1, as an example, it was shown that even in the absence of inner and outer electron donors the polymer with 85% of *iso*-specificity could be prepared (Table 5).

Under these conditions the consumption of aluminum-organic compound is less than in usual cases and the yield of polymer is higher.

Table 5. The Polymerization of 4-Methyl-1-pentene.[*]

Catalytic System	Concentration of		Al/Ti Ratio	Activity[a]	Isospeci-city
	Monomer (mol l^{-1})	Ti (g l^{-1}*10^{-5})	(mol mol^{-1})		(%)
TiCl$_4$/MgCl$_2$+AlR$_3$	8	6	800	53	41
TiCl$_4$/MgCl$_2$+AlR$_3$ +D1	8	6	800	62	58
TiCl$_4$/MgCl$_2$+AlR$_3$ +D1+D2	8	6	800	32	96
TiCl$_4$/MgCl$_2$/SiO$_2$ +ZrOC+AlR$_3$	2	3	80	138	85

[*] *ZrOC*, oligodienyl complex of Zr; *D1, D2,* inner and outer electron donors
[a] Activity of catalysts is expressed in kg of polymer per g of Ti per hour.

The usage of bifunctional catalysts in the polymerization of dienes leads to formation of both *cis-* and *trans* polybutadiene. In table 6 the microstructure of the obtained polybutadiene samples is shown. It is seen that by changing the composition of the catalyst and sequence of deposition of its components practically any ratio of *cis-* to *trans-*isomers of polybutadiene could be achieved.

Table 6. Polymerization of Butadiene on the Modified Titanium-Magnesium Catalyst (Temperature 50°C)

Catalysts (sequence of deposition)	Composition of Catalysts mol mol^{-1}		Conversion of butadiene	Microstructure of Polymer (%)		
	Ni/Ti	Al/Ti	(%)	1,4 -cis	1,4- trans	1,2-
SiO$_2$ + AlOC + Ti/Mg	-	1	97	2	93	5
SiO$_2$ + Ti/Mg + NiOC	1	20	79	28	66	6
SiO$_2$ + Ti/Mg + NiOC[a]	1	20	89	88	8	4
SiO$_2$ + NiOC + Ti/Mg	1	20	98	72	19	9
SiO$_2$ + NiOC + Ti/Mg	1	-	96	90	8	2
SiO$_2$ + NiOC + Ti/Mg	2	-	95	96	2	2
SiO$_2$ + NiOC + Ti/Mg	2	20	80	96	2	2

[a] temperature 20°C

The results of fractionation of the polymers and DSC data evidence that the polymers obtained are the blends of *cis* - and *trans-* isomers, but not the polymers with mixed structure.

According to some patent data, the addition of *trans-* isomer to butadiene rubber leads to considerable improvement of mechanical properties of vulcanized rubber.

In addition to the bifunctional catalytic systems their individual components have been studied as well.

For the first time it was shown [2,3] that Ti-Mg catalysts prepared according to our method in combination with triisobutylaluminum are very effective for *trans-*polymerization of butadiene and isoprene (Table 7), without contribution of oligomerization.

These catalysts are active in wide range of aluminum to titanium ratio (from 10 to 100). This indicates the high stability of the centers, which are responsible for *trans-*polymerization, and essentially differentiate the catalysts from the homogeneous catalyst TiCl$_4$+AlR$_3$.

Table 7. Microstructure of Polymers; Catalyst: TiCl$_4$/MgCl$_2$/SiO$_2$+AlR$_3$.

Polymer	Al/Ti (mol mol^{-1})	Temperature (°C)	Microstructure		
			cis-1,4	*trans*-1,4	1,2-[a]
Polybutadiene	30	50	6	88	6
Polybutadiene	100	50	5	92	3
Polybutadiene	20	50	2	93	5
Polybutadiene	30	30	2	93	5
Polyisoprene	30	30	2	93	5

[a], for polyisoprene – 3,4-

Some tests have been performed on co-polymerization of butadiene and isoprene with the aim of preparation of non-crystalline *trans*-copolymer. The microstructure, crystallinity and transition temperatures have been determined by IR-spectroscopy, X-ray analysis and DSC method respectively (Table 8).

Table 8. Butadiene - Isoprene Co-polymerization on Ti-Mg catalyst +AlR$_3$ (temperature 50°C)

Run	1	2	3	4
Composition of the monomer blend, wt.%:				
butadiene	100	80	60	
isoprene	-	20	40	100
Conversion of monomer, %	100	99	100	100
Composition of polymer, wt.%:	100	-	-	-
butadien			55	-
isoprene			45	100
Microstructure of -polymer, wt.%:				
butadiene part:				
1,4-*trans*	92	88	88	-
1,4-*cis*	5	8	4	-
1,2-	3	4	8	-
isoprene part:	-			
1,4-*trans*	-	95	97	97
3,4-	-	5	3	3
Crystallinity, %	60	25	7	81
Phase Transition temperatures, °C:				-
First transition (crystalline phase - mesophase)	72	42	32	-
Second transition (melting)	147	124	110	62

As it can be seen, the change in the composition of the monomer blend does not affect the microstructure of polymers obtained - the content of *trans*-units remains close to 90%. At the same time, the increase of isoprene units incorporated into polybutadiene chains leads to a drastically decrease of crystallinity and transition temperatures, indicating the formation of random copolymers. Thus these catalysts allow preparing both the crystalline and amorphous *trans*-copolymers.

Acknowledgements

The authors thank Prof E. Tinyakova, Prof. M. Gabutdinov, V. Cherevin, and C. Medvedeva for helpful discussions and Drs I. Gavrilenko, Yu. Bit-Gevorgizov, A Vakhbreit and S. Solodyankin for experimental assistance.

References

1   Mushina E et al (1998) Metal complex bicenter catalysts for olefin and diene polymerization. *Applied Catalysis A: General* **166**: 153-164.
2   Mushina E et al (1996) Polymerization of dienes with titaniummagnesium catalysts. *Polymer science* **38**: 270-273.
3   Antipov E et al (1997) Peculiarities of structure of some polydienes and polyolefins prepared with supported metal complex catalysts. *Polymer science* **39**: 430-439.

Long-Chain Branched Polyethene via Metallocene-Catalysis: Comparison of Catalysts

Esa Kokko*, Petri Lehmus[1], Anneli Malmberg[1], Barbro Löfgren, and Jukka V. Seppälä

Helsinki University of Technology, Department of Chemical Technology, P.O. Box 6100, FIN-02015 HUT, Finland
[1] Present Address: Borealis Polymers, P.O.Box 330, FIN-06101 Porvoo, Finland
E-mail: kokko@polte.hut.fi

Abstract. Metallocene catalysts have enabled the production of long-chain branched (LCB) polyethene at low pressure and temperature. The assumed LCB mechanism for the branched structure is the copolymerization of vinyl terminated macromonomers with ethene. In order to obtain LCB polyethene effectively, the employed metallocene-catalyst should be able to produce polyethene with vinyl terminals and effectively copolymerize the formed macromonomers with ethene. We present results of our recent investigation in which we have compared the properties of polyethenes polymerized with five conventional metallocene catalysts activated with methylaluminoxane (MAO); $Et[Ind]_2ZrCl_2$, $Et[H_4Ind]_2ZrCl_2$, $(n\text{-}BuCp)_2ZrCl_2$, $Me_2Si[Ind]_2ZrCl_2$ and Cp_2ZrCl_2. We have examined and discussed the relation between chain transfer mechanisms, hydrogen effect, copolymerization abilities, and rheological behavior of the polyethenes.

1 Introduction

The properties of polyethene are strongly influenced by its molecular weight, molecular weight distribution (M_w/M_n), and branching. The effect of branching on the properties of polyethene depends on the length and the amount of the branches. Short-chain branches (SCB), of less than approximately 40 carbon atoms, interfere with the formation of the crystal structure. SCB mainly influences the mechanical and thermal properties. As the branch length increases, the branches are able to form lamellar crystals of their own and the influence on the mechanical and thermal properties is diminished. The longer branches – whose length is comparable or longer than the critical entanglement distance of a linear polymer chain – have a tremendous effect on the melt rheological behavior. Even very small quantities of LCB alter the polymer processing properties significantly.

Conventional high-pressure low-density polyethene (PE-LD) grades have a broad M_w/M_n and the polymers contain LCB. This structure makes PE-LD easy to process. Though, the broad M_w/M_n causes the mechanical properties of PE-LD to be inferior to those linear low-density polyethene (PE-LLD) grades which have narrower M_w/M_n.

The highly active metallocene catalysts for the olefin polymerization [1,2] have enabled the production of linear, narrow M_w/M_n polyethenes with greatly improved mechanical properties. The processability of these polymers, however, is more difficult when compared to the grades with the broad M_w/M_n and LCB. The introduction of LCB in the metallocene-catalyzed polyethenes can help to improve the poor processability in some applications. [3] The combination of the good mechanical properties of metallocene-catalyzed polyethenes with the good processability of PE-LD is an obvious target. The catalytic approach to LCB polyethene has been met with great interest during the last years. [4-10]

In this work the formation of LCB with various metallocenes was studied. The work had three objectives: 1) An examination of polymerization behavior of different metallocene catalysts, 2) evaluation of the influence of the catalyst on LCB as seen in melt rheological behavior, and 3) evaluation the effect of polymerization conditions on LCB.

2   Polymerization Behavior of the Studied Catalysts

Long-chain branching in metallocene-catalyzed polymerizations is believed to take place via a copolymerization reaction in which a vinyl-terminated macromonomer is incorporated into a growing polyethene chain. [7-11] According to this LCB formation mechanism two factors are of primary interest; chain transfer mechanisms and comonomer response of metallocene catalysts. The structures of the studied metallocene catalysts are shown in Fig. 1.

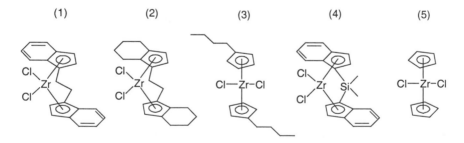

Fig. 1. The catalysts used in this work: (**1**) Et[Ind]$_2$ZrCl$_2$; (**2**) Et[H$_4$Ind]$_2$ZrCl$_2$; (**3**) (n-BuCp)$_2$ZrCl$_2$; (**4**) Me$_2$Si[Ind]$_2$ZrCl$_2$; (**5**) Cp$_2$ZrCl$_2$.

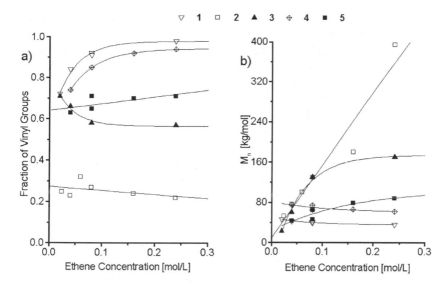

Fig. 2. a) The fraction of the vinyl end-groups of total unsaturations. **b)** Number average molecular weight of polyethenes produced with the studied catalysts.

2.1   Chain Transfer Mechanisms

Chain transfer to monomer, β-H elimination, and σ-bond metathesis result in a vinyl end-group in polyethene. [1,2,12] An isomerization reaction related to the chain transfer has been proposed [13] to be the origin of *trans*-vinylene end-groups found in metallocene-catalyzed polyethenes. In copolymerization, chain transfer after 1,2-insertion of comonomer results in a vinylidene unsaturation and chain transfer after 2,1-insertion results in an internal *trans*-vinylene. Chain transfer to an external chain transfer agent, i.e. hydrogen or aluminum results in a saturated end-group.

The studied catalysts appeared to produce polyethene with approximately one unsaturation in each chain. This suggested that chain transfer to MAO was negligible. Chain transfer to aluminum may have been present to some degree with 3/MAO and 5/MAO. The unsaturations consisted mainly of vinyl end-groups and *trans*-vinylene double bonds. In addition, a very small amount of vinylidene double bonds was found.

Fig. 2a shows the vinyl end-group fractions of the polyethenes produced with the studied catalysts. The remaining fraction of unsaturations in homopolyethene consisted almost exclusively of *trans*-vinylenes. Indenyl-ligand substituted **1** and **4** produced polyethene with the highest selectivity towards vinyl end-groups. Furthermore, the vinyl end-group selectivity decreased and *trans*-vinylene increased as ethene concentration (C_E) was decreased and the M_n was almost constant at the studied C_E range (Fig. 2b). These results indicate that chain transfer to a coordinated monomer was the dominating chain transfer mechanism. The same chain transfer mechanism has been observed for siloxy-substituted Et[Ind]$_2$ZrCl$_2$ derivatives. [14]

Polyethenes produced with **2** had the lowest vinyl selectivity. The selectivity was independent of C_E but M_n increased proportionally to C_E. Based on the end-group and GPC analysis, β-H elimination was highly favored with **2**. A moderate vinyl selectivity was obtained in **3** and **5** –catalyzed polymers. Also, C_E did not seem to have a clear influence on the end-group selectivity but did have a notable impact on M_n. We conclude that for **3** and **5** β-H elimination was favored at lower C_E but chain transfer to the monomer seemed to be more favored at higher C_E.

Hydrogen effect. The studied catalysts had marked differences in their sensitivity towards hydrogen insertion and in the resulting chain transfer. Table 3 shows the hydrogen sensitivity of the studied catalysts. The introduction of very small amounts of hydrogen resulted for each catalyst in a polymer structure without *trans*-vinylene double bonds. The vinyl bonds were also saturated to a large extent when **2**, **3**, or **5** were used and the molecular weight decreased drastically. This difference in the hydrogen insertion sensitivity was attributed to the dominance of different chain transfer mechanisms. [9]

Table 1. The influence of hydrogen on the molecular properties of homopolyethene. Polymerization conditions: T = 80°C; [C$_2$H$_4$] = 0.08 mol/L.

Catalyst	H$_2$ Feed	GPC analysis			End-Group Analysis by FTIR. C=C / chain.[a]		
		M_w	M_n	M_w/M_n	*trans-*		
	[mmol]	[kg/mol]	[kg/mol]		Vinylene	Vinyl	Vinylidene
1	0.0	86	39	2.2	0.07	0.92	0.01
1	1.3	46	25	1.9	0.00	0.6	<0.03
2	0.0[b]	1,000	400	2.5	0.76	0.22	0.02
2	0.4[b]	69	34	1.9	0.00	<0.05	<0.05
3	0.0	290	130	2.3	0.39	0.58	0.03
3	0.5	17	8	2.1	0.00	<0.05	<0.05
4	0.0	167	65	2.6	0.10	0.85	0.05
4	1.3	91	37	2.4	0.01	0.6	<0.05
5	0.0	112	46	2.4	0.29	0.65	0.06
5	1.3	18	9	2.0	0.00	<0.2	<0.05

[a] Each chain was assumed to contain one C=C bond when no hydrogen was used. When hydrogen was used, the amount of C=C bonds per chain was calculated from the reduction of M_n.
[b] [C$_2$H$_4$] = 0.24 mol/L.

2.2   Comonomer Response

Copolymerization abilities of the studied catalysts can be used to predict the reactivity of very long 1-olefins. In this work, ethene-co-1-hexene copolymers were polymerized and the reactivity ratios of the components were estimated. Already small variations in the catalyst ligand structures were found to cause rather significant differences in comonomer reactivities.

Table 2 shows the observed influence of 1-hexene on the properties of polyethene produced with different metallocene catalysts. At the same comonomer content, M_n decreased and end-group types of copolymers produced with **2** and **5** changed dramatically compared to **1** and **4**. A similar sensitivity to chain transfer after comonomer insertion was observed for **3** compared to **1** when 1-hexadecene was used as a comonomer. [9] The different sensitivities of the catalysts towards chain transfer after comonomer insertion were attributed to the dominance of the different chain transfer mechanisms; β-H elimination appears to facilitate chain transfer reaction after 1,2-insertion of comonomer. [9]

Table 3 shows the calculated reactivity ratios. The bridged indenyl-ligand substituted metallocenes **1** and **4** had the highest comonomer responses. The tetrahydroindenyl ligand substituted **2** had lower and the unbridged metallocenes **3** and **5** had the lowest comonomer responses.

Table 2. Influence of 1-hexene concentration on the molecular weight and end-group types of polyethene. Polymerization conditions: T = 80°C; $[C_2H_4] = 0.08$ mol/L.

Catalyst	$[C_6H_{12}]$ in Feed	Comonomer Content	GPC Analysis		End-Group Analysis by FTIR. C=C / chain.		
			M_w	M_n	trans-Vinylene	Vinyl	Vinylidene
	[mol/L]	[mol-%]	[kg/mol]	[kg/mol]			
1	0.00	0.0%	71	33	0.07	0.91	0.02
1	0.09	2.5%	60	28	0.26	0.67	0.07
1	0.19	4.6%	56	28	0.37	0.47	0.16
1	0.38	8.4%	50	25	0.42	0.31	0.28
2	0.00	0.0%	220	93	0.73	0.21	0.06
2	0.09	1.6%	146	72	0.53	0.12	0.35
2	0.19	3.0%	100	49	0.36	0.07	0.57
2	0.38	5.8%	53	29	0.28	0.03	0.69
4	0.00	0.0%	167	65	0.10	0.85	0.05
4	0.09	4.2%	86	45	0.42	0.38	0.20
4	0.19	7.0%	73	38	0.47	0.25	0.28
4	0.38	12.2%	60	31	0.50	0.13	0.37
5	0.00	0.0%	112	46	0.29	0.65	0.06
5	0.09	1.2%	63	27	0.16	0.31	0.53
5	0.19	1.9%	70	33	0.18	0.28	0.54
5	0.38	3.7%	38	17	0.10	0.12	0.77

Table 3. Calculated monomer reactivity ratios of the studied catalysts at 80°C.[a]

Catalyst	1	2	3	4	5
r_{ethene}	48±4	71±4		26±2	112±9
$r_{1\text{-hexene}}$	<0.02	<0.01		<0.05	<0.01
r_{ethene}	51±7[b]		160±13[b]		
$r_{1\text{-hexadecene}}$	<0.01				

[a] Based on monomer concentrations in the liquid phase.
[b] From ref. [9]

3   Influence of the Catalyst on Long-Chain Branching

LCB content of one branch point per 10,000 carbon atoms is enough to affect the processing properties of polyethene. At this concentration level LCB is difficult to detect by ^{13}C NMR spectroscopy or multidetector GPC. The most convenient method for detecting such a low LCB content is the use of rheological measurements, which are very sensitive in detecting differences in the molecular structure of polymers. Molecular weight, M_w/M_n, and LCB greatly influence the melt properties including melt viscosity, melt elasticity, and temperature dependence of melt viscosity. [7,15-19]

The differences in the vinyl end-group selectivity (Fig. 2a) and in the copolymerization ability (Table 2 and 3) suggested that **1** and **4** were more suitable catalysts for producing LCB polyethene than **2**, **3**, or **5** on the basis of the suggested LCB mechanism. It was found that catalyst, monomer, and hydrogen concentration had a major influence on LCB in produced polyethenes.

3.1   Catalyst Comparison

Fig. 3a shows melt viscosity curves of polyethenes produced with the selected catalysts. The measured complex viscosity (η^*) values of polyethenes produced with **1** and **4** were markedly higher than theoretical η_0 values [20] for linear polyethenes of similar M_w. The discrepancy was smaller for a **2**–PE and **5**–PE and the theoretical value matched very closely for a **3**–PE. The observed melt viscosity behavior for the **1**–PE and **4**–PE can readily be explained with a small content of LCB, which enhances melt viscosity at low shear rates. [16]

Magnitude of the elastic response, evaluated as G' (storage modulus) at constant G" (loss modulus) is affected by M_w/M_n and LCB. In these polymers, the differences in M_w/M_n were insignificant (M_w/M_n 2.2 - 2.5) and the differences in G' values at low frequency region, can be used in evaluating the presence of LCB.

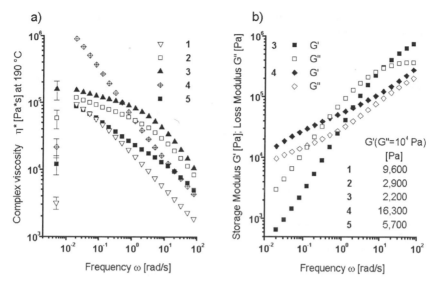

Fig. 3. a) Complex viscosity curves of homopolyethenes produced at $[C_2H_4]$ = 0.08 M. Single points shown at the left side of the graph are theoretical zero-shear viscosity (η_0) values expected for linear polymers of that M_w; η_0 was calculated using equation η_0 = $3.4 \times 10^{-15} \times M_w^{3.6}$ (Pa×s). [20] **b)** Dynamic modulus values of **3**–PE and **4**–PE and G'(G"$_{ref}$) of all polyethenes.

Fig. 3b shows the frequency dependency of the dynamic moduli for **3**–PE and **4**–PE. Catalyst **3** appeared to produce very linear polyethene in these polymerization conditions. This conclusion is based on the shape of the modulus curves being in line with the narrow M_w/M_n, the closely matched η^* and η_0 values (Fig. 3a), and the low E_a=29 kJ/mol. [9] In contrast, the dynamic modulus curves of **4**–PE (Fig. 3b) revealed the behavior of a network-like structure resulting from LCB. Similarly, **1**–PE had markedly enhanced G'(G"$_{ref}$) value.

3.2   Ethene Concentration Effect

In our earlier paper [9] we showed that decreasing C_E increased branching in **1** catalyzed polyethenes as evaluated by elevated E_a, contradiction of measured M_w, M_w/M_n and the η^*(0.02 rad/s), and increased G'(G"$_{ref}$) values. Also, an increase in polymerization time increased branching. Polymers produced with **2** and **3** had η^* values indicating a deviation from linearity only at low C_E.

Fig. 4 shows η^* curves of polyethenes produced with **5**. An increase in C_E resulted in an increase in the M_w. However, η^*(0.02 rad/s) was the highest for a polymer produced at C_E = 0.08 M. In spite of the increase in the M_w for a polymer produced at C_E = 0.16 M, η^*(0.02 rad/s) was lower for this polymer. The peculiarity can be explained with the decreasing LCB content with increasing C_E, as observed earlier for **1**.

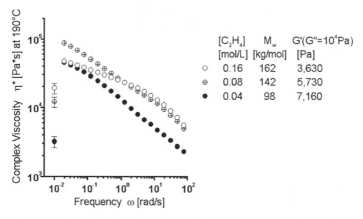

Fig. 4. Complex viscosity curves of polyethenes produced by **5** varying the ethene concentration.

3.3   Hydrogen Effect on LCB

Table 4 shows the effect of hydrogen on the melt rheological properties. The introduction of hydrogen suppressed LCB formation. For the polymer polymerized with **1** at $C_E = 0.08$ M the introduction of hydrogen decreased the values of E_a, G', and decreased the discrepancy between the theoretical and measured viscosity. The values obtained for **2**–PE indicate that the structure of this polymer was very linear with neither LCB nor high M_w tail. These findings were in accordance with the observed hydrogen effect on the end-group types (Table 1) and are in line with the assumed branching mechanism.

Table 4. Hydrogen effect on the melt rheological properties of polyethene. From ref. [9]

Catalyst	[C₂H₄]	H₂ Feed	M_w	Theoretical η_0[a]	$\eta*(0.02$ rad/s) at 190°C	G' at G"(10⁴ Pa)	E_a
	[mol/L]	[mmol]	[kg/mol]	[Pa×s]	[Pa×s]	[Pa]	[kJ/mol]
1	0.24	0.0	70	900	2,080	2,300	30
1	0.24	1.3	59	500	980	1,800	n.m.
1	0.08	0.0	98	3,200	94,000	9,600	42
1	0.08	0.5	65	700	5,930	5,400	38
2	0.24	0.4	69	900	1,200	1,200	27

[a] Theoretical η_0 was calculated using equation $\eta_0 = 3.4 \times 10^{-15} \times M_w^{3.6}$ [Pa×s]. [20]

4 Conclusions

The studied indenyl-ligand substituted metallocenes, Et[Ind]$_2$ZrCl$_2$ and Me$_2$Si[Ind]$_2$ZrCl$_2$ produce polyethene with a high selectivity towards vinyl terminations. Selectivity is reduced by decreasing the ethene concentration. Chain transfer to monomer is the dominating chain transfer mechanism. Cp$_2$ZrCl$_2$ and (n-BuCp)$_2$ZrCl$_2$ produce polyethene with a lower selectivity towards vinyl terminals. β-H elimination dominates at low monomer concentration and chain transfer to monomer seems to dominate at higher monomer concentration. Et[H$_4$Ind]$_2$ZrCl$_2$ produces polyethene with the lowest vinyl selectivity. β-H elimination is the prevailing chain transfer mechanism with this catalyst system in the studied monomer concentration range at 80°C.

Of the studied catalysts, Me$_2$Si[Ind]$_2$ZrCl$_2$ and Et[Ind]$_2$ZrCl$_2$ produce polyethenes with significantly modified rheological properties due to long-chain branching. At low monomer concentration, Cp$_2$ZrCl$_2$ produces slightly more branched polyethene than Et[H$_4$Ind]$_2$ZrCl$_2$ and (n-BuCp)$_2$ZrCl$_2$. The main factors that seem to influence the branching probability are the vinyl end-group selectivity and the copolymerization ability. In addition, the properties of the polymer slurry may influence the formation of long-chain branching. We shall investigate that in future.

The amount of long-chain branching can be increased by decreasing the ethene concentration. The introduction of hydrogen suppresses the branching. The findings support the view that long-chain branching takes place via the copolymerization reaction: i) The catalyst produces a vinyl terminated polyethene macromonomer and ii) reincorporates it into a growing polyethene chain.

5 Experimental

Materials. Metallocene catalysts, rac-[ethylenebis(indenyl)]zirconium dichloride (**1**), rac-[ethylenebis(4,5,6,7-tetrahydroindenyl)]zirconium dichloride (**2**), bis(n-butylcyclopentadienyl)zirconium dichloride (**3**), rac-[dimethylsilylbis-(indenyl)]zirconium dichloride (**4**), and bis(cyclopentadienyl)zirconium dichloride (**5**), and cocatalyst methylaluminoxane (10 w-%/w MAO in toluene) were obtained from Witco. Other materials used, handling, and purification procedures have been reported elsewhere. [9,14]

Polymerization. All polymerizations were carried out in a 0.5 dm^3 Büchi stainless steel autoclave operated semibatchwise at 80°C. The polymerization procedures for using **4** and **5** were identical as for **1**, **2**, and **3**, which have been reported earlier. [9] In a typical polymerization, 0.1-0.6 μmol of catalyst (depending on the polymerization rate) and 1.0 mmol MAO were dissolved in 330 mL of toluene. Polymerization rates and yields were kept low in order to avoid mass-transfer limitations.

Characterization. Molecular weights were determined by GPC (Waters 150C). End-group analysis was carried out using FTIR (Nicolet Magna 750) and comonomer content by ^{13}C NMR spectroscopy (Varian Gemini 2000). LCB in polymers was measured using a stress-controlled dynamic rheometer (Rheometric Scientific SR-500) using plate-plate geometry. The details of analysis are given elsewhere. [7,9,14]

Acknowledgments.

The authors wish to thank the National Technology Agency (TEKES) and the Neste Foundation for the financial support and Mr. Jari Koivunen for the assistance in polymerizations.

References

1   Sinn H, Kaminsky W (**1980**) *Adv Organomet Chem* 18: 99-149
2   Brintzinger H-H, Fischer D, Mülhaupt R, Rieger B, Waymouth RM (**1995**) *Angew Chem Int Ed Engl* 34: 1143-1170
3   Kim YS, Chung CI, Lai S-Y, Hyun KS (**1996***) J Appl Polym Sci* 59: 125-137
4   Hamielec AE, Soares JBP (**1996**) *Prog Polym Sci* 21: 651-706
5   Lai S-Y, Wilson JR, Knight GW, Stevens JC, Chum P-WS, US Patent 5,272,236, Dec. 21, **1993**
6   Brant P, Canich JAM, Dias AJ, Bamberger RL, Licciardi GF, Henrichs PM, Int Pat Appl WO 94/07930, 24 April **1994**
7   Malmberg A, Kokko E, Lehmus P, Löfgren B, Seppälä JV (**1998**) *Macromolecules* 31: 8448-8454.
8   Wang W-J Yan D, Zhu S, Hamielec AE (**1998**) *Macromolecules* 31: 8677-8683
9   Kokko E, Malmberg A, Lehmus P, Löfgren B, Seppälä JV (**2000**) *J Polym Sci Part A Polym Chem* 38: 376-388
10 Kolodka E, Wang W-J, Charpentier PA, Zhu S, Hamielec AE (**2000**) *Polymer* 41: 3985-3991
11 Shiono T, Moriki Y, Soga K (**1995**) *Macromol Symp* 97: 161-170
12 Siedle AR, Lamanna WM, Newmark RA, Schroepfer JN (**1998**) *J Mol Catal A Chem* 28: 257-271
13 Thorshaug K, Støvneng JA, Rytter E, Ystenes M (**1998**) *Macromolecules* 31: 7149-7165
14 Lehmus P, Kokko E, Härkki O, Leino R, Luttikhedde H, Näsman J, Seppälä JV (**1999**) *Macromolecules* 32: 3547-3552
15 Harrell ER, Nakajima N (**1984**) *J Appl Polym Sci* 29: 995-1010
16 Bersted BH (**1985**) *J Appl Polym Sci* 30: 3751-3765
17 Carella JM, Gotro JT, Graessley WW (**1986**) *Macromolecules* 19: 659-667
18 Vega JF, Santamaría A, Muñoz-Escalona A, Lafuente P (**1998**) *Macromolecules* 31: 3639-3647

19 Malmberg A, Liimatta J, Lehtinen A, Löfgren B (**1999**) *Macromolecules* 32: 6687-6696

20 Raju VR, Smith GG, Marin G, Knox JR, Graessley WW (**1979**) *J Polym Sci Polym Phys Ed* 17: 1183-1195

New Half-Sandwich Titanocenes for the Polymerization of Butadiene

Walter Kaminsky, Volker Scholz

Institute of Technical and Macromolecular Chemistry, University of Hamburg,
Bundesstrasse 45, 20146 Hamburg, Germany
E-mail: kaminsky@chemie.uni-hamburg.de

Abstract. Different half-sandwich titanium chlorides and –fluorides were synthesized and used as catalysts in combination with methyl aluminoxane for the polymerization of 1,3-butadiene. Highest activities were obtained by using 1,3-dimethylcyclopenta-dienyl-titanium compounds and an aluminum/titanium ratio of about 700. The polybutadiene shows a technical useful microstructure with about 80 % 1,4-cis, 1 % 1,4-trans and 19 % 1,2-linked units. The catalyst used and the polymerization temperature can influence the microstructure. The molecular masses of the polybutadienes are very high (> 1 million g/mol).

1. Introduction

Polybutadiene belongs to the most important rubber for technical purposes. In 1999, more than 2 million tons were produced worldwide, that is about 20 % of all synthetic rubbers [1]. One part is produced by using different types of Ziegler catalysts with titanium, nickel, cobalt or neodymium as transition metal and shows a high content up to 98 % of 1,4-cis linked diene units. The other part is catalyzed by lithium butyl showing a different microstructure of about 40 % cis-1,4, 49 % trans-1,4 and 11 % vinyl (1,2) linked units. The vinyl structures are useful for later vulcanization by sulfur. It is difficult to produce polybutadienes with microstructures having a high content of vinyl but low content of trans-1,4 structures.

Early investigations have disclosed that butadiene can be incorporated into a growing polymer chain by copolymerizating it with ethene using biscyclopentadienyl zirconium dichloride (Cp_2ZrCl_2) and MAO as cocatalyst in toluene [2]. The activity reaches values of 30 000 kg copolymer/mol Zr · h. With this catalyst butadiene is incorporated with 1,4-trans structures. The amount incorporated is low and remains under 5 mol%.

Homopolybutadiene was polymerized by $Co(acac)_2$/MAO or $Ni(acac)_2$(MAO in good yields [3]. With a $Ni(acac)_2$ concentration of $1,5 · 10-5$ mol/l, MAO = 10-3 mol/l by 50 °C from 20 g 1,3-butadiene, 9 g of the polymer are obtained. The polymer contains of about 90 % of 1,4-cis structures.

Porri and Olivera [4,5] demonstrated that a half sandwich complex biscyclopentadienyl titanium trichloride (CpTiCl$_3$) in combination with methylaluminoxane (MAO) is able to polymerize 1,3-butadiene to a rubber which contains high cis-1,4 and 1,2-structures but low trans 1,4-microstructures. An explanation for the differences between the mechanism of polymerization of butadiene and olefins is given by Porri. The first difference relates to the type of bond between the transition metal of the active species and the growing polymer chain. There is an η^3-allyl bond in diene polymerization and a σ-type bond in olefin polymerization. The other difference is the control of the insertion and the coordination of the monomer to the transition metal. We have reported [6-8] that substituted half sandwich complexes of titanium – especially if they are fluorinated – are more active for the polymerization of styrene than unsubstituted and corresponding chlorinated compounds. There is a similarity between the polymerization behavior of styrene and butadiene [8].

2. Results and Discussion

The half sandwich complexes of titanium, synthesized and used for the polymerization of 1,3-butadiene are shown in Fig. 1. The different methyl substituted cyclopentadienyl titanium compounds were fluorinated and also employed as catalysts.

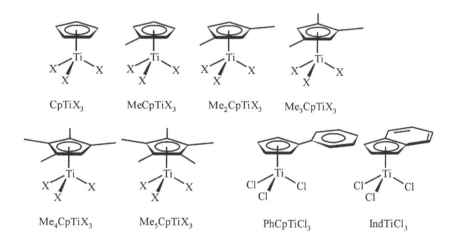

CpTiX$_3$ MeCpTiX$_3$ Me$_2$CpTiX$_3$ Me$_3$CpTiX$_3$

Me$_4$CpTiX$_3$ Me$_5$CpTiX$_3$ PhCpTiCl$_3$ IndTiCl$_3$

Fig. 1.

The polymerizations were carried out in a glass reactor filled with 100 ml toluene, 10g. 1,3-butadiene and 290 mg MAO. The concentration of the titanium complex was $5 \cdot 10^{-5}$ mol/l resulting in an Al/Ti-ratio of 1000. The polymerization was conducted for 20 minutes at 30 °C. The results are shown in Table 1.

Table 1. Activities of Titanium Complexes for the Polymerization of 1,3-Butadiene in 100 ml toluene, 10 g 1,3-butadiene, 0,29 g MAO, [Ti] = 5 · 10^5 mol/l, Al/Ti = 1000, T = 30 °C, polymerization time = 20 min.

Catalyst	Activity*	Catalyst	Activity
$CpTiCl_3$	260	$CpTiF_3$	260
$MeCpTiCl_3$	300	$MeCpTiF_3$	310
$Me_2CpTiCl_3$	750	Me_2CpTiF_3	605
$Me_3CpTiCl_3$	340	Me_3CpTiF_3	350
$Me_4CpTiCl_3$	165	Me_4CpTiF_3	350
$Me_5CpTiCl_3$	60	Me_5CpTiF_3	350
$IndTiCl_3$	310	$Cp*TiF_2 (OCOCF_3)$	330
$PhCpTiCl_3$	325	$Cp*TiF_2(OCOC_6F_5)$	340

* Activity : kg BR/mol Ti · h

At a polymerization temperature of 30 °C the chlorinated and the fluorinated complexes show nearly the same activity. Only the highly substituted fluorinated compounds (Me_4CpTiF_3, Me_5CpTiF_3) are significantly more active than the corresponding chlorinated ones. At higher polymerization temperatures a corresponding behaviour can be observed, however with increasing polymerization temperature also the activity of the complexes increase. The activities of the 1,3-dimethyl-cyclopentadienyl titanium trihalides are the highest and reach about 700 kg BR/mol Ti·h. It makes no difference if one of the fluorides is substituted by another ligand like perfluoroacetic or perfluorobenzoic acid ($Me_5CpTiF_2(OCOCF_3)$, $Me_5CpTiF_2(OCOC_6F_5)$).

As can be seen in Fig. 2, the activity reaches a maximum value for all catalysts after a short induction period of 5 to 10 minutes. After this, the activity decreases to a value being constant for a longer period of time of up to about 1 hour.

The substitution pattern influences the induction period. The most active compounds show the shortest induction period, whereas the less active ones need a clearly longer period.

There is a linear relation between the activity and the butadiene concentration in the starting phase of the polymerization (Fig. 3). The kinetic order of the butadiene concentration is 1. At constant Al:Ti ratio the polymerization rate is given by ·

$$r_p = k_p \cdot c_{cat} \cdot c_\beta$$

where $c\beta$ is the concentration of butadiene. Even at low Al:Ti ratios of 200 the activities are quite high (Fig. 3). The activity increases with an increasing Al:Ti ratio, reaches a maximum at an Al:Ti ratio of about 700 and decreases slowly with increasing Al:Ti ratios.

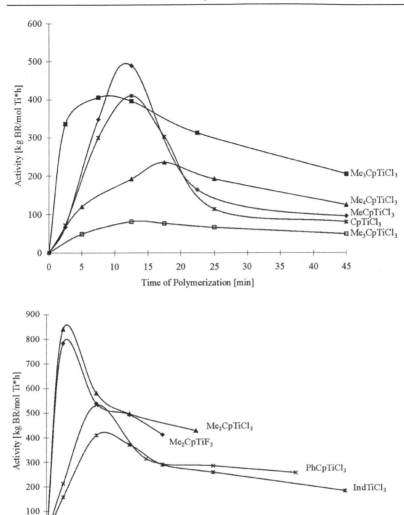

Fig. 2.

Surprisingly, high molecular weights are obtained for the polybutadienes produced with these new catalysts.

The di- and trimethylcyclopentadienyl titanium trichlorides give the highest molecular weights while the fluorinated compounds have significantly lower molecular weights, even if their activity is higher, as shown for the Me_4CpTiF_3 and Me_5CpTiF_3 complexes.

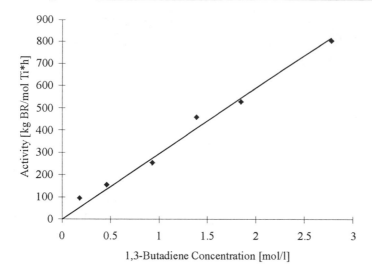

Fig. 3.

Table 2. Molecular Weights of the Polybutadienes Produced with Fluorinated and Chlorinated Catalysts.

Catalyst	$X = Cl$ Molar Mass $M_w[g/mol*10^6]$	$X = F$ Molar Mass $M_w[g/mol*10^6]$
CpTiX$_3$	1,2	0,97
MeCpTiX$_3$	1,6	1,22
Me$_2$CpTiX$_3$	3,1	1,28
Me$_3$CpTiX$_3$	3,6	1,25
Me$_4$CpTiX$_3$	3,3	1,5
Me$_5$CpTiX$_3$	2,6	1,4
IndTiCl$_3$	1,25	-
PhCpTiCl$_3$	0,86	-
CyCpTiCl$_3$	1,00	-
(Me$_3$Si,MeCp)TiCl$_3$	1,5	-

Most important for an industrial application are the microstructures and the glass transition temperatures. The glass transition temperatures range of –90,1 and -96,9 °C (Table 3). The polybutadienes produced with the most active catalysts have the highest content of 1,4-cis units and the lowest glass transition temperature.

Table 3. Microstructure and Glass Transition Temperatures of Polybutadienes produced with Chlorinated Catalyst Precursors

Catalyst	1,4-cis [%]	1,4-trans [%]	1,2 [%]	T_g [°C]
$CpTiCl_3$	81,7	1,1	17,2	-95,1
$MeCpTiCl_3$	81,9	1,1	17,0	-95,3
$Me_2CpTiCl_3$	85,8	0,5	13,7	-96,9
$Me_3CpTiCl_3$	83,8	1,1	15,2	-95,6
$Me_4CpTiCl_3$	80,0	1,7	18,3	-91,5
$Me_5CpTiCl_3$	74,8	2,6	22,6	-91,0
$IndTiCl_3$	74,3	4,2	21,5	-90,1
$PhCpTiCl_3$	80,9	2,1	16,9	-95,8
$CyCpTiCl_3$	82,6	0,8	16,7	-95,0

For all catalysts, the 1,4-cis structure units of the polybutadiene range between a content of 74 and 85,8 %, the 1,4-trans between 0,5 and 4,2 %, and the 1,2-units between 13,7 and 22,6 % (Table 4). The most active systems generate the polymer with the highest content of cis-1,4 and the lowest content of 1,4-trans and 1,2-units. The fluorinated compounds show a similar behavirour. In average, the analysis of the microstructure by infrared spectroscopy gives values of 82 % for 1,4-cis, 1,5 % for 1,4-trans and 16,5 % for 1,2-linked units. A mechanism for the formation of these microstructures is published by Porri.

Table 4. Microstructure and Glass Transition Temperatures of the Polybutadienes Produced by Fluorinated Catalysts

Catalyst	1,4-cis [%]	1,4-trans [%]	1,2 [%]	T_g [°C]
$CpTiF_3$	81,8	1,4	16,8	-95,0
$MeCpTiF_3$	81,9	1,2	16,9	-92,7
Me_2CpTiF_3	82,0	2,0	16,0	-95,0
Me_3CpTiF_3	84,0	1,1	14,9	-94,1
Me_4CpTiF_3	80,4	1,9	17,7	-89,9
Me_5CpTiF_3	74,6	2,8	22,5	-87,9

There is no dependence of the microstructure on the polymerization time (between 10 and 120 minutes the cis content is $81,8 \pm 0,3$ % for $MeCpCl_3$) and on the Al:Ti ratio (between Al:Ti = 500 and Al:Ti = 10 000 the cis content is about $80,7 \pm 1,2$ for $MeCpTiF_3$). Also, half sandwich titanium compounds can polymerize isoprene. The catalysts used and polymerization conditions are illustrated in Table 5.

Table 5. Homopolymerization of Isoprene. Polymerization Conditions: 50 ml toluene, 50 ml isoprene, [Ti] = 5*10^5 mol/l, Al/Ti = 200, T$_p$ = 30 °C, t$_p$ = 5-24 h

Catalyst	Activity [g IR/mol Ti*h]	T$_g$ [°C]
CpTiCl$_3$	28	- 52,0
Me$_5$CpTiCl$_3$	8	n.b.
CpTiF$_3$	840	- 50,3
MeCpTiF$_3$	250	n.b.
Me$_5$CpTiF$_3$	29	n.b.

As of steric effects the unsubstituted cyclopentadienyl compound is more active than the substituted ones. For isoprene the fluorinated compounds are much more active (up to a factor of 30) than the chlorinated ones. The glass transition temperature of the polyisoprenes is about –52 °C. The microstructure is similar to that of the polybutadienes.

It is possible to synthesize copolymers of 1,3-butadiene and isoprene by using a MeCpTiF$_3$/MAO catalyst. The copolymer has a glass transition temperature of –60 °C.

3. Acknowledgement

We thank the Bayer AG for supporting this research.

References

1 Nentwig
2 W. Kaminsky, M. Schlobohm,. Makromol. Chem. Macromol. Symp. 4, 103 (1986)
3 M. Schlobohm, Dissertation, University of Hamburg 1985
4 G. Ricci, S. Italia, A. Giarrusso, L. Porri, J. Organomet. Chem. 451, 67 (1993)
5 L. Oliva, P. Longo, A. Grassi, P. Ammendola, C. Pellecchia, Makromol. Chem., Rapid Commun. 11, 519 (1990)
6 W. Kaminsky, D. Arrowsmith, C. Strübel, J. Polym. Sci, A. Polym. Chem. 37, 2959 (1999)
7 W. Kaminsky, S. Lenk, Macromol. Symp. 118, 45 (1997)
8 W. Kaminsky, S. Lenk, V. Scholz, H.W. Roesky, A. Herzog, Macromolecules 30/25, 7647 (1997)
9 L. Porri, A. Giarrusso, G. Ricci in Metallocene-based Polyolefins, J. Scheirs, W. Kaminsky (eds.), Wiley Series, Vol. 2, Chichester 2000, p. 115

Metallocene Catalyzed Copolymerization of Propene with Mono- and Diolefins

Manfred Arnold,* Steffen Bornemann, Jana Knorr and Thomas Schimmel

Martin-Luther-University Halle-Wittenberg, Institute of Technical Chemistry and
Macromolecular Chemistry, 06099 Halle (Saale), Germany
E-mail: arnold@chemie.uni-halle.de

Abstract. To compare the copolymerization behavior of propene with
mono- and diolefins, the copolymerization of propene with 1-octene as well
as 1,7-octadiene and 1,9-decadiene was investigated. Aim of the
investigations was the synthesis of polymers with different chain structures.
The copolymerization of propene with 1-octene leads to copolymers with
hexyl side chains. In contrast to this, the result of the copolymerization of
propene with linear, nonconjugated diolefins is a copolymer containing both
linear side chains and cyclic units in one polymer chain. The result is a
cycloolefin copolymer containing free double bonds which can be used for
further modification reactions. A syndiospecific, an aspecific and two
isospecific metallocene catalysts in combination with methylaluminoxane
were used for propene/ 1-octene copolymerization reactions to detect
differences in polymerization behavior. The copolymerization of propene
with linear, nonconjugated diolefins was carried out only by a constrained
geometry catalyst. Furthermore, the interactions between comonomer and
catalyst were investigated. It can be shown that catalyst activity,
incorporation rate and reactivity parameters, respectively, as well as thermal
and mechanical properties of the copolymers are strongly influenced by the
type and amount of the comonomer or type of the catalyst used.

Introduction

In the early 90´s metallocene catalysts were firstly used in the commercial
production of LLDPE [1]. The basement for the industrial breakthrough of the
metallocenes was the possibility of the selective modification of the properties of
the ethene copolymers [2]. Disadvantages of the classical Ziegler-Natta catalysts,
e.g. nonuniform and insufficient comonomer incorporation, could be overcome
now. The combination of narrow molecular weight distribution and random
distribution of comonomers in the polymer chain is an important tool for materials
with tailor-made properties. Up to now, several industrial applications of ethene
copolymers started [3].

However, not only for ethene copolymers the use of metallocene catalysts has
importance for the development of new materials. The introduction of metallocene
catalysis in the propene homo- and copolymerization gives access to new

polypropene (PP) materials with a broader spectrum of material properties [4]. A replacing of conventional Ziegler-Natta catalysts by metallocene catalysts is possible [5-6].

Our group is active in the field of mono- and diolefin polymerizations for some years [7-9]. Furthermore, we investigated the synthesis of propene homo-, co- and graft copolymers [10-12], respectively.

The use of 1-octene in copolymerization with ethene [13-16] and propene [17,18] was described by several authors. Now we want to review some differences in the copolymerization behavior of propene with 1-octene by using several metallocene catalysts. Furthermore, we copolymerized propene with the nonconjugated diolefins 1,7-octadiene and 1,9-decadiene, respectively, to compare the influence of the chain structure on the properties of the propene copolymers obtained.

Experimental

Materials

The metallocene catalysts and MAO (10 wt.-% solution in toluene) were donated by BASF AG and Witco GmbH. Propene was dried by passing through a molecular sieve 3 A column. The comonomers were commercially obtained from Fluka and purified by distillation over calcium hydride. Toluene (BSL GmbH Schkopau) was refluxed over sodium/ benzophenone.

Copolymerization procedure

All copolymerizations were carried out as described [19] at 30 °C in a 0,5 l steel reactor. To calculate the solubility of propene in toluene, we used a method described elsewhere [20]. Catalyst and cocatalyst amounts used were different. When IpFluCp or CBT was used, a one thousandfold MAO surplus was given. In the case of MBI [Al] : [Zr] = 2,000:1 and in the case of EBI Hf [Al] : [Hf] = 830:1.

Copolymer characterization

Molecular weight and molecular weight distribution (MWD) of the copolymers were determined by size-exclusion chromatography (SEC) at 25 °C in tetrahydrofurane or at 135 °C in 1,2,4-trichlorobenzene. The ^{13}C NMR spectra of the obtained copolymers were recorded on a Varian 500 spectrometer operating at 250 or 400 MHz. $CDCl_3$ or C_6D_5Br were used as solvents according to the solubility of the products. The composition of the copolymers were calculated using the diad sequence distribution according to the literature [21]. The thermal behavior of the copolymers was investigated using differential scanning calorimetry (DSC). To characterize the mechanical properties of the obtained copolymers plates with a thickness of 1 mm were produced. From these plates dumb-bell objects submitted for testing were stamped. Afterwards, the mechanical behavior was studied at a tensile-strength test machine (EN ISO 527-1).

Results and discussion

The incorporation of 1-octene into the PP chain leads to a polymer chain containing hexyl side chains. In contrast to this the incorporation of a diolefin yields in a totally changed structure. Due to the possibility of cyclization, both side chains and cyclic units can be obtained.

Cyclopolymerization was observed for the first time more than 40 years ago [22]. The homopolymerization of nonconjugated diolefins gives access to cyclopolymers from linear monomers is also described [23-25].

To compare the copolymerization behavior of propene with 1-octene and 1,7-octadiene we copolymerized the monomers by using the aspecific metallocene dimethyl silyl(tetramethylcyclopentadienyl) ^{t}butyl amidotitanium dichloride (**CBT**) and the syndiospecific metallocene [2,4-cyclopentadien-1-ylidene(isopropylidene)-fluoren-9-ylidene]zirconium dichloride (**IpFluCp**). Recently, we described the synthesis of propene/1-octene copoylmers with the isospecific metallocene catalysts *rac*-[(ethylene)bis(η^5-inden-1-ylidene)]hafnium dichloride (**EBI Hf**) [26] and *rac*-[(dimethylsilylene)bis(2-methylbenzo(e)indenyl)]-zirconium dichloride (**MBI**) [19], respectively. For a comparison some data in the figures are taken from that publications.

Some years ago, Dow Chemical developed the so called Insite$^®$ process for the production of polyethene and polyolefin elastomers (POE) using constrained geometry catalysts like CBT [27-29]. Due to the open structure of the complex it is not possible to polymerize propene stereospecifically. In contrast to this, the propene polymerization with catalysts with C_s symmetry, e.g. IpFluCp, leads to PP with syndiotactic structure [30]. For some years, syndiotactic PP is produced by metallocene catalysis in a technical scale [31].

In Table 1 the results of the propene/1-octene copolymerization experiments are listed. For the copolymerizations a catalyst concentration of [CBT] = 4.10^{-5} mol/L and [IpFluCp] = 2.10^{-5} mol/L was used.

In contrast to the propene/1olefin copolymerization only a few number of publications dealing with the copolymerization of propene and nonconjugated diolefins can be found [32-35]. Table 2 summarizes some results of the copolymerization of propene with the linear nonconjugated diolefin 1,7-octadiene and 1,9-decadiene. For copolymerization reactions we used the constrained geometry catalyst **CBT**. A catalyst concentration of 3.10^{-5} mol/L was used.

Table 1. Propene/1-octene copolymerization using different metallocene catalysts/ MAO

metallocene catalyst	1-octene in monomer feed (mol-%)	1-octene in copolymer[1] (mol-%)	M_n[2] (kg/mol)	M_w/M_n	T_g[3] (°C)
Me₂Si TiCl₂ **(CBT)**	0	0	134	1.4	-3
	10	6	44	1.7	-7
	25	12	34	1.6	-11
	40	21	46	1.6	-16
	60	40	24	1.4	-37
	75	68	20	1.3	-50
	90	89	13	1.7	-60
	100	100	11	1.3	-68
ZrCl₂ **(IpFluCp)**	0	0[4]	95	2.2	-3
	10	5[5]	115	3.0	-7
	20	10	112	2.0	-9
	40	25	101	3.0	-25
	50	32	100	2.0	-29
	60	46	107	1.9	-35
	80	71	100	1.7	-59
	90	85	63	2.0	-57
	100	100	46	1.9	-60

Table 2. Propene/diolefin copolymerization using the constrained geometry catalyst **CBT** and MAO. [CBT] = 3.10^{-5} mol/L

diolefin	diolefin content in the feed (mol-%)	Diolefin[1] In the copolymer (mol-%)	c.u.[1] (mol-%)	n.c.u.[1] (mol-%)	M_n[2] (kg/mol)	M_w/M_n	T_g[3] (°C)
	0	0	0	0	135	1.4	-3
1,7-octadiene	10	15	14	1	47	5.4	2
	20	23	20	3	34	2.8	7
	30	46	41	5	39	2.4	16
	60	64	50	14	14	2.7	14
	90	84	72	12	5,5	2.6	20
	100	100	n.d.	n.d.	27	1.5	20
1,9-decadiene	20	30	20	10	109	10.3	11
	40	48	29	19	37	2.9	30
	60	66	46	20	40	6.5	71
	80	70	41	29	25	7.4	59
	100	100	61	39	13	3.1	13

1) diolefin content, content of cyclized diolefin units (c.u.) and content of noncyclized diolefin units (n.c.u.) determined by ^{13}C NMR spectroscopy
2) determined by size exclusion chromatography
3) determined by differential scanning calorimetry
4) [IpFluCp] = 4.10^{-5} mol/L; double melting point 131 and 119°C, crystallinity 14%
5) Melting point 86°C, crystallinity 1%

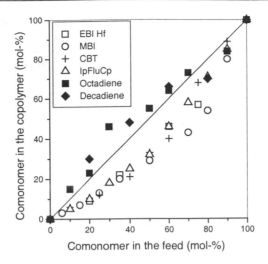

Fig. 1. Copolymerization diagram for the copolymerization of propene with 1-octene, 1,7-octadiene and 1,9-decadiene, respectively, using different metallocene catalysts in combination with MAO. The data for MBI and EBI Hf are taken from literature (19,26).

In Figure 1 the copolymerization diagram for the copolymerization of propene with 1-octene, 1,7-octadiene and 1,9-decadiene is shown. The data for the experiments with the two isospecific catalysts **EBI Hf** and **MBI** are taken from our publications [19,26]. With the four different metallocene catalysts used it is possible to synthesize copolymers with every desired composition. In general, a comparable incorporation of the 1-octene into the polymer chain can be observed. That finding is different to published data, in which an influence of the stereospecifity of the catalyst type on the incorporation rate was observed for propene/1-hexene copolymers [36]. Only at higher olefin content in the feed some differences in the copolymerization behavior of the metallocenes can be found. The highly isospecific **MBI** is the one with the lowest 1-octene incorporation at higher 1-octene concentration, followed by the second isospecific catalyst **EBI Hf**. The syndiospecific and aspecific metallocenes are able to incorporate under comparable polymerization conditions a higher amount of the 1-olefin into the copolymer.

If we have a look at the propene/diolefin copolymerization we can find a totally changed polymerization behavior. We found a very high incorporation rate for the diolefin. Only at very high diolefin content in the monomer feed the insertion of the propene is favored.

The r-values for the propene copolymers calculated by the method of Kelen and Tüdös [37] are listed in Table 3.

The incorporation of 1-octene leads to a rapid decrease of M_n when the **EBI Hf** catalyst was used. In principle, the incorporation of higher amounts of the 1-octene leads to a decreasing M_n. Under the polymerization conditions used here the incorporation of 1-octene yields in a increasing M_n if the **MBI** catalyst was used. Generally, we advance for M_n of propene/1-octene copolymers the following order:

IpFluCp > MBI > EBI Hf > CBT. For poly(1-octene) we found the following molecular weights: MBI > IpFluCp ≈ EBI Hf > CBT.

The MWD is ranging between 1.5 and 3.5. That´s typical for metallocene catalysis. Polymers with relatively broad MWD were obtained from **IpFluCp** and **EBI Hf**. However, with higher amount of the 1-octene in the copolymer the MWD is reduced. Obviously a correlation between catalyst activity, molecular weight and MWD is existing. A very homogeneous MWD without any dependence from the 1-octene content we can consider when the **CBT** catalyst was used.

Table 3. Copolymerization parameters for propene copolymerizations

comonomer	metalloce ne catalyst	r_1	r_2	$r_1 \cdot r_2$
1-octene	EBI Hf	2.10	0.55	1.16
	MBI	2.46	0.30	0.74
	CBT	1.80	0.41	0.74
	IpFluCp	2.17	0.65	1.16
1,7-octadiene	CBT	0.89	1.07	0.95
1,9-decadiene	CBT	0.33	0.43	0.14

The thermal behavior of the copolymers was investigated by DSC. Normally, the applications of a polymeric material are limited by melting point T_m and glass transition temperature T_g [39]. In Figure 2 the influence of the 1-octene content on the thermal properties of the copolymers is illustrated. Only isotactic or syndiotactic propene homo- and copolymers are semicrystalline materials in which T_m can be detected. The result of the incorporation of 1-olefins is a propene copolymer with reduced T_m and crystallinity. This tendency is comparable for all of the catalyst systems used here.

Crystallinity of polymers was calculated from DSC measurements. Data for calculations were taken from literature for iso- [40] or syndiotactic PP [41].

The phenomenon of a double peak T_m of syndiotactic PP was investigated by several authors [42,43]. In our own DSC measurements we also observe two T_m for the syndiotactic PP synthesized with **IpFluCp**. Already, the incorporation of only 5 mol-% 1-octene leads to a polymeric material with drastically decreased T_m and crystallinity. For copolymers with more than 5 mol-% no T_m could be detected.

A comparison of the thermal properties of the copolymers obtained from the isospecific metallocene catalysts **MBI** and **EBI Hf** shows similar melting behavior. By use of the **MBI** a PP with T_m of more than 150°C can be synthesized. The crystallinity of the copolymers prepared by **MBI** is higher than those reached by the other catalysts. Copolymers with more than approx. 20 mol-% 1-octene incorporated are found to be amorphous materials.

The T_g of PP is reported as approximately −10°C [44]. This value limits the application possibilities of PP at lower temperatures. By the incorporation of linear 1-olefins the T_g can be lowered. In Figure 3 the T_g of the propene/1-octene copolymers as a function of the 1-octene content is shown. For a comparison, the T_g values for propene/diolefin coplymers are added.

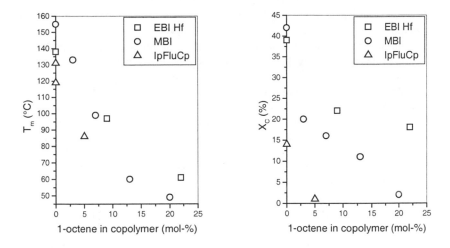

Fig. 2. Melting points T_m (left) and crystallinity X_c (right) of propene/1-octene copolymers determined by DSC measurements

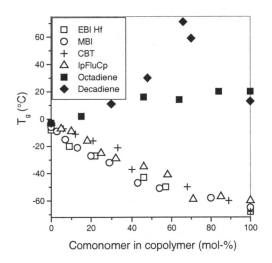

Fig. 3. Glass transition temperatures T_g of propene copolymers determined by DSC measurements (heating rate 10 K/min)

In all cases we detect a strong influence of the 1-octene content on the T_g. A straight decrease of the T_g with increasing comonomer content can be observed. The value for the T_g of the poly(1-octene) measured by us is approx. $-65°C$. In literature a T_g region for poly(1-octene) from -65 up to $-60 °C$ is published [45].

In contrast to the propene/1-octene copolymers the incorporation of 1,7-octadiene or 1,9-decadiene leads to an increasing T_g.

We assume that three different effects can describe the values of propene/diolefin copolymer T_g:
-    the size and content of the incorporated cyclic units in the copolymer
-    the ratio between cyclic (c.u.) and noncyclic units (n.c.u.)
-    crosslinking.

An increasing T_g was found in copolymerization of cyclic olefins with ethene [2,46] and propene [47]. A comparison of the published data demonstrates that the effect of the ring size is very important for the increase of the T_g. Now, we can confirm that observation. The insertion of 1,9-decadiene by cyclization leads to a ring involving nine C-atoms. Compared to this, the incorporation of 1,7-octadiene forms a cycloheptane unit. The result of the smaller rings of the propene/1,7-octadiene copolymers is a lower increase of the T_g.

The attaching of alkyl side branches into a polymer chain leads to a lowering of the T_g [48]. In propene/diolefin copolymers the ratio between cyclic units c.u. and noncyclic units n.c.u., respectively, seems to be also important for an estimation of the value of the T_g. The amount of the n.c.u. in a propene/1,9-decadiene copolymer is higher than in a comparable propene/1,7-octadiene copolymer. The n.c.u./c.u. value could be used to describe the decreasing values for the T_g of propene/1,9-decadiene copolymers with high content of the diolefin. Apparently, the high amount of the side branches leads to a reduction of the T_g.

Furthermore, the curve of the T_g for propene/1,9-decadiene seems to be influenced by a crosslinking. In propene copolymerization with nonconjugated diolefins the crosslinking cannot totally avoided. The insertion of a pendent double bond of a polymer chain into a second chain can be detected by SEC measurements. A low crosslinking yields in polymers with a broader MWD (see Table 2). Crosslinking reduces the molecular mobility of a polymer chain segments resulting in an increasing T_g. For propene/1,9-decadiene copolymers with a content of approx. 70 mol-% 1,9-decadiene incorporated a much broader MWD compared to the poly(1,9-decadiene) was detected. Combined with the high amount of noncyclic units in the copolymer this effect could be the reason for the T_g values observed.

The insertion of 1-olefins into the PP chain seriously influences the mechanical properties of the plastic materials. To determine the mechanical properties a tensile-strength test was carried out at propene/1-octene copolymers. We used copolymers synthesized by the **MBI** catalyst.

As described before, the incorporation of a small amount of the 1-olefin leads to totally changed thermal behavior. Such an influence can also be observed when the mechanical properties are investigated. The incorporation of a few mol-% 1-octene yields in a disturbance of the crystallinity resulting in a depressed T_m. Furthermore, the reduced crystallinity is important for the stretchability of the copolymers. As shown in Figure 4 the rising amount of 1-octene leads to materials with high stretchability.

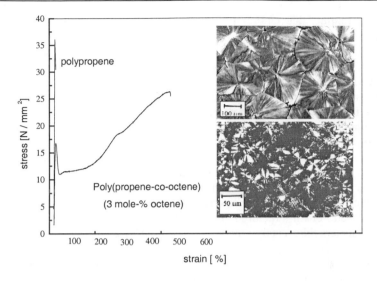

Fig. 4. Stress-strain curves of a PP and a propene/1-octene copolymer containing 3 mol-% 1-octene. Copolymers were synthesized by using the MBI catalyst.

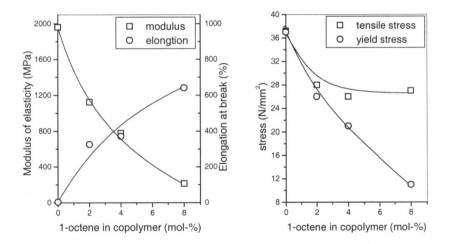

Fig. 5. Mechanical properties of propene/1-octene copolymers as a function of 1-olefin content in the copolymer. Copolymers were synthesized by using the MBI catalyst.

In Figure 5 some mechanical properties of propene/1-octene copolymers are illustrated. PP synthesized by the **MBI** catalyst is a polymeric material with a modulus of elasticity of approx. 1950 MPa, a tensile stress at yield of 37 N/mm^2 and an elongation at break of only 3 %. The copolymerization of propene with linear 1-olefins, e.g. 1-octene, allows the tailoring of material properties. Especially the elongation at break can be improved clearly. However, the improvement of the stretchability is combined with a reduction of the modulus of elasticity. The modulus of elasticity describes the stiffness of a polymeric material. A lowering of the modulus means that the products are more and more softer. It can be shown that the reduction of the yield stress as a function of the 1-octene content is not combined with a decreasing tensile stress.

Having a look at the results above, we can compare the copolymerization of propene with 1-octene, 1,7-octadiene and 1,9-decadiene as followed:

1. The copolymerization behavior in propene/1-octene copolymerization is not strongly influenced by the type of the metallocene used.
2. The use of different metallocene catalysts results in a broad variety of material properties. Amorphous or semicrystalline polymers with syndiotactic, isotactic or atactic chain structure can be prepared. The type of the catalyst controls parameters like catalyst activity, molecular weight and thermal properties, respectively.
3. The incorporation of a low amount of an α-olefin into the poly(propene) chain yields in a totally changed mechanical behavior. The modulus of elasticity is reduced while a higher elongation at break was measured.
4. In propene/diolefin copolymerization another copolymerization behavior was found. Compared to 1-octene the incorporation rate for the diolefin is higher.
5. The incorporation of a diolefin yields in a totally changed chain structure. Consequently, a conflicting thermal behavior was observed. The incorporation of the 1-octene leads to a decreasing T_g. Due to the cyclic units in a propene/diolefin copolymer the T_g is increasing. Obviously, the size of the rings is more important for the value of the T_g than the content of the cyclic units.

Acknowledgement

We wish to express our appreciation to BASF AG, Witco GmbH and BSL GmbH Schkopau for donating the homogeneous catalysts, the MAO and the toluene, respectively. Furthermore, we thank Dr. T. Lüpke and Dr. G.W.H. Hoehne for some DSC measurements, Dr. E. Brauer for high-temperature SEC measurements as well as Dr. R. Thomann for light microscopy. This project was supported by the Deutsche Forschungsgemeinschaft.

References

1   Kissin Y V (**1996**) in: Kroschwitz I (ed) Kirk-Othmer: Encyclopedia of chemical technology, 4th Ed., vol 17, John Wiley & Sons, New York, p. 756
2   Kaminsky W, Arndt M (**1997**) *Adv Polym Sci* 127: 143
3   McKnight A L, Waymouth R M (**1998**) *Chem Rev* 98: 2587
4   Blanco A (**2000**) *Plastics Eng,* May 2000, p. 40
5   Grasmeder J R (**1997**) Metallocene technology: One-day seminar, paper 2: 1

6   Kristen M O (**1997**) *New Adv Mater* 1997/2 and 3: 1
7   Arnold M, Wohlfarth L, Schmidt V, Reinhold G (**1991**) *Makromol Chem* 192: 1017
8   Arnold M, Reußner J, Wohlfarth W (**1992**) *Angew Makromol Chem* 196: 37
9   Arnold M, Reußner J, Utschick H, Fischer H (**1994**) *Macromol Chem Phys* 195: 2653
10   Arnold M, Knorr J, Köller F, Bornemann S (**1999**) *J M S - Pure Appl Chem* A36: 1655
11   Arnold M, Henschke O, Köller F (**1996**) *Macromol Reports*, A33(Suppls. 3&4): 219
12   Henschke O, Neubauer A, Arnold M (**1997**) *Macromolecules* 30: 8097
13   *Quijada R, Galland G B, Mauler R S (**1996**) Macromol* Chem Phys 197: 3091
14   Schneider M J, Suhm J, Mülhaupt R, Prosenc M H, Brintzinger H H (**1997**) *Macromolecules* 30: 3164
15   Xu G, Ruckenstein E (**1998**) *Macromolecules* 31: 4724
16   Wang W J, Kolodka E, Zhu S, Hamielec A E (**1999**) *J Polym Sci, Part A: Polym Chem* 37: 2949
17   Schneider M J, Mülhaupt R (**1997**) *Macromol Chem Phys* 198: 1121
18   Jüngling S, Mülhaupt R, Fischer D, Langhauser F (**1995**) *Angew Makromol Chem* 229: 93
19   Arnold M, Bornemann S, Köller F, Menke T J, Kressler J (**1998**) *Macromol Chem Phys* 199: 2647
20   Wohlfarth C, Finck U, Schultz R, Heuer T (**1992**) *Angew Makromol Chem* 198: 91
21   Uozumi T, Soga K (**1992**) *Makromol Chem* 193: 823
22   Butler G B Cyclopolymerization and Cyclocopolymerization, Marcel Dekker, Inc., New York, **1992**
23   Marvel C S , Stille J K (**1958**) *J Am Chem Soc* 80: 1740
24   Resconi L, Coates G W, Mogstad A, Waymouth R M (**1991**) *J M S: Chem* A28: 1225
25   Naga N, Shiono T, Ikeda T (**1999**) *Macromol Chem Phys* 200: 1466
26   Arnold M, Henschke O, Knorr J (**1996**) *Macromol Chem Phys* 197: 563
27   Sylvest R T, Lancaster G, Betso S R (**1996**) Metallocenes´96, Proc Int Congr Metallocene Polym, 2nd Ed:197
28   Stevens J C (**1996**) *Stud Surf Sci Catal* 101: 11
29   Rotman D (**1997**) *Chem Week* 159: 23
30   Ewen J A, Elder M J, Jones R L, Haspelagh L, Atwood J L, Bott S G, Robinson K (**1991**) *Makromol Chem, Macromol Symp* 48/49: 253
31   Shamshoum E, Schardl J (**1998**) in: Benedikt G M, Goodall B L (eds.), Metallocene catalyzed polymers: materials, properties and markets, Plastics Design Library, Norwich, NY, p. 359
32   Naga N, Shiono T, Ikeda T (**1999**) *Macromolecules* 32: 1348
33   Pietikainen P, Starck P, Seppala I V (**1999**) *J Polym Sci, Part A: Polym Chem* 37: 2379
34   Sernetz F G, Mülhaupt R, Waymouth R M (**1997**) *Polym Bull* 38: 141

35   Lee D H, Yoon K B, Park J R, Lee B H (**1997**) *Eur Polym* J 33: 447
36   Uozumi T, Soga K (1992) Makromol Chem 193: 823
37   Kelen T, Tüdös F, Földes Berezhnikh T (1975) J Polym Sci, Polym Symp 50: 109
38   Brintzinger H H, Fischer D, Mülhaupt R, Rieger B, Waymouth R M (1995) Angew Chem Int Ed Eng 107: 1255
39   Mathot VBF (ed) (1994) Calorimetry and thermal analysis of polymers, Hanser, Munich
40   Wunderlich B (1980) Macromolecular Physics, Crystal melting vol. 3, Academic Press, New York
41   Haffka S, Könneke K (1991) J Macromol Sci B30: 319
42   Thomann R, Kressler J, Mülhaupt R (1998) Polymer 39: 1907
43   Rodriguez-Arnold J, Zhang A, Cheng S Z D, Lovinger A J, Hsieh E T, Chu P, Johnson T W, Honnell K G, Geerts R G, Palackal S J, Hawley G R, Welch M B (1994) Polymer 35: 1884
44   Mark H F (ed) (1988) Encyclopedia of polymer science and technology, 2nd ed., vol. 13, Wiley, New York, p. 481
45   Krentsel B A, Kissin Y V, Kleiner V I, Stotskaya L L (1997) Polymers and copolymers of higher α-olefins, Hanser, Munich, p. 111
46   Huang B, Tian J (1996) in: Salamone J C (ed) Polymeric materials encyclopedia, vol. 6, CRC Press, Boca Raton, p. 4199
47   Henschke O, Köller F, Arnold M (1997) Macromol Rapid Commun 18: 617
48   Schneider H A (1996) in: Salamone J C (ed) Polymeric materials encyclopedia, vol. 4, CRC Press, Boca Raton, p. 2785

A Comparison of the Behavior of Nickel/MAO Catalytic Systems in the Polymerization of Styrene and 1,3-Cyclohexadiene

Riccardo Po, Nicoletta Cardi, Maria Anna Cardaci, Roberto Santi

EniChem S.p.A., Centro Ricerche Novara "Istituto Guido Donegani"
Via Giacomo Fauser 4, 28100 Novara, Italy
E-mail: riccardo.po' @enichem.it

Abstract. Several nickel complexes activated with MAO have been tested in the polymerization of styrene and cyclic conjugated diolefins. The effect of the ligands, temperature, solvents, metal concentration has been investigated. The most active complex is nickel bis(acetylacetonate), but polystyrene was obtained in fairly good yields also with other catalysts ($NiCl_2$, Cp_2Ni, etc.). The prepared polystyrene samples have been examined from the point of view of steric microstructure. On the contrary, the only satisfactory catalyst for 1,3-cyclohexadiene polymerization is $Ni(acac)_2$. Larger cyclic diolefins do not polymerize. Attempts to prepare copolymers of styrene and 1,3-cycloexadiene failed and only pure 1,3-cyclohexadiene homopolymer was recovered, suggesting that the active species for the two monomers have a very different structure.

1. Introduction

In the last years, nickel based catalysts have gained increased interest in homogeneous coordinated polymerization of ethylene, α-olefins and cyclic olefins [1]. Also the polymerization of non polar and polar monomers has been reported to be viable through these systems, and recently a neutral single-component class of catalyst has been discovered that do not require activation by any cocatalyst [2].

Nickel bis(acetylacetonate) ($Ni(acac)_2$) is perhaps the most simple compound of these families; it is unable to polymerize olefins, but it has also some advantages, because of the greater flexibility with other monomers. Indeed, $Ni(acac)_2$/MAO (MAO=methylaluminoxane) system is highly active in the polymerization of styrene [3-6]. A mixture of atactic and partially isotactic polystyrene is produced, and the isotacticity was found to increase by adding Lewis bases to the system [5].

Diolefins [7], such as butadiene [7,8] and isoprene [7,9], were also found to polymerize readily with the same catalytic system. The structural units of the resulting polymers have mainly *cis*-1,4 stereochemistry, while 1,2 units are almost absent, except for 4-methyl-1,3-pentadiene [7]. Number average molecular weights are in the range 13000-20000, with polydispersity ratios (M_w/M_n) between 2 and 2.3. Kinetic studies have been carried out [8,9] to evaluate the activation energies of the process and the reactivity ratios with styrene [9].

Since only few papers have been published on the use of nickel-based catalysts in cyclic conjugated olefins polymerization [10-14], it seemed to us particularly interesting to test Ni(acac)$_2$ and other "simple" nickel complexes in the polymerization of 1,3-cyclohexadiene (CHD), 1,3-cycloheptadiene and 1,3-cyclooctadiene, and to try to copolymerize these diolefins with styrene. Aim of this work is to report some results on these polymerizations.

2. Experimental Part

Materials

Styrene (EniChem), α-methylstyrene (Fluka), 1,3-cyclohexadiene (Aldrich), 1,3-cycloheptadiene (Fluka) and 1,3-cyclooctadiene (Aldrich) were purified by contact with CaH$_2$ and treatment with basic alumina prior to polymerization. Toluene (Carlo Erba) was distilled over sodium/benzophenone; dioxane (Carlo Erba) and 1,2-dichloroethane (Carlo Erba) were dried over activated basic alumina. Nickel derivatives were either commercial products or prepared according to known procedures [15-20] and analyzed by NMR spectroscopy and elemental analysis. Methylaluminoxane (10 wt-% solution in toluene; Witco GmbH) was mantained under vacuum at 60°C for 2 h to distill the free AlMe$_3$ and dissolved again in anhydrous toluene prior to polymerization to give 1.55-1.65 M solutions. Triethylaluminum (TMA, 2M in toluene) and triisobutylaluminum (TIBA, 1M in toluene) were purchased from Aldrich. Dimethyl-*p*-tolylammonium tetrakis-(pentafluorophenyl) borate (AFB) was kindly supplied by Witco GmbH.

Polymerization

A typical polymerization was carried out as follows. Into a reaction tube kept in nitrogen atmosphere, toluene (10 mL), the monomer (styrene; 5 mL, 44 mmol), bis(cyclopentadienyl) nickel (8.3 mg, 44 µmol) and MAO (1.33 mL of 1.648 M solution, 2.2 mmol) were added in that order. The tube was immersed in an oil bath at 70°C, and kept under stirring for 4 h. The reaction was stopped by the addition of 200 mL of ethanol and 0.5 mL of concentrated HCl. After filtration and washing with ethanol, the product was dried at 60°C for 16 h. A small amount of 3,5-di-*tert*-butyl-4-methylphenol as stabilizing agent was added to the ethanol when the recovery of 1,3 cyclic diene polymers was performed.

Characterization

Molecular weights and polydispersity indexes were determined by gel permeation chromatography (GPC) in tetrahydrofuran solution at 30°C using a Waters 600E chromatograph. Atactic polystyrene standards were used for calibration.

^{13}C-nuclear magnetic resonance (^{13}C-NMR) spectra were recorded in 1,2-dichlorobenzene using a Bruker AM300 spectrometer (75.43 MHz). The sequence distribution analysis of polystyrene samples was performed through deconvolution of the multiplets attributed to the aromatic quaternary carbon (~146 ppm) and to the CH$_2$ carbons (~45 ppm).

Differential scanning calorimetric (DSC) curves were registered by using a Perkin Elmer DSC 7 calorimeter. Indium was used for temperature calibration. Heating and cooling rates 20°C/min and 10°C/min, respectively, were adopted.

Infrared spectra were registered with a Perkin Elmer FTIR 1800 spectrophotometer on autosupported pellets kept under vacuum.

Gas chromatographic-mass spectrometric (GC-MS) analysis was performed with a Finnigan TSQ-700 mass spectrometer coupled with a Varian 3400 gas chromatograph.

3. Results and Discussion

Polymerization of styrene

The results of the styrene polymerization tests carried out using different nickel derivatives and different cocatalysts are summarized in Table 1. In several cases the reaction mixtures are not homogeneous, as, for instance, for NiCl$_2$; in other cases it is difficult to understand whether the reaction medium is homogeneous or not, because of the black color of the mixture, but dark organic-insoluble residues were isolated after polymer recovery and redissolution.

The precursors that exhibit higher activities are nickel chloride and nickel bis(acetylacetonate), followed by bis(cyclopentadienyl)nickel. Phosphane ligands decrease the activity of NiCl$_2$; the decrease is much more marked with phosphines than with phosphites. The ionic substituent seems to play a minor role, as evidenced by the fact that bis(diphenylmethylphosphine)nickel dichloride and dithiocyanate have a similar activity. Nickel(0) tetrakis(triethylphosphite) is almost uninfluential on the polymerization (it affords the same amount of polymer as MAO alone), while Ni[(C$_6$H$_5$)C$_2$S$_2$]$_2$ leads to poisoning.

Weight average molecular weight values are between 9000 and 24000, with polydispersity indexes greater than two. The polystyrene obtained with nickel tetrakis(triethylphosphite) (M$_w$=95000) is probably actually formed through a ionic process initiated by MAO [21]. This is confirmed from the fact that MAO alone afford atactic polystyrene in 2.8% yield.

Table 1. Polymerization of styrene with nickel catalysts. [styrene] = 2.6 mol/L in toluene, [styrene]/[Ni] = 1000, T = 70°, t = 4 h.

Catalyst precursor	Cocatalyst	[Al]/[Ni]	[B]/[Ni]	Yield (wt%)	M_w	M_w/M_n
None	MAO	-	-	2.8	48000	5.5
Ni[P(OC$_2$H$_5$)$_3$]$_4$	MAO	50	-	2.9	95000	3.4
Ni(acac)$_2$	MAO	50	-	95	19000	2.5
Ni(acac)$_2$	AFB+TIBA	10	1	>99	n.d.	n.d.
Ni(acac)$_2$	AFB+TIBA	50	1	83	n.d.	n.d.
Ni(acac)$_2$	AFB+TIBA	50	5	90	n.d.	n.d.
Cp$_2$Ni	MAO	50	-	46	18000	2.1
NiCl$_2$	MAO	50	-	>99	19000	2.0
L$_2$NiX$_2$						
X=Cl, L=o-phenanthtroline	MAO	50	-	28	24000	3.2
X=Cl, L=(C$_2$H$_5$O)$_3$P	MAO	50	-	65	18000	2.3
X=Cl, L=(C$_2$H$_5$)$_3$P	MAO	50	-	24	18000	2.2
X=Cl, L=(C$_6$H$_{11}$)$_3$P	MAO	50	-	19	19000	2.5
X=Cl, L=[(CH$_3$)$_2$CH]$_3$P	MAO	50	-	11	9000	4.1
X=Cl, L=(C$_6$H$_5$)$_3$P	MAO	50	-	5.3	15000	1.6
X=Cl, L=(C$_6$H$_5$)$_2$(CH$_3$)P	MAO	50	-	13	14000	2.6
X=SCN, L=(C$_6$H$_5$)$_2$(CH$_3$)P	MAO	50	-	9.5	18000	8.9
Ni[(C$_6$H$_5$)$_2$C$_2$S$_2$]$_2$	MAO	50	-	0	-	-

Perfluorinated arylborates salt (AFB)/TIBA cocatalyst is as efficient as MAO. α-methylstyrene is not able to polymerize under the adopted conditions.

For Ni(acac)$_2$/MAO system, the effect of reaction parameters was studied in more detail. The catalyst activity and the polymer molecular weight decrease in the temperature range 25-90°C [6]. Polymer yields are comparable when the reaction is carried out in bulk, in toluene or in 1,2-dichloroethane for [styrene]/[Ni]=1000 (97%, 95% and 94%, respectively), while a decrease of a few percent is observed for a [styrene]/[Ni] ratio of 5000 (93%, 81%, and 77%, respectively). In dioxane, yields are much lower (less than 20%), but the polymer has two molar mass peaks, the first one located at M_w=10000, the second one at M_w=140000; this sample is also the sole exhibiting a melting endotherm (T_m=211°C, ΔH=9.4 J/g). All the other samples appear to be amorphous.

NMR analysis demonstrate that the crude reaction products display an overall prevailing isotactic structure [6], and diads and triads distributions show a deviation from a bernoullian statistic. As evidenced by GPC measurements, at least two active species are likely to exist; this was supported by the data of Porri and coworkers, that were able to obtain polystyrene samples separable into two fractions [5] by using nickel catalysts activated by modified MAO. Therefore, it seems reasonable to suppose, in a first approximation, that the polymer products consist of two different species:

- an atactic polystyrene, produced by MAO [21] or by soluble nickel species;
- a partially isotactic (bernoullian) polystyrene, produced by isospecific nickel sites.

These sites may be associated to solid or nanocolloidal [22] nickel particles formed in the polymerization mixture [6], that could be responsible of the isospecific monomer insertion (site controlled).

Table 2. Calculated composition and stereoregularity of isotactic polymer fraction of the crude polymerization products.

Catalyst	Solvent	Fraction of isotactic polymer	Stereoregularity of isotactic fraction (% of *meso* diads)
Cp$_2$Ni	Toluene	0.70	81
(o-phenanthroline)NiCl$_2$	Toluene	0.77	79
[(C$_2$H$_5$O)$_3$P]$_2$NiCl$_2$	Toluene	0.90	73
(C$_6$H$_{11}$)$_3$P]$_2$NiCl$_2$	Toluene	1.00	75
NiCl$_2$	Toluene	0.84	74
Ni(acac)$_2$	Toluene	0.67	80
Ni(acac)$_2$	Dioxane	0.42	92
Ni(acac)$_2$	1,2-dichloroethane	0.68	78

Under these assumptions, the amount and stereoregularity degree of the isotactic fractions can be evaluated [6]. A linear correlation between the bernoullian parameter [23] (that is an estimate of the deviation from the ideal bernoullian behavior) and the fraction of isotactic polymer was found [6]. As shown in Table 2, most polymer samples contain isotactic fractions having 73-83% of isotacticity (as expressed as percentage of *meso* diads), with the exception of the polymer obtained in dioxane that exhibits a 92% of *meso* diads in the isotactic fraction; accordingly, as already said, this sample is the only one exhibiting a melting temperature. Ni(acac)$_2$ and Cp$_2$Ni have a lower fraction (<0.7) of more isospecific sites (>80%), while the other examined catalysts, all containing heteroatoms other than oxygen, behave in the opposite way; [(C$_6$H$_{11}$)$_3$P]$_2$NiCl$_2$ affords only the partially isotactic polymer. It is possible that phosphines or o-phenantroline poison the aspecific MAO initiating sites or favor the formation of nanocolloidal species. Worth to be remarked, the thorough elimination of free TMA from MAO and the addition of Lewis bases (amines) to Ni(acac)$_2$/MAO increase the isotacticity of the final polystyrene [5].

Polymerization of 1,3-cyclohexadiene and other conjugated cyclic diolefins

For a long time 1,3-cyclohexadiene (CHD) polymerization was found to proceed with difficulty, only affording low amounts of low molecular weight products. An improvement in anionic polymerization was recently discovered [24], that leads to high molecular weight amorphous products containing both 1,4 and 1,2 units. Nickel catalysts have been used to prepare highly stereoregular *cis*-1,4 polymers. The catalytic system (π-C$_4$H$_7$NiCl)$_2$/chloranil produces *cis*-1,4-PCHD (*cis*-1,4 > 90%; T$_m$ = 270°C) [10]; Cp$_2$Ni/MAO affords a fully *cis*-1,4 polymer with moderate activity [13]. The preparation of copolymers with other cyclic [12] or linear [13] dienes is also reported.

Nickel bis(acetylacetonate)/MAO is a very active catalyst for CHD polymerization (Table 3) [14]. All the other tested nickel complexes, including Cp_2Ni, are much less active or inactive at all. The polymer yields can be increased by adding an aluminum alkyl, especially TIBA (which role in the formation of active species is however not yet understood), in partial subtitution of MAO. Considering the strong reducing properties of aluminum alkyls, this seems to be in disagreement with the fact that in 1,3-butadiene polymerization, the prevention of reduction of nickel by BF_3 contributes to increase the catalytic activity [25].

Table 3. Polymerization of cyclic conjugated diolefins with nickel catalysts. [monomer] = 3.0 mol/L in toluene, [monomer]/[Ni] = 2000, [Al]/[Ni]=50, T = 50°, t = 30 min.

Monomer	Catalyst	Cocatalyst	Yield (wt%)
CHD	$Ni(acac)_2$	MAO	27
Cycloheptadiene	$Ni(acac)_2$	MAO	0
Cyclooctadiene	$Ni(acac)_2$	MAO	0
CHD	$Ni(acac)_2$	MAO+TMA (1:1)	30
CHD	$Ni(acac)_2$	MAO+TIBA (1:1)	42
CHD	$Ni(acac)_2$	TIBA	0
CHD	Cp_2Ni	MAO	1.1
CHD	$NiCl_2$	MAO	0.5
CHD	$[(C_2H_5O)_3P]_2NiCl_2$	MAO	0
CHD	$[(C_2H_5)_3P]_2NiCl_2$	MAO	0.3
CHD	$[(C_6H_5)_3P]_2NiCl_2$	MAO	0

1,3-cycloheptadiene and 1,3-cyclooctadiene do not give any polymer product. Unreacted monomer is recovered at the end of the reaction, as shown by GC-MS analysis. 1,3-cyclooctadiene also inhibits the polymerization of CHD: a 0.5% of pure PCHD was isolated after the polymerization of a 1:1 mixture of monomers. The reason for this behavior could be attributed to the different conformations of CHD (planar) and its larger homologues ("oyster"-shaped). Nickel active centers are possibly wrapped up by the larger cycles to form stable complexes and prevented to be approached by other monomer molecules.

The effect of temperature and catalyst concentration was studied on the $Ni(acac)_2$/MAO system. The polymer yield increase remarkably in the 0-90°C temperature range (Table 4).

Table 4. Effect of reaction temperature on the polymerization of 1,3-cyclohexadiene. [CHD] = 3.0 mol/L in toluene, [CHD]/[Ni] = 2000, [MAO]/[Ni]=50, t = 30 min.

Temperature (°C)	Yield (wt%)
0	4.9
25	17
50	27
90	57

Higher conversion are also obtained by increasing the nickel concentration or the [MAO]/[Ni] ratio. For instance, for a MAO/nickel molar ratio equal to 100, the polymer yields have the following values: [CHD]/[Ni]=10000, yield=0.9%;

[CHD]/[Ni]=5000, yield=24%; [CHD]/[Ni]=2000, yield=78%; [CHD]/[Ni]=1000, yield=96%;. By halvening the [MAO]/[Ni] ratio, the yields are <0.1%, 0.7%, 27%, 75%, respectively.

Poly(1,3-cyclohexadiene) is insoluble in all the common organic solvents: it even starts to precipitate from the polymerization mixture after few seconds from the addition of the catalyst. A good solubilty was only found to exists in naphthalene at very high temperatures (>160°). Thus, the characterization through the usual solution techniques (GPC, NMR) is rather difficult to carry out, and the data acquisition is still in progress.

DSC measurements show that most of the polymers have a melting temperature about 320-322°C. These values are a little higher for the polymer prepared at 0°C (328°C) and for that obtained with Cp_2Ni (327°C), indicating in these last cases a greater degree of stereoregularity. PCHD begin to be degraded around 300-310°C, just below the melting temperature. For this reason, a correct evaluation of the melting enthalpies is not possible. Degradation involves a partial crosslink of double bonds, as shown by the decrease of $v(C=C)_{cis}$ and $\omega(=CH)_{cis}$ bands in IR spectra of samples after treatment at temperatures above 280°C for prolonged times.

Copolymerization of styrene and 1,3-cyclohexadiene

In table 5 some catalysts are compared in the polymerization of styrene and CHD under the same conditions. As can be seen, the yields of polystyrene are much more high than those of poly(1,3-cyclohexadiene), the differences being particularly evident for $NiCl_2$. Because $Ni(acac)_2$ is the most active catalyst among those tested, and, as previously reported, is capable to afford copolymers of linear conjugated dienes and styrene [9], attempts to prepare styrene/CHD copolymers have been performed.

Table 5. Comparison of styrene and 1,3-cyclohexadiene in the polymerization with nickel/MAO catalysts. [monomer] = 2.8-3.0 mol/L in toluene, [monomer]/[Ni] = 2000, [MAO]/[Ni]=50, T = 50°, t = 30 min

Catalyst	Polymer yield (wt%)	
	Styrene	1,3-cyclohexadiene
$Ni(acac)_2$	46	27
$NiCl_2$	39	0.5
Cp_2Ni	9.5	1.1
$[(C_6H_5)_3P]_2NiCl_2$	14	0.3

The results are reported in Figure 1. By calorimetric analysis it was found that in all cases (except, obviously, when CHD is absent) the reaction products consist of the 1,3-cyclohexadiene homopolymer. Therefore, styrene behaves as an inert diluent in the same way as toluene, and conversions should be more correctly considered by referring them to the amount of polymerized CHD, excluding styrene from the computation.

The so calculated conversions show that the recovered polymer amount is roughly constant below [CHD]/[Ni]=1500, and decrease at higher [CHD]/[Ni] ratios (see the values reported above the experimental points in Figure 1).

From these results it can be speculated that the catalytic site structure is very different for the two monomers, and that the active species for styrene polymerization has a low stability in the presence of 1,3-cyclohexadiene.

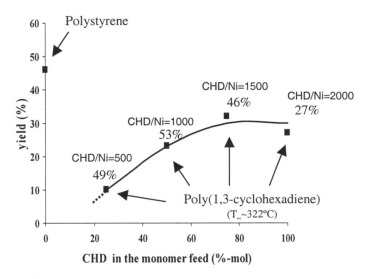

Fig. 1. Copolymerization of styrene and CHD. Yields are referred to the sum of the monomers; the percentage above the experimental points are the conversion of CHD to polymer. Catalyst: Ni(acac)$_2$/MAO, [monomers]/[Ni]=2000, [MAO]/[Ni]=50, T = 50°, t = 30 min.

4. Conclusions

The following conclusions can be drawn from the present work:
- The ability of nickel complexes/MAO to promote the isospecific polymerization of styrene can be modulated by appropriate ligands on the metal;
- Best activity is exhibited by Ni(acac)$_2$ and NiCl$_2$. Molecular weights are rather low (M$_w$~20000);
- Activity decrease with reaction temperature, dilution, and solvent basicity;

- Atactic polystyrene is formed along with partially isotactic polystyrene. Two catalytic sites are believed to exist; the isospecific one is probably heterogeneous;
- $Ni(acac)_2$ is the most active catalyst for 1,3-cyclohexadiene polymerization; Cp_2Ni has a much lower activity, while nickel chlorides, including those bearing phosphane ligands, afford only traces of polymer;
- Monomer conversion increase with the temperature, the nickel concentration and the Al/Ni ratio; addition of TIBA also increases the catalytic acivity;
- Poly(1,3-cyclohexadiene) cannot be easily characterized due to its low solubility. Melting temperatures of the polymers are remarkably high (up to 328°C);
- Larger cyclic conjugated diolefins (1,3-cycloheptadiene, 1,3-cyclooctadiene) polymerization is not activated by the reported nickel complexes;
- The attempted copolymerization of styrene and 1,3-cyclohexadiene failed, suggesting that the active species structure is significantly different for the two monomers.

The technical assistance of Dr. A.Guarini, Dr. A.M.Romano, Dr. M.Salvalaggio, Dr. S.Spera, Mr. F.Toscani, Dr. C.Zannoni is gratefully acknowledged.

References

1   S.D. Ittel, L.K. Johnson, M. Brookhart, *Chem. Rev.*, **100**, 1169 (2000)
2   T.R. Younkin, E.F. Connor, J.I. Henderson, S.K. Friedrich, R.H. Grubbs, D.A. Bansleben, *Science*, **287**, 460 (2000)
3   N. Ishihara, M. Kuramoto, M. Uoi, *Macromolecules*, **21**, 3356 (1998)
4   P. Longo, A. Grassi, L. Oliva, P. Ammendola, *Makromol. Chem.*, **191**, 237 (1990)
5   G. Lopes Crossetti, C. Bormioli, A. Ripa, A. Giarrusso, L. Porri, *Macromol. Rapid Commun.*, **18**, 801 (1997)
6   R. Po, N. Cardi, R. Santi, A.M. Romano, C. Zannoni, S. Spera, *J. Polym. Sci., Part A, Poly. Chem.*, **36**, 2119 (1998)
7   L. Oliva, P. Longo, A. Grassi, P. Ammendola, C. Pellecchia, *Makromol. Chem., Rapid Commun.*, **11**, 519 (1990)
8   K. Endo, Y. Uchida, Y., Matsuda, *Macromol. Chem. Phys.*, **197**, 3515 (1996)
9   K. Endo, K. Masaki, Y. Uchida, *Polymer J.*, **29**, 583 (1997)
10 B.A. Dolgopolsk, S.I. Beilin, Y.V. Korshak, L.M. Chernenko, L.M. Vardanyan, M.P. Teterina, *Eur. Polym. J.*, **9**, 895 (1973)
11 D.L. Gin, V.C. Conticello, R.H. Grubbs, *J. Am. Chem. Soc.*, **116**, 10507 (1994)
12 J.P. Claverie, D.L. Gin, V.C. Conticello, P.D. Hampton, R.H. Grubbs, *Polym. Prep. (Am. Chem. Soc., Div. Polym. Chem.)*, **33**, 1020 (1992)
13 P. Longo, C. Freda, F. Grisi, O. Ruiz de Ballestreros, *Macromol. Rapid Commun.*, in press
14 R. Po, R. Santi, M.A. Cardaci, *J. Polym. Sci., Part A, Polym. Chem.*, **38**, 3004 (2000)
15 A.R. Pray, *Inorg. Synth.*, **5**, 153 (1957)
16 A.E. Arbusov, V.M. Zoroastrova, *Dokl. Akad. Nauk. SSSR*, **84**, 503 (1952)
17 G. Garton, D.H. Henn, H.M. Powell, L.M. Venanzi, *J. Chem. Soc.*, 3625 (1963)
18 S.A. Broomhead, F.P. Dwyer, *Austr. J. Chem.*, **14**, 250 (1961)

19   G. Schrauzer, H.N. Rabinowiz, *J. Am. Chem. Soc.*, **90**, 4297 (1968)
20   M. Meier, F. Basolo, *Inorg. Synth.*, **28**, 104 (1990)
21   N. Cardi, R. Fusco, L. Longo, R. Po, S. Spera, G. Bacchilega, *Proceedings of International Symposium on Ionic Polymerization*, Paris (1997), page 378
22   A. Duteil, G. Schmid, W.Meyer-Zaika, *J. Chem. Soc., Chem. Commun.*, 31 (1995)
23   F.A. Bovey, *High Resolution NMR of Macromolecules*, Academic Press, New York (1972)
24   I. Natori, S. Inoue, *Macromolecules*, **31**, 4687 (1998), and references cited therein
25   G. Kwang, Y. Jang, H. Lee, *Polymer J.*, **31**, 1274 (1999)

6. Catalyst Heterogenization and Particle Microreactor Effects

Activity Limits of Heterogeneous Polymerization Catalysts

Timothy F. McKenna, Roger Spitz*

C.N.R.S. - L.C.P.P./C.P.E., Bât. F308,43 Bd du 11 Nov. 1918, B.P. 2077,
69616 Villeurbanne, France
E-mail: spitz@lcpp.cpe.fr

Abstract. Given the high activities of heterogeneous catalysts used in the polyolefins industry, it is reasonable to expect that the reactions might be controlled by mass or heat transfer. It is very difficult to establish the limit of the activity which can be reached without encountering any significant effect of diffusion limitations on the polymer properties using the commonly accepted models of olefin polymerization. Such models seem to restrict the activity far below the observed values (for realistic values of model parameters), typically much less than 10 kg of polyolefin per g catalyst per hour, and forbid the reaction rate to reach high values in a few minutes. In this paper, we intend to give some simple experimental arguments to show that limiting values of the intrinsic activity can reach much higher values, thus justifying the efforts devoted by chemists to improving the performance of heterogeneous catalysts. Examples will be restricted to slurry polymerization where mass transfer is dominant.

Introduction

As opposed to many other industrial catalytic processes, the lifetime of the catalysts used in olefin polymerization is restricted to the time they pass in the reactor, and never exceeds a few hours (this is to avoid the need for huge, inefficient reactors). These catalysts are not recovered and recycled, and their true cost should therefore be defined in terms of the amount of monomer that they convert. For this reason, as much polymer as possible must be produced in the shortest possible time: high rates must be obtained immediately upon injection of catalyst into the reactor, and maintained for as long as possible. Catalyst chemistry must therefore be conceived so as to favor these activities, and should be based on: choice of efficient catalysts, optimization of the dispersion of the active species (in the particles and in the reactor), perfection of the activation of the active centers, control of dormant species, mastering of deactivation, etc. Most of these issues are common to all families of catalysts used in olefin polymerization, including group IV metallocenes, even if the true design of the active species depends on the catalyst used. If the chemist has been able to prepare highly active species, and if the mechanical properties of the solid catalyst allow the reaction to proceed, it is possible that the observed reaction rate would be limited by some sort of diffusion process. This is doubly true as very highly polymerization rates are being reported - sometimes over 100 kg of polymer are produced per gram of catalyst per hour! Some diffusion problems involve transport of monomer at the reactor scale, and are generally not too difficult to solve. In the case of homogeneous catalytic processes, heat and mass transfer problems can be solved by dilution, with catalyst levels being brought below 1 micromole per liter. The same is true at the reactor scale for heterogeneous catalysts, however at the single particle scale one cannot solve the problem of monomer diffusion limitations by diluting the active sites once the catalyst has been made (although understanding the importance of this factor will help at the catalyst design stage). It is important to understand and/or eliminate diffusion resistances at this scale since important properties like the MW, MWD, and copolymer composition are (very) sensitive to local concentrations. Most of the attempts that have been made at modeling heat and mass transfer in highly active modern catalytic systems demonstrate more the limits of existing models than of the reaction itself: it appears that when performed correctly and in the right conditions, the reaction proceeds better than predicted by the models. Also, cases where mass transfer resistances have been reported in the literature often correspond to anomalous cases, e.g. large catalyst particles, unsuitable porosity, abnormal mechanical resistance of the support, but also in cases where the activities are much too low to suggest transfer resistance. Despite these shortcomings, a certain number of points have been demonstrated. Problems in gas phase processes, where heat transfer limitations dominate, differ greatly from those in liquid phase systems, where single particle monomer transfer limitations are more important. It is possible to circumvent heat transfer problems using a well-controlled combination of phases (e.g. super-condensed cooling) to push back the limits somewhat. Critical particle sizes have been defined for different, competing processes, which means that we cannot decrease the size of the particles indefinitely. However, even if we can identify these sizes and the parameters of the diffusion process, it is clear that mass transfer is a complex process. If we are to understand the limits set on catalyst activity by mass transfer, it is necessary to improve our description of the physical processes taking place, and to include a description of the morphology of the growing particles.

Polyolefins are catalytically produced using heterogeneous catalysis, generally organo-metallic compounds (sometimes derived from inorganic precursors like $TiCl_4$) on highly porous, solid supports such as magnesium chloride or silica. These catalysts can be used in either gas phase fluidized bed reactors (FBR), liquid-solid slurry, or in the three phase liquid pool process (only polypropylene). The polymers generally form an heterogeneous phase as they are (usually) insoluble in the monomer or monomer/diluent mixture, and the size of the polymer particles grows continuously throughout the polymerization. For polymerization to take place, monomer must circulate through narrow pores and penetrate through a thin film of polymer to reach the active sites. This is the first reason for which one might suspect that diffusion limits exist during the polymerization. The second reason is that, as a opposed to many other catalytic processes used for instance in the petrochemical industry, the rate of reaction is generally very high. Typical reaction rates for recent catalysts are on the order of 10,000 to 50,000 g/(g.h). High activities are necessary in the case of polyolefin production because the reactor residence time defines the size of the polymerization plant, and polymer production rates must be high for both economic reasons and to reduce the concentration of catalytic residue to avoid undesirable colors, odors or toxic effects. The specific cost of the catalyst decreases as the volume of polymer produced on it increases. This need for high reaction rates make it thus reasonable to suppose that one could encounter mass transfer resistance, especially in slurry and liquid pool reactions (less in the gas phase). Furthermore, since the polymerization is highly exothermic, with ΔH_p (heat of polymerization) being on the order of 100 kJ/mol, it is very important to understand and optimize heat transfer in order to avoid meltdown and reactor runaway which are occasionally observed in gas phase. A third group of observations is often interpreted by the chemists as proving the existence of diffusion control of the reaction: the molecular weight distributions are never narrow and copolymer compositions are not homogeneous in all processes using heterogeneous catalysts. The reaction kinetics are generally not stationary, the reactions are often not first order with monomer concentration, and different unexpected activation effects (by comonomers, by hydrogen) are observed which were not easy to explain immediately from a chemical point of view, suggesting that not only chemical effects are involved.

From the point of view of chemical engineering, the description of the heat and mass transfer, even if slightly complicated by the particle growth, does not appear to be very unusual. The reactants diffuse from the continuous phase, through the boundary layer surrounding the particles, through the pores of the catalyst to the active sites where they build a solid polymer. The heat produced by the reaction is evacuated in the opposite direction. Immediately after the reaction starts the original particle support fragments due to the hydraulic forces created by the formation of the polymer, but nevertheless retains its original form due to the adhesive forces of the polymer molecules, and one generally observes that the particle size distribution at the beginning and end of the reaction are of the same width. As such, the reaction begins with small particles (10-50 μm in diameter), and finishes one to two hours later with the same number of particles with a diameter on the order of 500 - 1000 μm. The original work in the area of modeling of mass and energy transport phenomena on Ziegler type catalysts was done in particular by Chiovetta et al. [1,2], and by the research group of Ray at the University of Wisconsin [3,4]. These models were based on what is now called the MultiGrain Model (MGM). In this, and similar models (e.g. polymer flow), a growing particle is assumed to be a pseudo-homogeneous medium in which mass transport in the entire polymer/catalyst particle is characterized by a single diffusivity, and the characteristic length scale for diffusion is the particle radius. More recently, Fink et al.[5] used microscopy studies to justify the addition skin effects on the surface of the particle. The same references also provide data for different diffusion constants and the way the porosity and tortuosity of the catalyst modifies the diffusion properties. The common feature of all these models is the prediction of limits in activity which are due, in liquid phase, to the consumption of a large part of the monomer flux into the particle in its outer shell, and to the overheating and loss of control of the reaction in gas phase. Even if these predictions where more or less in agreement with the catalytic activities developed in the early 80s, they no longer satisfactorily describe recent performance levels obtained with modern catalysts.

In this paper, we discuss a series of experimental results which show that polymerization can occur at very high rates without evidence of diffusion limits. We restrict the discussion to slurry polymerization, where it is safe to assume that only mass transfer limits. The experiments reported are laboratory scale semi-batch experiments. The activities which will appear are not all in the range of the highest reported activities, but, considering the experimental pressures (or the corresponding monomer concentrations and keeping in mind that the rate of diffusion is proportional to concentration) they are generally equivalent to very high activities reported in the literature. For instance, the catalyst considered below in a 4 bar propene polymerization in heptane slurry has an initial activity of 6 kg/g.h at 70 °C and 4 bars, and the activity rises to 50 kg/g.h at the same temperature in a batch of liquid propene.

Expected kinetic profiles

Before discussing this in detail, we should point out that the observation of high reaction rates in slurry phase at the laboratory scale can be difficult for reasons linked to a different type of mass transfer resistance. If the catalytic activity is high, the polymerization may be limited by the dissolution of the monomer in the diluent used for polymerization. This reactor-side mass transfer limitation might increase during polymerization, as the monomer exchange rate at the gas-liquid interface can slow down if polymer plugs the interface, or if the viscosity of the diluent increases.

On the contrary, during polymerization, the size of the polymer particle increases and mass transfer rates to the particle will increase. This is because the flux of monomer into the particle is proportional to the surface of exchange i.e. R^2, R being the radius of the growing particle, whereas the rate of monomer consumption is proportional to the mass of catalysts in the particle. Mass transfer therefore becomes less and less difficult as the polymerization progresses, and the polymerization rate, if controlled by diffusion, is expected to increase with polymerization times. Typical behaviors corresponding to the 2 types of diffusion control are presented in Figure 1.

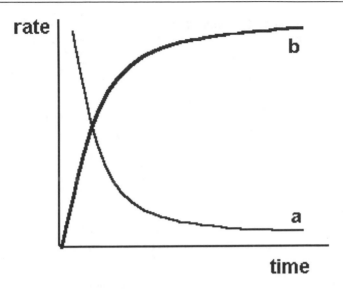

Fig. 1. Scheme representing the "observed" rates of reaction for a catalyst with a constant intrinsic activity in the presence of (case a) a reactor-side diffusion limitation (a diffusion barrier at the gas-liquid interface limiting the dissolution of the monomer) or (case b) a diffusion barrier in the growing polymer particle .

The limitation of monomer diffusion in a particle results in a gradient of monomer concentration. The monomer concentration decreases from the surface to the center of the particle . The effect shown in Figure 2 will depend on the particle size, and on how the catalytic activity varies as a function of time and reaction conditions (activation, deactivation, etc...). In the example shown in Figure 2, the maximum intrinsic activity cannot be reached in a brief time as most of the active centers are in contact with a reduced concentration of monomer.

The other consequences are that the molecular weight will be reduced and the molecular weight distribution will be predicted to be broader than under the conditions of a chemical control of the reaction.

High activities at the beginning of the polymerization

All reaction-diffusion models predict that mass transfer limitation inside the particle will limit the activity at the beginning of the polymerization. To illustrate this effect, let us examine the polymerization presented in Figure 3.

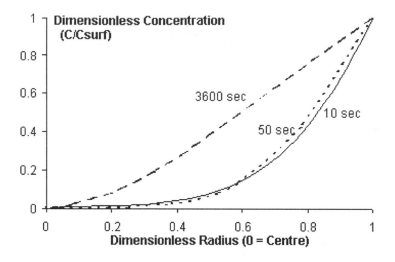

Fig. 2. Predicted values of the monomer concentration inside a particle of catalyst as function of polymerization time. The particle diameter is 25 microns at the beginning of the polymerization; the maximum activity of the catalyst in the absence of diffusion limitations is 25 kg/g.h, reached in about 100 s. Polymerization conditions : heptane slurry, 80 °C, 8 bars total pressure (6 bars of ethylene and 2 bars of hydrogen).

The polymerization starts with an extremely high activity which was measured to be greater than 20 kg/g.h after 2 min. This result is not easy to explain on the basis of existing models. In a second step, after 50 min., the activity increases up to 130 kg/g.h. At this time, the catalyst is diluted by 20 000 times its weight of polymer and the diffusion barrier has became more negligible.

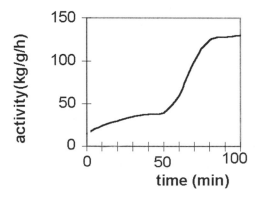

Fig. 3. Polymerization of ethylene in heptane slurry at 80 °C and 8 bars total pressure (hydrogen pressure fixed at 3 bars at the beginning of polymerization). Catalyst polymer particles diameter 25 micrometers

The activity measured at an early stage of the reaction, before the active sites become significantly diluted by the polymer, is an excellent indicator of the importance of the diffusion barrier, which is why we focus our attention on this point. Diffusion barriers are also sensitive to the initial size of the catalyst particle: initial activity varies as R^3, where as the flux of monomer entering the particle varies as R^2. The two parameters (activity, size), can be combined in the same experiments, using a catalyst with a broad distribution of particles (1-80 microns) and an activity reaching 20 000 g/g/h in less than 3 min. We already mentioned that the observation of such high initial activities contradict the predictions of the models in the literature. These experiments give us direct access to the effect of the diffusion barriers. The molecular weight and molecular weight distributions are expected to be sensitive to the ratio of monomer to hydrogen which is used as a transfer agent. Since hydrogen is consumed at a rate about 1000 times less than that of the monomer, and its diffusivity is much higher, we would expect that the diffusion barrier will be negligible for hydrogen. We compared the MW and MWD in small and large polymer particles in different conditions of polymerization leading to initial activities ranging from a few kg/g.h to 25 kg/g.h. The MWD is the same in all fractions and experiments. The MW values are not exactly the same for small and large particles, probably due to slight differences in the formation of the active centers according to the particle size. But the ratios do not change for experiments corresponding to extremely different activities. Another set of experiments is obtained by changing the diffusion constant in the medium. This is simply done by changing the viscosity of the diluent by addition of soluble oligomer. Enough oligomer was added to change the viscosity by a factor of ten, and no change in MW and MWD was observed with respect to low viscosity experiments.

In all these experiments, we verified that there was no significant change in the number of particles in the medium. There is no agglomeration, nor particle rupture during the polymerizations. Polymer particle size only increases as expected with the monomer conversion.

Lowering activities

The experiments presented above were intended to demonstrate that systems employing highly active catalysts do not show diffusion limitations. Completely different experiments can also be performed. A simple method consists in changing the level of activity by chemical poisoning. As an example, we will consider the poisoning of a polymerization of propene in heptane slurry at 70 °C with 4 bars total pressure including 0.1 bars hydrogen. In these conditions, a convenient choice of a magnesium chloride supported titanium catalysts leads to activities reaching 6 kg/g.h after much less than 1 min. This activity is in fact very high considering the low monomer concentration in solution according to the low pressure. The activity decreases slowly with polymerization time. Adding a small quantity of a poison, ethylene glycol dimethoxy ether, to the catalytic system allows us to decrease the activity by as much as a factor of 10. Note we used a compound which, contrary of others like CO or allene, cannot be inserted at all. Enough polymer is produced to allow characterization, and no change is observed concerning tacticity and MWD. MW decreases which is opposite to the effect expected from a change in diffusion control: at low activity, the monomer concentration in the particles is expected to be higher than at high activity because if there were mass transfer limitations at higher activity, they would be reduced at low activity. The observed variation must be due to a change in the chemistry of the system because of the poison. The polymer is produced under chemical control of the reaction both at low and high activity.

Heterogeneous metallocene catalysts

Metallocene catalysts were first used industrially in processes operating in solution, but they being developed more and more for supported systems [6]. One of the most important properties of these new catalysis is that there are often single site systems, leading to the production of uniform (co)polymers. If the reaction occurs with a diffusion control of the monomers, the copolymers produced in the outer shell of the particles and the copolymers produced close to the center of the particles will not have the same composition. The overall copolymer will appear to be a mixture which can be separated by the usual fractionation methods like TREF. Many industrial companies have turned to gas phase polymerization which is expected not to give too much mass transfer limitations. We have done some experiments using silica supported catalysts in heptane slurry [7]. Since the particles were designed for the gas phase, they are rather large, typically more than 40 microns in diameter. Despite this fact, moderately high activities (the activities ranges from 1 to more than 10 kg/g.h) have been observed and the polymer retains properties which are very close to a single site system: Mw/Mn not larger than 2.5; the ethylene-hexene-1 copolymers cannot be fractionated by solvents in fractions differing significantly in comonomer composition. Experiments covering a wide range of comonomer compositions and temperatures have been performed, ranging from 60 to 95 °C, and from 0 % hexene to an hexene to ethylene molar ratio of 4 in the reactor. The polymers differ in melting point (90-130°C), as well as in crystallinity, but no change in the homogeneity of the polymers could be clearly demonstrated. The result is especially surprising since some of the high temperature, high hexene content experiments the polymer is soluble in the polymerization medium and in the experiments were performed in slurry.

Conclusion

All the experiments presented in this paper suggest that, with a convenient choice of the catalyst properties and polymerization conditions it may be possible to polymerize without reaching the diffusion limitations predicted by existing models for a wide range of polymerization rates in conditions similar to those used in industrial processes. This does not mean that there is no diffusion barrier at all. We often meet limitations due to an incorrect development of the polymer particles for instance. It is also necessary to distinguish between the different polymerization processes. All the examples are presented in slurry polymerization in the presence of an inert diluent. In this particular case, limitations are expected to concern only mass transfer and to manifest themselves inside the particle. The presence of inert diluent molecules hinders monomer diffusion, and the case of the polymerization of propene in liquid pool would, in principle, be more favorable for mass transfer. We have not discussed gas phase polymerization either. The effect of a limitation of heat transfer is often obvious, e.g. when the polymer melts, but is sometimes more discrete but detectable when the melting is limited to the inner part of the particle.

It will be necessary, in the future, to develop models that agree much better with experimental observations. The modeling of the mass transfer of monomer(s) inside the growing polymer particle must be improved [8,9]. On the one hand, some simplifications can be introduced in the models, taking in account the fact that, below a defined dimension, certain barriers can be neglected. This typical dimension is generally in the range of 1 micrometer for the highest activities which have to be considered. On the other hand, the characteristic length scale for which diffusion takes place inside the particle has also to be considered. This depends on the particle size, but also on the porous morphology of the swollen particle in polymerization conditions, which will of course add some complexity to the models [10]. The synergistic effect of a fast reaction on the diffusion process should also be investigate, especially if convection processes are not negligible.

Acknowledgment

The authors thank Elf-Atochem for the financial support of the work and Fanny Barbotin, Jérôme Dupuy, Nathalie Verdel, Virginie Mattioli and Valérie Benoit for their contributions. One of the authors (TM) wishes to acknowledge the financial support of the European Commission. Portions of this work were funded by BRITE-EURAM project CATAPOL BE 3022.

References

1 Laurence, R. L. and M. G. Chiovetta, in Polymer Reaction Engineering: "Influence of Reaction Engineering on Polymer Properties," K.H. Reichert and W. Geisler (eds.), Hanser Publishers, Munich (1983).
2 Ferrero, M. A. and M. G. Chiovetta, Polym. Eng. Sci., 27, 1436, (1987).
3 Floyd, S., K. Y. Choi, T. W. Taylor and W. H. Ray, J. Appl. Polym. Sci., 32, 2935 (1986).
4 Hutchinson, R. A. and W. H. Ray, J. Appl. Polym. Sci., 34, 657 (1987).

5   Przybyla C., Weimann B. and Fink G. in "Metalorganic Catalyst for Synthesis and Polymerization", W. Kaminsky ed. Springer Berlin 1999 p.333.

6   Hlatky G. G. , Chem. Rev. **100**, 1347 (2000)

7   Spitz R., Verdel N., Pasquet N., Dupuy J., Broyer J. P.and Saudemont T. in Metalorganic Catalyst for Synthesis and Polymerization", W. Kaminsky ed. Springer Berlin 1999 p.347.

8   McKenna T. F., Dupuy T. F.and Spitz R. , J. Applied Polym Sci 63 315 (1997)

9   McKenna T. F, Cokljat D., Spitz R., Schweich D., *Catalysis Today,* **48(1-4)**, 101 (1999).

10   Kittilsen P. , McKenna T. F, Svendsen H. , proceedings of the "1st European Conference on the Reaction Engineering of Polyolefins" Lyons, july 2000, p.21.

Reversibly Crosslinked Polystyrene as a Support for Metallocenes
Part I: Covalently Bonded Metallocenes and Functionalized Supports

Markus Klapper, Matthias Koch, Martin Stork, Nicolai Nenov, Klaus Müllen

MPI für Polymerforschung, Ackermannweg 10, 55128 Mainz, Germany
E-mail: Klapper@mpip-mainz.mpg.de

Abstract: Via radical polymerization and followed by polymer analogous transformations cyclopentadienyl functionalized polystyrenes are generated. The pending cyclopentadienyl functions are used for the attachment of metallocenes and, furthermore, for the reversible formation of a network generated by a Diels Alder reaction. The resulting catalyst beads can be applied in the olefin homo- and copolymerization. A high activity as well as productivity is achieved. Films prepared by the obtained polyolefins exhibit a high transparency. These results suggest a fragmentation of the polymeric support, similar as in the silica case, induced by the reversibility of the Diels-Alder reaction. Additionally, by the incorporation of dyes into the polymeric support via copolymerization colored polyolefins can be generated.

Introduction

Metallocenes have several advantages over conventional Ziegler-Natta catalysts. They allow the tailoring of polyolefin properties by the design of the ligands around the metal center, and through their single active site, they produce polymers with narrow distributions of molecular weight and comonomer insertion. Their main disadvantage is their solubility, which makes they as not suitable for many industrial processes. In homogeneous polymerization one has to deal with large quantities of solvents as well as a dustlike, light polyolefin product. The heterogeneous Ziegler-Natta systems can be used in slurry as well as in solvent free gas phase processes.

It is thus obvious that metallocenes need to be heterogenizied before they can be used in existing Ziegler type production. [1,2] The most common used support is silica, but some polymeric carriers are also reported in the literature.

The fragmentability of the support is a crucial point when choosing materials for the immobilization of metallocenes. In the ideal case, the active centers are distributed evenly over the whole carrier particle. During the polymerization the particle breaks up and the inner centers can be reached by the monomer [3,4]. In this way the overall productivity of the catalyst is higher than for unfragmentable supports. Additionally, as the support always remains in the product, the smaller the fragments are, the less is their influence on the products properties, especially the transparency. Zeolites or divinylbenzene crosslinked polystyrene beads have low fragmentation abilities; for this reason productivity ratios are quite low, compared with silica systems, which fragmente well. The unfragmented carrier is visible in the product for example as inhomogenities when films are made out of it.

On the other hand the product is expected to have a high bulk density and being ideally to be formed into spherical particles. The size and shape of the product particles are determined by size and shape of the catalyst particles. Homogeneous polymerization products are dustlike and light. This makes processing difficult and can lead to severe 'reactor fouling', when the polyolefin sticks to the reactor walls and the stirrer. Similar products are obtained by immobilization on small particles or soluble supports. The active centers of a catalyst system have to be hold together during the polymerization in such a way that particles can be formed.

A good support therefore has to satisfy two main criteria. The primary particles have to have a certain size (usually about $50\mu m$) and strength to achieve high density product particles (preferred diameters are in the range of 1mm and above). On the other hand, the carrier must be able to break apart during the polymerization for a high productivity and homogeneous distribution of the support in the product.

Reversible Polymer Networks as Supports

As already said, the use of polymers as supports has some advantages over inorganic materials. They can be easily functionalized to satisfy certain requirements and it can be expected that an organic support will incorporate better into the product polymer than an inorganic one.

Organic supports for metallocenes are mainly based on divinylbenzene crosslinked polystyrenes [5,6], which are hindered in fragmentation. Often the metallocenes are synthesized directly on the support, producing multi-site systems in most of the cases. As metallocenes supported on soluble polymers can only produce dust or very small particles, a polymer network has to be used. For the need for fragmentation this network should be able to break up during the polymerization.

Therefore, we have developed a process which allows for an easy synthesis of a metallocene containing polymer under homogeneous conditions and a reversible formation of the network. [7] Via radical polymerization we generated a chloro-methylenestyrene/styrene copolymer 1 which can be directly functionalized with cyclopentadienyl units (Fig. 1). This approach has the advantage that one functionality can be used for both the synthesis of the catalyst as well as for the formation of the network. One part of the pendant cyclopentadienyl units is converted to the catalytically active metallocene 4, the other undergoes upon heating the crosslinking Diels-Alder reaction (Fig.2).

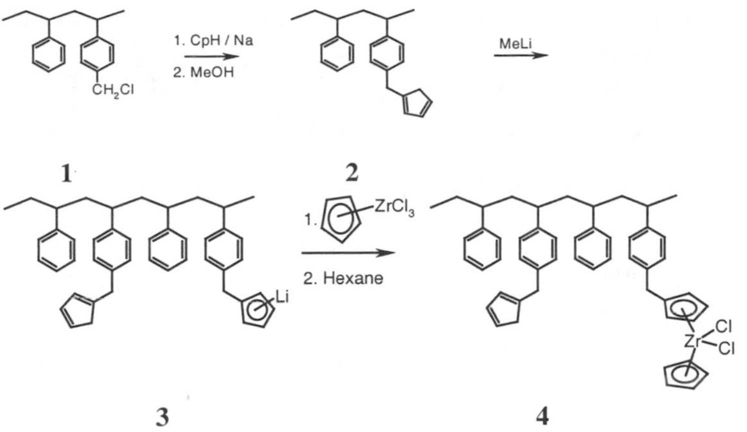

Fig. 1. Synthesis of a polymer-supported zirconocne catalyst

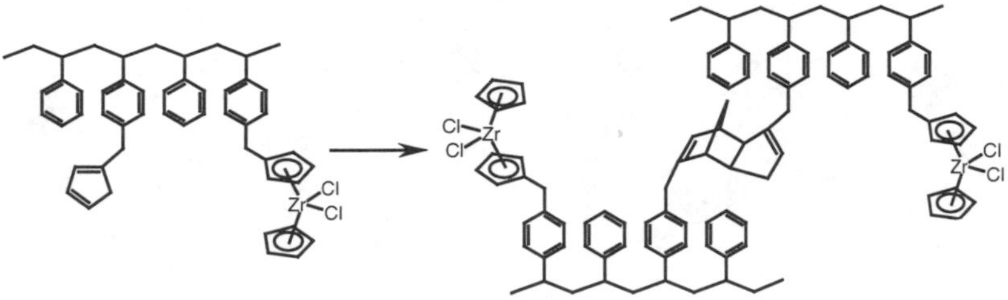

Fig. 2. Crosslinking reaction via Diels-Alder reaction

The catalyst beads so obtained can be directly applied in the polyethylene polymerization.

The polymerizations were carried out in a 1l stainless steel autoclave at 70°C in 400 ml isobutane and 40 bar pressure. The activity strongly depends on the MAO (methylaluminoxane) concentration as it is shown in table 1.

The product beads are spherical with a diameter of 0.1-3 mm depending on the reaction conditions. The SEM pictures of the metallated catalyst particle and of the product bead suggest that each polyolefin particle originates from one catalyst bead (Fig.3). [7] In comparison to the literature using other organic supports high productivities were achieved. (Table 1). The higher activities of our catalysts (exp A and B) in comparison to the ones supported on Merryfield resins suggest that the fragmentation of the network, as shown in Fig. 4 gives more and more access to the inner catalytically active centers inside of the network. All metallocenes can take part in the polymerization process. In contrast, in highly crosslinked materials as is the case for the divinylbenzene crosslinked materials, it is very likely that only the outer centers are active. Access to the centers inside of the network is blocked by growing polymer chains after a short time.

Table 1. Results of ethylene polymerization (650 μmol cyclopentadienyl units and 100 μmol zirconocene per g polymer)

Run	Mole Ratio Al/Zr	Activity [kg (PE) / molZr h]	M_w [g / mol]	M_w/M_n	T_m [°C]
1	300	500	-	-	134.9
2	500	1450	706000	3.4	135.2
3	1000	4250	600000	3.0	135.4
4	1500	5300	634000	3.1	137.8
5	2000	6660	635000	2.9	136.7

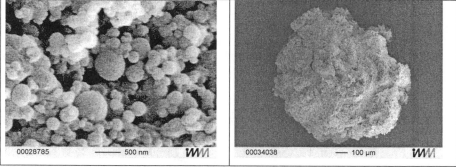

Fig. 3. Scanning electron microscope image of a catalyst particle (left side), scanning electron microscope picture of a product bead (right side)

Table 2. Comparison of polymerization results with different polymeric supports

	Zr-content μmol/g	Al/Zr	activity kg PE / molZr h	productivity g PE / g cat h	Tm °C
A	200	2000	11500	2300	136.9
B	200	1000	6700	1350	136.8
Soga [5]	7,1	14000	3192	23	139.2

Furthermore, these catalysts can also be applied in copolymer synthesis as we demonstrated for the ethylene/hexene copolymerization. (Table 3). The results are quite similar as expected from the case of silica. A decrease of Tm and of the degree of crystallinity is observed.[9]

Table 3. Copolymerisation of ethylene / 1-hexene (Loading: 200 μmol [Zr] / g [cat], [Al]:[Zr] = 1000, 90 min, solvent: 400 ml isobutane, 2 ml TiBA)

Run	Hexene [ml]	Activity [Kg [PE] / mol[Zr] h]	Productivity [g [PE] / g [cat] h]	T_m [°C]	Xc [%]
1	0	6800	1400	136.2	58
2	5	4500	900	132.7	58
3	15	4800	960	131.8	53
4	30	3300	670	129.5	52
5	400	11800	2400	129.0	49

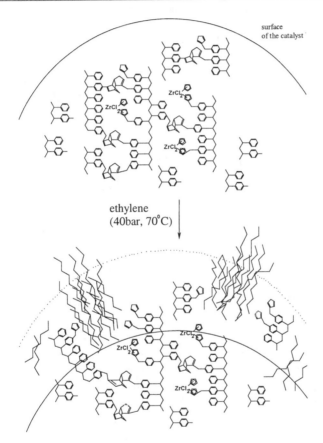

Fig. 4. Schematic description of the polymerization process on a fragmentable polymer bead

Dyeing of Polyolefins

It is quite difficult to incorporate dyes into polyolefins with a very homogeneous distribution. In most cases the pigments are clearly visible as spots in the material. Therefore, to get a perfect distribution of the dyes we developed a process to attach dyes directly onto the supports. This is possible for our organic materials as well as for silica supports. For such an approach it is essential to find dyes which are stable against MAO and do not interfere with the metallocenes. As we could show perylene dyes (e.g. **5**) fulfil this prerequirement (Table 4). [8] Depending on the type of support we used two different techniques for the dyeing of the support. For the inorganic materials, we just treated a solution of the dye with silica. Due to an adsorption process the dye is strongly physically linked to the support.[9]

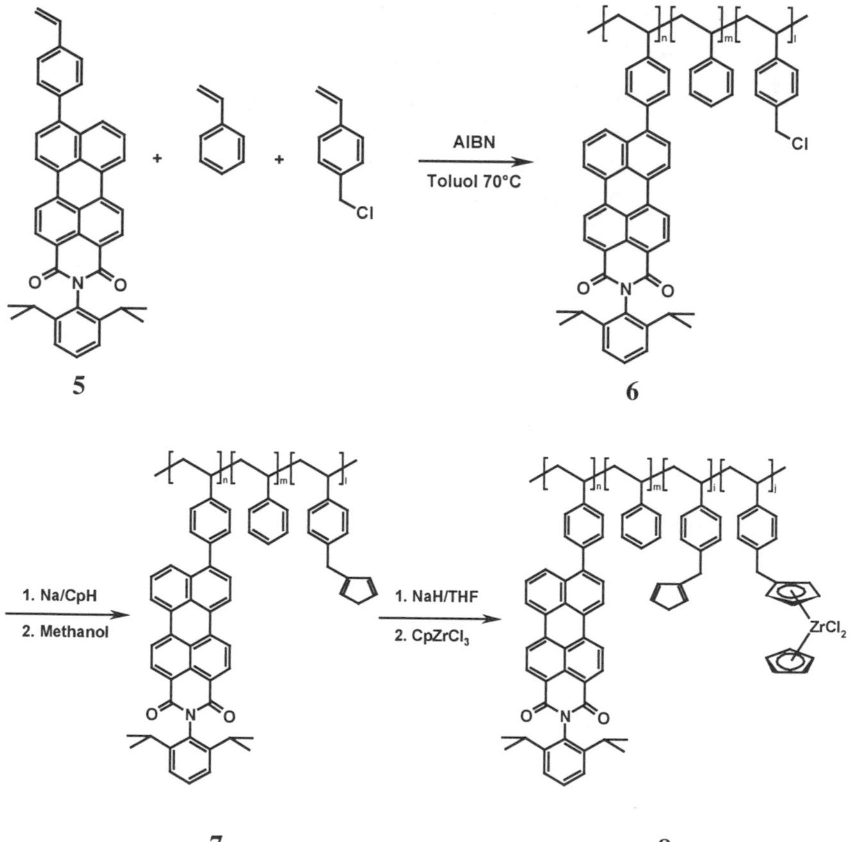

Fig. 5. Synthesis of polymeric support containing a fluorescent dye

For the organic supports we copolymerized a styryl functionalized dye with styrene and chloromethylstyrene followed by the polymer analogous build up of the zirconium catalyst. This procedure has the advantage that leaching of the catalyst can be completely excluded.

Table 4. Polymerization conditions and results of supports containing dyes (GPC-Analysis: Waters150C, o-DCB, 135°C, Standard: polystyrene)

	Metallocene	Supported on	[Al]:[Zr]	Activity Kg [PE] / mol [Zr] h	M_n g / mol	D
PE -1	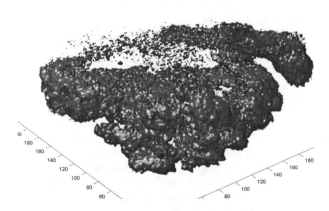	Silica	1000	5700	120000	4,0
PE- 2	**8**	Polymer	1500	5000	388000	2,8

The incorporation of a fluorescent dye allows not only for the dyeing of polyolefins but also for an easy and fast investigation of the fragmentation behavior via confocal fluorescence microscopy and a tracing of the catalyst. This technique allows for the 3-dimensional imaging of particles with a resolution of about 30-400 nm depending on the applied Laser wave length.

In a cooperation with W. Trabesinger, B. Hecht and U. Wild (ETH Zürich) we recorded 3D fluorescence images of several polyolefin product beads. [9,10]

Fig. 6. Fluorescence image of a PE-bead (PE-1) recorded by confocal microscopy

One can expect that that due to the fragmentation of the fluorescent tagged support a homogeneous distribution of the fluorescent dye in the product is observable. If a fragmentation would not occur only single spots of fluorescent areas in the polyethylene beads would be detectable. And indeed, as shown in Figure 6 (image of PE-1) a homogeneous distribution of the fluorescent dyes can be achieved. Even in the outer spheres of the product particles the fluorescent dye is detectable. Similar images are recorded from PE-2 obtained from the catalyst supported on a polymer containing the covalently attached dye.

It should be mentioned at this point that the concept of fragmentation is as well backed by the investigation of the properties of films prepared from polyethylene.

Films of PE generated with fragmentable supports show a much higher transparency than ones obtained from PE made with metallocenes supported on Merryfield resins. This might be due to the much higher activity of our catalysts which dilutes the polystyrene beads much more, but is, more likely to be, due to the prevention of scattering centers by the fragmentation of the supporting particles below the wave length of visible light

Experimental

Polymerisation of ethylene (General procedure)

To a Büchi stainless steel reactor (1 litre) equipped with a magnetic stirrer 0.25 ml of tri-iso-butylaluminum is introduced. Isobutane (400 ml) is injected using a pressure burette followed by saturation of ethylene at 40 bar pressure at 70°C. The silica or the polymeric supported catalyst is treated with 3.3 ml (10 %-w/w-solution in toluene) of MAO for 30 min and the preactivated catalyst is injected into the reactor using a pressure gate without adding further MAO. In order to quench the polymerization, the product slurry is drained and the polymer is dried at 70°C in vacuum.

Conclusion

We can demonstrate that using reversibly crosslinked polystyrene as a support has some important advantages over the already applied Merryfield resins. The catalyst can be prepared in a homogeneous way. High activities and productivities in ethylene polymerization can be achieved. The approach is very flexible in controlling the degree of zirconocene loading. and crosslinking. Spherical polyethylene beads can be isolated. Furthermore, the incorporation of dyes in a polymeric backbones was presented. This allows on one hand for the dyeing of polymers and on the other for the visualization of catalyst particles and for the study of the fragmentation process.

However, it should be mentioned that due to the fact that the catalyst is generated by a polymer analogous reaction on the polymeric backbone the numbers of applicable catalysts is restricted. To overcome this drawback we improved our concept of fragmentable supports towards non-covalently linked catalysts (Figure 7). [11] This is described separately in a second contribution in this book. [12]

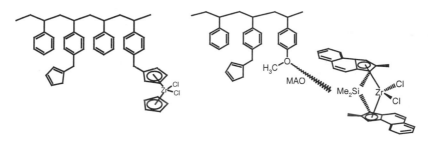

Fig. 7. Polymeric supports with covalently and non-covalently bonded metallocenes

Acknowledgement:

Financial support from BASF AG, Ludwigshafen is gratefully acknowledged.

References

1   F. Langhauser, J. Kerth, M. Kerstin, P. Kölle, D. Lilge, P. Müller: *Angew. Makromol. Chem.* 233, 155 (**1994**).
2   M. R. Ribeiro, A. Deffleux, M. F. Portela,: *Ind. Eng. Chem. Res.* 36, 1224-1237 (**1997**).
3   B. Steinmetz, B. Tesche, C. Przybyla, J. Zechlin, G. Fink: *Acta Polym.* 48, 392-399 (**1997**).
4   F. Bonini, V. Fraaije, G. Fink: *J. Polym. Sci. A* 33 2393-2402 (**1995**).
5   H. Nishida, T. Uozumi, T. Arai, K. Soga: *Macromol. Rapid. Commun.* 16, 821-830 (**1995**).
6   S. B. Roscoe, J. M. J. Fréchet, J. F. Walzer, A. J. Dias: *Science* 280, 270-273 (**1998**).
7   M. Stork, M. Koch, M. Klapper, K. Müllen, H. Gregorius, U. Rief, *Macromol. Rapid Commun.* 20, 210-213 (**1999**)
8   Martin Stork, Andreas Herrmann, Tanja Nemnich, Markus Klapper, Klaus Müllen, *Angew. Chem Int. Ed. Engl.* accepted
9   Martin Stork, Dissertation Johannes Gutenberg-Universität, Mainz **2000**
10   W. Trabesinger, B. Hecht, U. P. Wild M. Stork, A. Herrmann, M. Klapper, K. Müllen in preparation
11   M. Koch, M. Stork, N. Nevov, M. Klapper, K. Müllen: *Macromolecules*, accepted
12   Matthias Koch, Markus Klapper, Klaus Müllen, "Reversibly Crosslinked Polystyrene as a Support for Metallocenes Part II: Non-Covalent Bonding for the Immobilization"

Reversibly Crosslinked Polystyrene as a Support for Metallocenes
Part II: Non-Covalent Bonding for the Immobilization

Matthias Koch, Markus Klapper, Klaus Müllen

MPI für Polymerforschung, Ackermannweg 10, 55128 Mainz, Germany
E-mail: koch@bessel.mpip-mainz.mpg.de

Abstract: Polymeric supports suitable for various kinds of metallocenes are presented. The metallocene is non-covalently bound to the support by nucleophilic functionalities. Cyclopentadiene functions act as crosslinkers forming a reversible network. The network can break up during olefinpolymerization and high productivities are achieved.

Introduction

Part I showed the principles of reversibly crosslinked polystyrene supports. The system consisting of a bis-cyclopentadienyl zirconocene synthesized directly on the polymeric support works well for the polymerization of ethene. Stereocontrol in propene polymerisations requires more sophisticated metallocenes. Typically bridged systems are used. To obtain a single site catalyst a separate metallocene synthesis and purification is desirable.

Here we present a functionalized polymeric support that is suitable for the immobilization of various metallocenes.

Reversible Polymer Networks as Supports

As already said, the use of polymers as supports has some advantages over inorganic materials. They can be easily functionalized to satisfy certain requirements and it is to be expected that an organic support incorporates better into the product polymer than an inorganic one.

Organic supports for metallocenes are mainly based on divinylbenzene crosslinked polystyrene [1, 2], which are hindered in fragmentation. Often the metallocenes are synthesized directly on the support, producing multi-site systems in most cases. The use of stereospecific *ansa* metallocenes for propene polymerization needs a separate metallocene synthesis and purification. The immobilization can be done non-covalently, analogous to silica supports. As metallocenes supported on soluble polymers can only produce dust or very small particles, a polymer network has to be used. Due to the need for fragmentation this network should be able to break up during the polymerization.

A polystyrene bearing cyclopentadiene functionalities can be thermally crosslinked via the well known Diels-Alder reaction, which is also thermally reversible. In a first approach we used the cyclopentadiene functions not only for crosslinking but also for the synthesis of the metallocene directly on the support by reacting them with cyclopentadienylzirconium dichloride [3]. In this way an easily synthesizable catalyst system for the polymerization of ethene with high productivities was obtained.

Propene polymerization has higher demands on the metallocenes used, as stereocontrol has to be taken into account. Some attempts to synthesize bridged metallocene systems on inorganic and organic supports are known in the literature, but none of them is a single site catalyst. Therefore a different strategy was used.

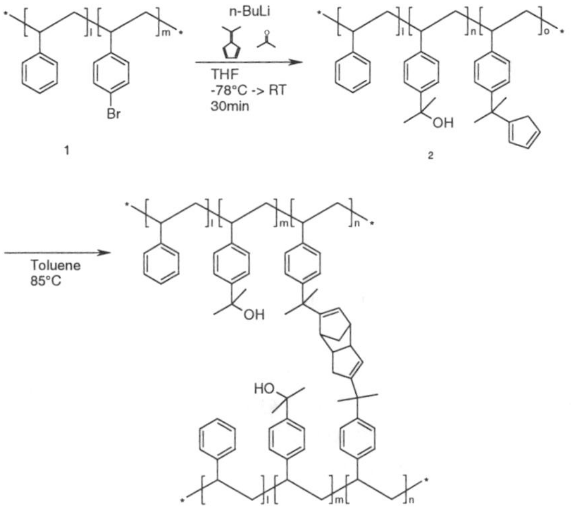

Scheme 1

The support we describe here is also based on cyclopentadienyl containing polystyrene, but bearing additional 'anchor groups' for the non-covalent bonding of separately synthezised metallocenes.

Starting with a copolymer (1) from styrene and p-bromostyrene obtained via radical polymerization, this support precursor is functionalized with cyclopentadienyl and hydroxy functions in a one-pot synthesis (2). A lithiation of the bromofunctions is followed by the simultaneous addition of dimethylfulvene and acetone. The product is crosslikable by heating a solution of it (3) and can be further functionalized [4].

Metallocene Bonding through MAO (4)

A copolymer containing 50% p-bromostyrene is reacted with acetone and dimethylfulvene in the ratio 2:1, yielding a polymer containing about 34% hydroxy and 16% cyclopentadiene functions. This polymer is crosslinked by heating a solution in toluene to 80°C, forming a highly swollen gel. This gel is treated with methylalumoxane (MAO) for the deactivation of the hydroxy functions, analogous to common silica supports. Afterwards, the metallocene (preferably a solution in MAO) is added to form the complete catalyst system. Upon addition of hexane the polymer separates from the solvent; the supernatant solution is removed of, the support is washed with additional hexane and dried in vacuum.

Metallocene Bonding through Ether Functions and MAO (5)

A copolymer containing 30% p-bromostyrene is reacted with acetone and dimethylfulvene in the ratio 1:2, yielding a polymer containing about 10% hydroxy and 20% cyclopentadiene functions. The hydroxy functions are reacted with tosylated polyetheneoxide, giving a support bearing polyethene sidechains. This support does not have the acidic hydroxy functions and a deactivation with MAO before adding the metallocene is not neccessary. After crosslinking, the support can be directly treated with the metallocene / MAO solution. It precipitates immediately, having the color of the metallocene and the supernatant solution is almost colorless.

The polyether chains give the support a higher affinity to the metallocene / MAO complex even when a lower molar content of bromostyrene is used in the initial polystyrene. Additionally one avoids the deactivation step - the support can be treated directly with the metallocene.

Metallocene Bonding and Activation through Boron Functions (6)

Instead of the reaction with tosylated polyetheneoxides, a functionalization with a boron compound is also possible. The hydroxy functionalized crosslinked support is treated with N,N-dimethylaniline and boron tris-pentafluorophenyl. In this way one obtains a covalently attached activator in the form of a boron-anilinium salt. When the support is treated with a metallocene dissolved in triisobutylaluminum (TIBA), the protonated N,N-dimethylaniline reacts with the metallocene forming cationic activated species. The cation is bound to the support via Coulomb interactions with the anionic boron. A leaching of the metallocene during the polymerization is impossible, eliminating the production of dustlike polyolefin and risk of reactor fouling.

Additionally, the activator can be used in equal molar amounts to the metallocene and no MAO is needed, reducing the amount of inorganic compounds in the product, which is important especially for the production of highly transparent films.

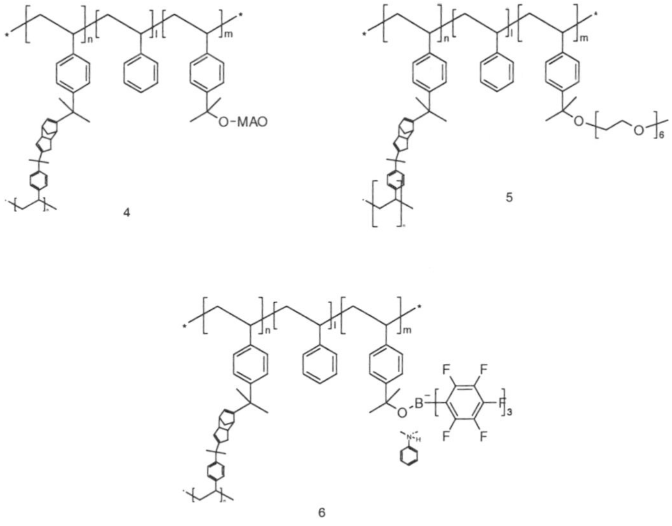

Scheme 2

Polymerizations

Polymerizations were carried out in a Büchi 1l stainless steel autoclave. Propene was polymerized in hexane at 4bar and 50°C, ethene at 40bar and 70°C in isobutane. The catalyst was transferred into the reactor through a pressure gate. 2ml triisobutylaluminum (TIBA) were used to scavenge impurities. Results are given in tables 1 and 2.

Activities are influenced highly by the amount of the co-catalyst MAO, values are only comparable between runs with similar MAO amounts. The polymerizations at 4bar were done with additional MAO activation as they showed only low activities without it. Therefore activities are not comparable between propene and ethene polymerizations. The metallocene used was (2-methyl benzindenyl)dimethylsilyl zirconium dichloride (MBI) for the MAO activated systems and (phenyl indenyl)dimethylsilyl zirconium dichloride (PI) in case of the boron based systems.

Activites range between 2600 and 8400 kg PP / mol Zr h bar and 200 and 970 kg PE / mol Zr h bar respectively which is in the range of silica supported MBI and much higher compared to other catalysts on polymeric supports. The catalysts made from the divinylbenzene crosslinked support (runs 1 and 9) are much less active than the reversibly crosslinked systems. They are comparable to other catalysts using this type of support.

Runs 4 and 5 compare a non crosslinked polymer with one containing no cyclopentadienyl. Similar activities give evidence that the cyclopentadiene groups have no influence on the catalysts activities. The activities of the non crosslinked systems are 30% higher than the comparable run 7, using a crosslinked support.

Table 1. Propene Polymerization Results

run	support	Zr [μmol / g cat.]	MAO / Zr	amount [mg]	Yield [g]	activity [kg PP / mol Zr h bar]	productivity [g PP / g cat.]
1	4*	33	800	93	12	2000	250
2	4	33	400	76	14	2600	340
3	4	33	840	60	45	8100	1400
4	5*	40	890	70	65	8400	1900
5	5*	40	890	70	62	8400	1900
6	5	40	630	80	11	1400	300
7	5	40	840	70	44	6300	1300
8	none		500	1μmol	44	22000	

Conditions: 400ml hexane, 2ml TIBA, 4bar propene, 50°C, 30min.

* run 1: 4% divinylbenzene crosslinked support, run 4: support not crosslinked, run 5: support containing no cyclopentadiene.

Morphology

The polyolefin is produced in spherical particles of about 1mm diameter. Polypropene bulk densities (around 200 g/l) are lower than polyethene ones (250-300 g/l), which might be due to the lower polymerization pressure and the additional MAO activation. If additional MAO is used in the polyethene polymerizations, a sharp increase of initial activity can be observed, followed by a rapid activity loss after a few minutes. In this case the products are very lightweight with popcorn like structures.

Table 2. Ethene Polymerization Results

run	support	Zr [μmol / g cat.]	MAO / Zr	amount [mg]	yield [g]	activity [kg PE / mol Zr h bar]	productivity [g PE / g cat.]
9	4*	33	300	70	9	100	130
10	4	29	375	55	32	500	600
11	4	33	300	75	51	500	700
12	5	43	180	90	32	200	350
13	5	33	300	90	71	600	900
14	6	48		25	23	970	1900

Conditions: 400ml isobutane, 5ml TIBA, 40bar ethene, 70°C, 60min.
* run 9: 4% divinylbenzene crosslinked support.

None of the catalysts produced any reactor fouling. Dustlike product, which is due to leaching of the metallocene / MAO from the support, resulting in a homogeneous polymerization, was observed only in very small quantities with the MAO activated systems and mainly with support 4. System 6 produced no dust at all. In this case the active species is formed only together with the covalently bound boron compound. A possible leaching of the metallocene results in a deactivation, therefore the metallocene is only able to polymerize when it is heterogenized. To prove this, we charged the reactor with metallocene solution and added the carrier. Polymerization started immediately and neither dust nor reactor fouling was observed.

Runs 4 and 5 are made with non crosslinked supports. Though the activities are higher, the product is similar to the one obtained from homogeneous polymerization. A large amount stuck to reactor walls and stirrer (reactor fouling). It shows that soluble polystyrene (single polymer chains) is no suitable support for metallocenes, even when they are used in a non solvent like hexane.

To investigate the fragmentation of our carrier, we produced polyolefin films in a heated press. About 100mg polymer were heated to 160°C for 10min and a pressure of 20kPa was applied for 1min. The samples were cooled rapidly in cold water. While the products from runs 1 and 9 showed inhomogenities, all other films looked similar to the one made from homogeneously produced polypropene.

Conclusion

The metallocenes are attached to the polymeric support non-covalently either via the interaction between nucleophilic groups and MAO or via the Coulomb interactions between the cationic zirconocene species and the anionic support.

The support is a thermally reversible network, that brakes up under the conditions of olefin polymerization, allowing a homogeneous distribution of the support througout the the product and higher productvities than with unfragmentable supports. The support incorporates well into the product as no visible defects were found in transparent polyolefin films.

Experimental

The zirconocene was donated by BASF-AG. MAO (10wt.-% in toluene) was obtained from Witco, Germany. All other chemicals were purchased from Aldrich Chem. Corp.. Styrene and its derivatives were distilled prior to use. Hexane, THF and toluene were dried by distillation from sodium / potassium alloy. All experiments were carried out under argon using standard Schlenk techniques.

The initial styrene / p-bromostyrene copolymers are made by radical polymerization. Styrene, p-bromostyrene are dissolved in toluene (10ml / 10g monomers) together with AIBN (50mg / 10g monomers). The mixture is degassed by several freeze-pump-thaw cycles and heated to 70°C for 24h. It is dilluted with dichlorormethane (5 times the volume), the product is precipitated from methanol and dried in vacuum at 60°C. Depending on the further functionalization, different molar ratios of styrene and p-bromostyrene are used (table 3).

The functionalization was done by dissolving 1g of that polymer in 100ml THF and adding 1.1eq n-buthyllithium 1.6M inhexane at −78°C. The mixture is stirred for 10min and 0.8eq Acetone 1M in THF are added simultaneously with 0.4eq dimethylfulvene 1M in THF. After further stirring for 20min the mixture is allowed to warm up to room temperature. The product is precipitated from methanol and dried in high vacuum at room temperature. Table 3 lists the exact quantitiies used.

Table 3. Synthesis of the Support Polymers

S	BS	BS / g	n-BuLi 1.6M	acetone 1M	fulvene 1M
4.10	7.23 (50mol-%)	3.48	2.4	2.8	1.4
7.29	5.49g (30mol-%)	2.35	1.6	1.9	.9
9.89	0.92g (5mol-%)	0.46	0.3	.4	.2

All supports are crosslinked by heating a solution of 200mg in 2ml toluene to 85°C for 2d. This solution is directly used for further functionalization.

The OH / MAO based support is obtained by deactivation of the hydroxy groups with MAO and subsequent mixing of the support with a solution of MBI in MAO (0.01M). The support precipitates together with the metallocene, is washed with hexane and dried in vacuum.

For the introduction of PEO sidechains the hydroxyfunctionalized, crosslinked polymer is reacted with 1.1eq n-buthyllithium and 1.2eq tosylated PEO at –78°C for 10min and is then allowed to warm up to room temperature.

By recation of the crosslinked polymer with 1.2eq N,N-dimethylaniline and 1.2eq boron tris pentafluorophenyl the covalently bound boron activator is formed. The product is precipiitated from hexane and dried in vacuum. For the immobilization, 4.8ml of a solution of MI in TIBA (0.001M) is added to a suspension of 100mg of the support in hexane. A successful activation results in a color change; for the metallocene used here it was from orange to blue. This suspension is transferred to the reactor.

Acknowledgement

Financial support from BASF AG, Ludwigshafen is greatfully acknowledged.

References

1   H. Nishida, T. Uozumi, T. Arai, K. Soga: *Macromol. Rapid. Commun.* 16, 821-830 (**1995**).

2   S. B. Roscoe, J. M. J. Fréchet, J. F. Walzer, A. J. Dias: *Science* 280, 270-273 (**1998**).

3   M. Stork, M. Koch, M. Klapper, K. Müllen, H. Gregorius, U. Rief: *Macromol. Rapid Commun.* 20 210-213 (**1999**).

4   M. Koch, M. Stork, N. Nevov, M. Klapper, K. Müllen: *Macromolecules*, accepted

An Intrinsic Defect of Slow Monomer Diffusion Theory for Explaining Broad MWDs

Tominaga Keii

Japan Advanced Institute of Science and Technology, Asahidai, Tatsunokuchi, Ishikawa 923-1292, Japan

Abstract: The theory of slow monomer diffusion to catalyst sites through polymer matrix explains broad molecular weight distribution(MWD) of polymer produced with heterogeneous Ziegler catalysts as well as rate-decay during polymerization. The explanation for MWD at pseudo-stationary polymerization with hydrogen dominant transfer can not be applied for polymerization with monomer dominant transfer, which is an intrinsic defect of the theory. The theory should predict that the narrowest MWD(M_w / M_n =2) for the latter case and reduction of the monomer contribution by changing concentration of monomer or by hydrogen-addition broaden MWD, which gives the basis of useful method to evaluate the theory itself. The accumulated experimental results of propene polymerizations with traditional and supported catalysts refute the above prediction.

Introduction

For explaining broad molecular weight distributions (MWDs) of polymers produced with heterogeneous Ziegler catalysts many rival theories have been proposed and discussed[1]. In the previous paper the two groups of theories ;site activities are non-uniform and site activities depend on chain-length of growing polymer, have been compared with the experimental results of MWDs during the transition from living to stationary polymerization[2].

The remaining group of theories is that of monomer diffusion, as a rate-controlling mechanism, is discussed in this paper. The concept of slow monomer diffusion has been introduced for explaining of rate-decay during polymerization by Pasquon[3] who based upon a "solid-core" model; diffusion of monomer through polymer layer covered the catalyst surface is rate-controlling. Buls and Higgins[4] pointed out the possibility that slow monomer diffusion can explain broad MWDs of produced polymers in which catalyst was torn and dispersed. Applying the kinetic theory of porous catalysts[5], Schmeal and Street[6],

Merril[7] and Ray[8] developed quantitative theories of broad MWDs as well as rate-decay of heterogeneous polymerization with modified models ; "polymeric core", "expansion", "flow" and "multi-grain" ones. Those models are corresponding to observed physical states of growing polymer particles, in which catalyst particles collapsed are dispersed. On the other hand, some experimental results have been referred as evidences against for the diffusion control theory. Soga[9] found that the rate-decay of the propene polymerization with $TiCl_3/Al(C_2H_5)_3$ continues during intermission of polymerization when monomer is eliminated. Hsu[10] and the present author[11] reported that the constant of second order decay in propene polymerization with $MgCl_2$-supported $TiCl_4/Al(C_2H_5)_3$ depends on the concentration of $Al(C_2H_5)_3$. These authors have supposed some chemical deactivation as the reason of the rate-decay. Concerning the possibility of the diffusion theory for explaining broad MWDs, the present author[2] noted that MWDs of polymers produced in the living stages of propene polymerization with $MgCl_2$-supported catalyst are broad even when volume of produced polymers(≈ 0.1 of the catalyst-volume) seems to be not sufficient to form such a "polymeric core".

In this paper discussions are limited on the pseudo-stationary polymerization and focused to disclose an intrinsic defect of the theory of slow monomer diffusion itself, being compared to the experimental results accumulated comprehensively with the propene polymerization with the traditional and supported Ziegler catalysts[12],[13].

Implications of the theory of slow monomer diffusion

The implications of the theory of slow monomer diffusion have been generally summarized and discussed by Schmeal and Street[14], with the expressions for the Q value, or ratio of weight-average to number-average MW; $Q_0 = FX$ for polymers at living stages and $Q_\infty = 2FGX$ for polymers terminated dominantly by hydrogen at pseudo-stationary stages. The factors are greater than unity: F depends on particle geometry and the nature of diffusion and reaction; G depends on environmental history of polymer particle and X depends on the distribution of polymer particle sizes.

Usual kinetic measurements of slurry polymerizations are carried out in semi-batch reactor under constant pressures of monomer and hydrogen and then at pseudo-stationary stages of polymerization the main factor is F which is described with the Thiele modulus, $\phi = r_0\sqrt{k_p C^* / D}$ (where r_0 is the particle radius, k_p rate constant of propagation reaction, C^* site concentration, and D is diffusivity of monomer in polymer), as bellow.

The mass balance of monomer in the spherical shell between r and $r+dr$, within a particle of radius r_0 under quasi-stationary state may be represented by

$$\frac{1}{\eta^2}\frac{d}{d\eta}\left(\eta^2\frac{d[M]_\eta}{d\eta}\right)-\phi^2[M]_\eta=0 \tag{1}$$

where η is dimensionless radius(r/r_o), $[M]_\eta$ the monomer concentration in the spherical shell. With the boundary condition that

$$[M]_\eta=[M] \quad \eta=1 \quad \text{and} \quad d[M]_\eta/d\eta=0 \quad \eta=0 \tag{2}$$

where $[M]$ is the monomer concentration in solvent, the solution of the differential equation (1) is expressed by

$$[M]_\eta=[M]\left(\frac{\sinh(\phi\eta)}{\eta\sinh(\phi)}\right) \tag{3}$$

We introduce here "Effectiveness Factor" E, the ratio of the observed rate of polymerization to the intrinsic rate of polymerization $R_p^{\sim}$ or the initial rate of polymerization,

$$E=\frac{R_p}{R_p^0}=\frac{3}{\phi^2}(\phi\coth(\phi)-1)=\frac{\langle[M]_\eta\rangle_\eta}{[M]} \tag{4}$$

where $\langle\;\rangle_\eta$ means averaging through polymer particle. Although effect of geometry of polymer particle, not sphere but cylinder or flat plate is slightly different, the essential feature of E is same as that $E\sim1$ for $\phi<3$ and E decreases inversely with increasing ϕ[15].

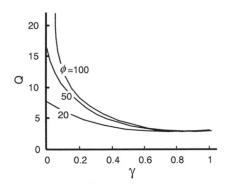

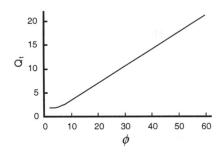

Fig.1 Polydispersity and Contribution of transfer by monomer at various values of Thiele modulus ϕ.

Fig.2 Polydispersity at $\gamma=0$ and Thiele modulus φ

Effects of transfer reactions

The quasi-stationary stages of polymerizations with decaying rate correspond to the later stages where MWD of polymers is stationary, i.e. transfer reactions occur simultaneously on the polymerization centers. Even in the absence of hydrogen, transfer reactions occur: by spontaneous termination, by co-catalyst(alkyl aluminum) as well as by monomer. The effects of the transfer reaction by monomer on MWD are very important for evaluating the theory of slow monomer diffusion, as describe below.

Number average degree of polymerization and Polydispersity

It has been experimentally established that the stationary number average MW or degree of polymerization $\overline{P_n}$, of polymer produced with heterogeneous Ziegler catalysts is generally represented by

$$\overline{P_n} = \frac{k_p[M]}{k_{t,s} + k_{t,a}[A]^{0.5} + k_{t,m}[M] + k_{t,h}[H]^{0.5}} \tag{5}$$

where $k_{t,s}$, $k_{t,a}[A]^{0.5}$, $k_{t,m}[M]$ and $k_{t,h}[H]^{0.5}$ are the rates of transfer reactions per site: spontaneous transfer, transfer by alkyl aluminum, transfer by monomer and transfer by hydrogen, respectively[16]. According to the above theory of slow monomer diffusion, Eq.(4), the rates of propagation and transfer by monomer are expressed by $k_p E[M]$ and $k_{t,m} E[M]$, respectively. The transfer reaction by monomer affects and changes k_p of Thiele modulus to $(k_p + k_{t,m})$, which may be practically ignored because of that $k_{t,m}/k_p \sim 10^{-3}$. Denoting the contribution of transfer by monomer to the overall transfer by γ

$$\gamma = \frac{k_{t,m} E[M]}{k_{t,s} + k_{t,a}[A]^{0.5} + k_{t,m} E[M] + k_{t,h}[H]^{0.5}} \tag{6}$$

the equation (5) can be represented by

$$\overline{P_n} = (\frac{k_p}{k_{t,m}})\gamma \tag{7}$$

or

$$\frac{1}{\overline{P_n}} = (\frac{k_p}{k_{t,m}}) \left\{ 1 + \frac{k_{t,s} + k_{t,a}[A]^{0.5} + k_{t,h}[H]^{0.5}}{k_{t,m}E}(\frac{1}{[M]}) \right\} \tag{8}$$

The polydispersity M_w/M_n at stationary state can be represented by

$$Q_\infty = 2\left\langle \frac{(k_p[M]_\eta)^2}{k_{p,s} + k_{t,a}[A]^{0.5} + k_{t,m}[M]_\eta + k_{t,h}[H]^{0.5}} \right/_\eta$$

$$\times \frac{k_{t,s} + k_{t,a}[A]^\alpha + k_{t,m}E[M] + k_{t,h}[H]^{0.5}}{k_p^2 E^2[M]^2} \tag{9}$$

or

$$Q_\infty = 2\left\langle \frac{([M]_\eta / E[M])^2}{1 - \gamma + \gamma([M]_\eta / E[M])} \right\rangle_\eta \qquad (10)$$

As illustrated in Fig.1, 2 the polydispersity-value is always 2 at $\gamma=1$ independent of Thiele modulus ϕ and increases to its largest value at $\gamma=0$ which depends on Thiele modulus as follows.

$$Q_\infty = \frac{2\left\langle [M]_\eta^2 \right\rangle_\eta}{E^2 [M]^2} \cong \frac{\phi}{3} = \frac{1}{E} \qquad (11)$$

Observed MWDs in pseudo-stationary polymerization.

The experimental results obtained with the propene polymerization with $MgCl_2$-suppored $TiCl_4$-$C_6H_5COOC_2H_5$/$Al(C_2H_5)_3$ catalyst[12][13] are mainly summarized and discussed in connection to the theory of slow monomer diffusion. As shown in Fig.3 and 4, the kinetic order of rate-decay is second order with respect to rate, excepting the beginning stages where the order seems to be third. The polymerization rate at time τ can be expressed by

$$R_p^\tau / R_p^0 = 1/(1 + k_d \tau) = E_\tau \qquad (12)$$

where $R_p^{\vee}$ is the initial polymerization rate of kinetic control and k_d is constant of second order rate decay.

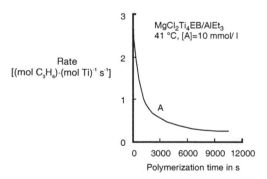

Fig.3 Rate-decay of propene polymerization with $MgCl_2$-suppored catalyst.

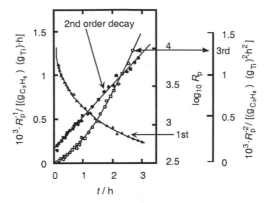

Fig.4 Kinetic order of rate-decay.

The number average molecular weight and the polydispersity are kept constant from 5s to 3h(Tab.1)[17]. These results show that the rate-decay continues for long time, i.e. $E\tau$ decreaces, while broad MWD is kept constant. Basing upon the theory of slow monomer diffusion, this situation should be interpreted as that in the case with $\gamma =1$ or $\gamma =0$. The latter case has been took into accounted as dominant hydrogen transfer case by Schmeal and Street[14].

Tab. 1. MWD of polypropylene after different polymerization times with the $MgCl_2$-supported $TiCl_4$-$C_6H_5COOC_2H_5$/$Al(C_2H_5)_3$ catalyst

Polym. time in s	$M_n \sim 10^{-4}$	M_w / M_n
5	4.21	5.2
15	4.68	5.3
45	4.78	5.2
240	3.97	5.2
2400	4.37	5.5
10800	4.59	6.5

Polym temp. = 41 °C, [M]= 0.45mol/ l, [Al]=10 mmol/ l

For the problem of time-invariant broad MWDs the theory of slow monomer diffusion may be rescued by introducing further sophisticated models. Then, the discussions are limited to pseudo-stationary polymerizations at a constant polymerization time $\tau =3h$.

The contribution of transfer by monomer to the overall transfer reaction, γ, can easily be estimate, on the basis of eq.(8), from the observation of $\overline{M_n}$ at various values of [M], as illustrated in Figs.5 and 6[18].

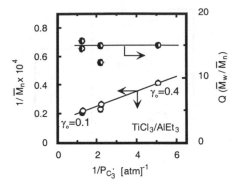

Fig.5. Dependency of $\overline{M_n}$ of propene concentration at 41°C, [A]=20mmol/l. TiCl₃/Al(C₂H₅)₃ catalyst.

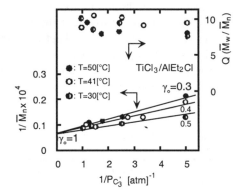

Fig.6. Dependency of $\overline{M_n}$ of propene concentration at 41°C, [A]=20mmol/l. TiCl₃/Al(C₂H₅)₂Cl catalyst.

Denoting the contribution of the control system at constant concentrations of [M]₀ and [A]₀ in the absence of hydrogen by γ_0, we can shift its value widely by changing the concentrations [M] and [A] as well as by the addition of hydrogen. The effect of hydrogen has been summarized by

$$\overline{M_n}\Big/\overline{M_n}\,H = {\gamma_0}\Big/{\gamma_H} = 1 + \alpha\sqrt{P_H} \tag{13}$$

with $0.40 \le \alpha \le 1.1$ (cmHg$^{-0.5}$) for the propene polymerizations with traditional and MgCl₂-supported Ziegler catalysts[18]. In Fig.7,8. it is shown that the value of γ_0 is large close to 1 and its reduction to 0.4 on the addition of hydrogen does not affect Q-value in the case with MgCl₂-supported catalyst[13]. The similar results received with the traditional TiCl₃ catalyst systems have been obtained(Fig.9)[18].

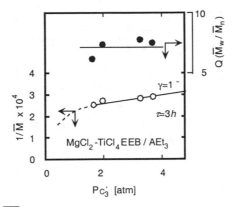

Fig.7 Dependency of $\overline{M_n}$ on [M] in the case with $MgCl_2$-supported catalyst.

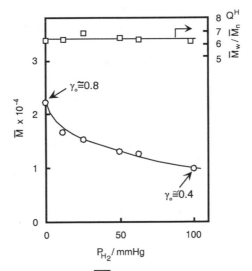

Fig.8 Effects of hydrogen addition on $\overline{M_n}$ and Q ($\gamma_0 = 0.8 \rightarrow 0.4$)

Discussion

It has been confirmed that concentration of hydrogen is kept constant during polymerization[13], which may guarantee that the diffusion of hydrogen through polymer-matrix is fast. It is supported also with the experimental fact that on the addition of hydrogen to polymerization system, after the co-catalyst was eliminated from solvent, the rate of polymerization decreases rapidly[19], i.e. the response of active sites to hydrogen addition is rapid. Addition of hydrogen, however, affects

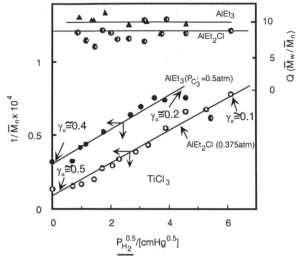

Fig.9 Effects of hydrogen addition on $\overline{M_n}$ and Q in the cases with TiCl$_3$, ($\gamma_0 \cong 0.5 \rightarrow 0.1$)

the polymerization up and / or down its rate. Similarly changes of the concentration of aluminum triethyl cause changes of the polymerization rate as well as of rate-decay constant. Because of that these are effects on the concentration of active sites, the above discussion of the effect of hydrogen on pseudo-stationary systems may be not suffered. On the other hand, the monomer concentration affects only the rates of propagation and transfer, excepting the initiation reaction stages of rate-increase[20]. Then, the existence of the contribution of transfer by monomer is very important for evaluating the theory of slow monomer diffusion. As described above, for the case of dominant transfer by monomer, such as the propene polymerization with MgCl$_2$-supported TiCl$_4$ catalyst, the theory of slow monomer diffusion can not explain any broad MWD. No response of Q-value to the change of γ, as proved experimentally, does disclose the intrinsic defect of the theory of slow monomer diffusion.

Acknowledgement

The author sincerely thanks Drs. H. Nakatani, T. Shimizu, JAIST, for preparing the manuscript.

References

1   Zucchini, U. and Cecchin, G., *Adv. Polym. Sci.* **51**, 103(1983)
2   Keii,T. and Matsuzawa,T.,"Metalorganic Catalysts for Synythesis and Polymerizatiopn." ed. W. Kaminsky,(1999),p38, Springer.

3   Pasquon,I., Dente,M. and Marduzzi, F., *Chim. Ind.* (Milan) **41**, 387(1959)
4   Emig,G. and Dittmeyer,R. "Handbook of Heterogeneous Catalysis", vol.3(1997) , Wiley-VHC, p1214
5   Buls, V. W. and Higgins, *J. Polym. Sci.* **A-1,8**,1025,1037(1970)
6   Schmeal,W.R. and Street,J.R. *AIChE J.,***17**,1188(1971)
7   Singh, D., Merril, R. P.: *Macromolecules* **4**, 599(1971)
8   Ray, H. et al, *I.&.C. Prod. Res. Dev.*,**19**, 372(1980)
9   Keii, T., Soga, K. and Saiki, N., *J. Polym. Sci.*,**C16**, 1507(1967)
10   Hsu, C. C. et al., " Transition Metal Catalyzed Polymerizations" Ed. Quirk,R.P.(1988) p136
11   Keii, T., "Catalyst Design for Tailor-made Polyolefins" Ed. Soga, K. and Terano, M.(1994) p1
12   Keii, T. et al., *Makromol. Chem.* **183**, 2285(1982)
13   Keii, T. et al., *Makromol. Chem.* **185**, 1537(1984)
14   Schmeal, W. R. and Street, J. R., *J. Polym. Sci. :Polymer Physics Edition*, **10**, 2173(1972)
15   Eming G. and Dittermeyer, "Handbook of Heterogeneous Catalysis", ed. Ertel .G. (1997) vol.3, p1209
16   Kissin, Y. V. "Isospecific Polymerization of Olefins",(1985), Springer-Verlag, p14
17   Suzuki, E., Tamura, M., Doi Y., and Keii, T., *Macromol. Chem.* **180**, 2235(1979)
18   Echigoya, S., MS Thesis, Tokyo Inst. Technology(1978), Keii, T., "Transition Metal Catalyzed Polymerization ", Ed. Qurk, R. P. (1988) p84
19   Kohara, T., PhD Thesis, Tokyo Inst. Technology(1979)
20   Keii, T., "Kinetics of Ziegler-Natta Polymerization",(1972), p21

Prepolymerization and Copolymerization Studies Using Homogeneous and Silica-supported Cp$_2$ZrCl$_2$/MAO Catalyst Systems

Peter J T Tait* and Johannes A M Awudza

Department of Chemistry, UMIST, Manchester, M60 1QD, UK.
E-mail: Dorothy.lovett@umist.ac.uk

Abstract. The "comonomer effect" associated with catalyst pre-polymerization and ethene/α-olefin copolymerization has been studied using the homogeneous and silica-supported Cp$_2$ZrCl$_2$/MAO catalyst systems, noting in particular the influence of catalyst heterogenisation. The "comonomer effect" was studied as a function of the type and concentration of the α-olefin used for the catalyst prepolymerization prior to ethene homopolymerization and in ethene/α-olefin copolymerization. 1-Butene, 1-hexene, 4-methylpentene-1 (4-MP-1) and 1-octene were used as comonomers for these prepolymerization and copolymerization studies. The concentrations of the active species responsible for these polymerizations were determined using a tritiated alcohol technique. The results are discussed taking into account the various processes that take place in ethene polymerization in the presence of a comonomer.

1. Introduction

It has been well established that both catalyst prepolymerization prior to ethene polymerization and ethene/α-olefin copolymerization invariably result in rate enhancement when these polymerizations are carried out using conventional Ziegler-Natta catalyst systems [1-3]. However, when an ethene/α-olefin copolymerization is carried out using metallocene-based catalyst systems, either a moderate rate enhancement [4-10] or a rate depression [11,12] is observed. Several reasons, both chemical and physical in nature, have been proposed to account for the "comonomer effect" observed both after catalyst prepolymerization and during copolymerization. These include modification of the catalytic centres, increased numbers of active centres, more rapid diffusion of monomer through amorphous polymer or combinations of these [1,2,13].

In earlier publications [4,11], we reported on rate enhancement or rate depression resulting from ethene/α-olefin copolymerizations using the homogeneous and silica-supported Cp$_2$ZrCl$_2$/MAO catalyst systems under different polymerization conditions. In the present study, the investigation has been extended to include the "comonomer effect" associated with catalyst prepolymerization prior to ethene homopolymerization or ethene/α-olefin copolymerization. 1-Butene, 1-hexene and 4-MP-1 were used for the prepolymerizations whilst 1-butene, 1-hexene, 4-MP-1 and 1-octene were the comonomers employed for the copolymerization studies.

2. Experimental

2.1 Materials

Ethene (99.92 %) and 1-butene (99.4 %) were purchased from the British Oxygen Company Ltd (BOC). These gases were purified by passage through preactivated molecular sieves (types 4A and 13X). 4-MP-1 was kindly donated by ICI (now BASF) as a pure liquid. The 4-MP-1 was stored over sodium wire for at least one week, distilled over the wire in a nitrogen atmosphere and stored over molecular sieves (type 4A) for at least 24 h before use. 1-Hexene (97 %) and 1-octene (98 %) were purchased from Aldrich Chemical Company Ltd. These monomers were purified prior to use by storage over preactivated molecular sieves (type 4A). Nitrogen gas ("white spot" grade, 99.99%) was obtained from BOC and purified by passage through phosphorus pentoxide, potassium hydroxide, self-indicating silica and columns of preactivated molecular sieves (types 4A and 13X). Silica EP10 was kindly donated by Crosfield Catalysts and was claimed to have a surface area of 291 m^2g^{-1}, a pore diameter of 269 Å, a pore volume of 1.8 cm^3g^{-1} and a median particle size of 100 μm. The silica was calcined at 260 °C under vacuum before use for the preparation of the supported Cp_2ZrCl_2 catalyst.

2.2 Catalysts

Bis(cyclopentadienyl)zirconium dichloride (Cp_2ZrCl_2), was obtained from the Aldrich Chemical Company, and claimed to be of 98 % purity. It was used without further purification. Methylaluminoxane (MAO) was purchased from Witco as a 30 % (w/w) solution in toluene and used as received.

The silica-supported Cp_2ZrCl_2 catalyst was prepared using MAO-modified silica according to the procedure outlined below:

2.3 Polymerization Procedure

Polymerizations were carried out in a glass reactor at 1 atm and 70 °C using toluene as the diluent. Details of the procedure have been described in previous publications [2,14].

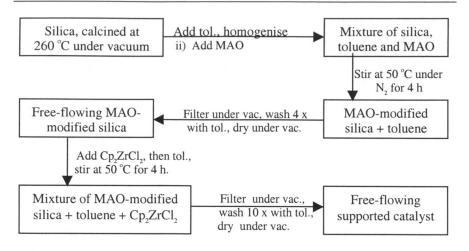

Scheme 1. Preparation of Supported Cp$_2$ZrCl$_2$ using MAO-modified Silica

Prepolymerization was carried out in a modified form of that described previously [2]. Polymerization was commenced, in the usual manner, but using an •-olefin and allowed to continue for a specified period of time (between 5 and 30 min). After this period the •-olefin was evacuated from the reactor until there was a solvent reflux. This refluxing was carried out above the boiling point of the •-olefin, if liquid, in order to ensure complete removal of the monomer. The system was then purged with dry oxygen-free nitrogen. The solvent refluxing and nitrogen purging were carried out for 5 min. This was followed by the addition of the •-olefin if required (for copolymerizations) and then the addition of ethane before recommencing polymerization. After the desired polymerization period, which was timed from the admission of ethene, the reaction was terminated and the polymer worked up in the usual manner.

Prepolymerization was carried out in order to develop maximum catalyst potential whilst retaining better control of catalyst and polymer morphology due to control of initial catalyst fragmentation during the slower prepolymerization step [1,2].

2.4 Active Centre Determination

The tritiated alcohol technique was used to determine the concentration of active centres in this work. The details of the procedure have been published elsewhere [14,15].

3. Results

3.1 Effect of Type and Concentration of Comonomers

In order to compare the influence of the presence of a comonomer on the ethene polymerization rate , the term comonomer effect factor (CEF) is used. CEF = $R_{p(av)}^{*}/R_{p(homopolymerization)}$, where $R_{p(av)}^{*}$ = average rate of polymerization for prepolymerization, copolymerization or prepolymerization + copolymerization

Figure 1 shows a comparison of the effects of 10 min prepolymerization for the homogeneous Cp_2ZrCl_2/MAO catalyst system using different concentrations of 1-butene, 1-hexene and 4-MP-1 prior to ethene homopolymerization on the comonomer effect factor. It can be seen from this figure that the value of the CEF is dependent on the type as well as the concentration of the α-olefin used for the prepolymerization. It is also evident that rate enhancement occurred with each of the prepolymerizing monomers, albeit at different concentrations. Taking into account the low concentrations of 1-butene used, the maximum CEF values follow the order CEF_{max} (4-MP-1) < CEF_{max} (1-hexene) < CEF_{max} (1-butene). This is the reverse order of the steric interference between the prepolymerized α-olefin monomer and the incoming ethene molecule in the subsequent ethene homopolymerization step.

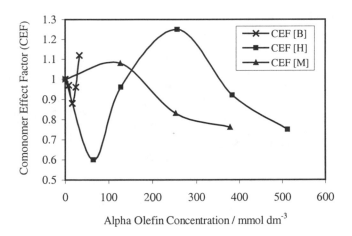

Fig. 1. Ethene Polymerization Using the Homogeneous Cp_2ZrCl_2 / MAO Catalyst System Following 10 min Catalyst Prepolymerization: Variation of the Comonomer Effect Factor with the Type and Concentration of α-Olefin Used for Prepolymerization <u>Experimental Conditions</u>: Temperature = 70 °C; Ethene Pressure = 1 atm; Diluent = 250 cm³ Toluene; 1 h Ethene Polymerization; [Zr] = 4.83 x 10⁻³ mmol dm⁻³; [Al] = 26.58 mmol dm⁻³; [Al] / [Zr] = 5.5 x 10³; B = 1-Butene; H = 1-Hexene; M = 4-MP-1.

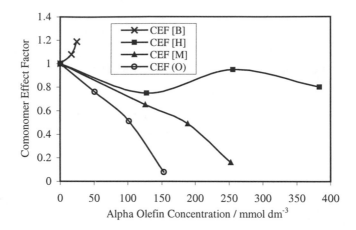

Fig. 2. . Variation of CEF with the Type and Concentration of α-Olefin in Ethene / α-Olefin Copolymerization Using the Homogeneous Cp₂ZrCl₂ / MAO Catalyst System. Experimental Conditions: Same as in Figure 1. Duration of Copoly-merization = 1 h. B = 1-Butene; H = 1-Hexene; M = 4-MP-1 and O = 1-Octene.

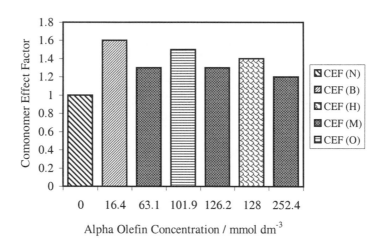

Fig. 3. Variation of the CEF with the Type and Concentration of α-Olefin in Ethene/α-Olefin Copolymerization Using the Silica-supported Cp₂ZrCl₂ / MAO Catalyst System. Experimental Conditions: Temperature = 70 °C; Ethene Pressure = 1 atm; Diluent = 250 cm³ Toluene; Duration of Copolymerization = 1 h; [Zr] = 9.66 x 10⁻³ mmol dm⁻³; [Al] = 53.16 mmol dm⁻³; [Al] / [Zr] = 5.5 x 10³; B = 1-Butene; H = 1-Hexene; M = 4-MP-1, O = 1-Octene, N = No comonomer (Homopolymerization).

Figures 2 and 3 show the variation of the CEF with the type and concentration of the α-olefins used in ethene/α-olefin copolymerizations employing the homogeneous and silica-supported Cp_2ZrCl_2/MAO catalyst systems respectively. In the case of the copolymerizations using the silica-supported Cp_2ZrCl_2/MAO catalyst system, only the concentration of 4-MP-1 was varied.

From Figure 2, it is evident that for the homogeneous Cp_2ZrCl_2/MAO catalyst system, whereas rate enhancement occurred in the ethene/1-butene copolymerizations, rate depression occurred during the copolymerizations carried out using 1-hexene, 4-MP-1 and 1-octene as the comonomers. The extent of the rate depression generally increased with an increase in the concentration of the comonomer. The CEF values follow the order CEF (1-butene) > CEF (1-hexene) > CEF (4-MP-1) > CEF (1-octene), which is also the trend of the level of steric interference (due to alkyl chains) between the inserted comonomer to the incoming ethene molecules. It is worth mentioning that we have already published observations of rate enhancement when the ethene/4-MP-1 copolymerization was carried out at a lower temperature (50 °C) [11, 16] and also at a [MAO]/[Zr] molar ratio ten times that used in the present study [4].

In contrast to the general rate depression observed when the homogeneous Cp_2ZrCl_2/MAO catalyst system was used for the copolymerizations, there was moderate rate enhancement in all the copolymerizations carried out using the silica-supported Cp_2ZrCl_2/MAO catalyst system (Figure 3). Variations in the type and concentration of the comonomer used did not significantly affect the CEF values. The latter observation is also in stark contrast to the significant changes in the CEF values with changes in the type and concentration of the α-olefin when the homogeneous catalyst system was employed in the copolymerizations.

Table 1. Influence of Polymerization Procedure on Kinetic Parameters in Polymerizations Employing the Homogeneous or Silica-supported Cp_2ZrCl_2/MAO Catalyst Systems

Polymerization Procedure	Yield / g	$R_{p(av)} \times 10^{-3}$ / g Polym (mmol Zr h atm)$^{-1}$	$R_{p(max)} \times 10^{-3}$	Decay Index	CEF
Homogeneous Catalyst System:					
Homopol	30.83	25.5	55.9	5.9	1.00
Copol	20.05	16.6	31.5	3.5	0.65
Prepol	33.13	27.5	47.5	4.6	1.08
Prepol + Copol	37.09	30.7	48.0	4.7	1.20
Silica-supported Catalyst System:					
Homopol	8.73	3.6	4.7	1.8	1.00
Copol	11.59	4.8	5.4	1.6	1.33
Prepol	12.73	5.3	7.1	1.5	1.47
Prepol + Copol	12.34	5.1	7.7	2.0	1.42

Experimental Conditions: Duration of Prepolymerization = 10 min; Duration of homopolymerization or copolymerization = 1 h; 126.2 mmol dm^{-3} of 4-MP-1 was used in prepolymerizations or copolymerizations; Decay Index = $R_{p(max)}$ / $R_{p\ (60\ min)}$. Homopol = Homopolymerization; Prepol = Prepolymerization; Copol = Copolymerization. Other conditions are the same as in Figures 2 and 3.

Table 1 collates the variation of some kinetic parameters with the polymerization procedure employed (i.e., homopolymerization, prepolymerization, copolymerization or prepolymerization followed by copolymerization) for polymerizations carried out using both the homogeneous and the silica-supported Cp_2ZrCl_2/MAO catalyst systems. It can be seen that when the homogeneous catalyst system was used for these polymerizations, the polymerization rates, the decay indices and the CEF values were all dependent on the polymerization procedure. With the use of the homogeneous catalyst, rate enhancement occurred when the ethene homopolymerization and ethene/4-MP-1 copolymerization were preceded by a catalyst prepolymerization using 4-MP-1. Ethene/4-MP-1 copolymerization using the non-prepolymerized catalyst, however, resulted in a rate depression.

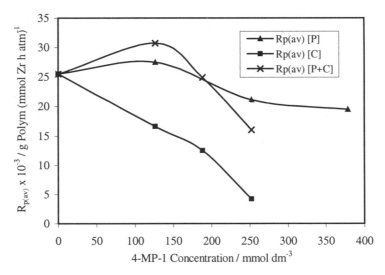

Fig. 4.. Variation of $R_{p(av)}$ with 4-MP-1 Concentration in Ethene Polymerization Following Catalyst Prepolymerization and in Ethene/4-MP-1 Copolymerization (With and Without Catalyst Prepolymerization). <u>Experimental Conditions</u>: Same as in Table 1 for the homogeneous catalyst. P = Prepolymerization;   C = Copolymerization

In order to further investigate the behaviour of the homogeneous Cp_2ZrCl_2/MAO catalyst system in the presence of a comonomer, the prepolymerizations, copolymerizations and prepolymerizations followed by copolymerizations were carried out using varying amounts of the 4-MP-1. Figure 4 shows plots of the variation of $R_{p(av)}$ with 4-MP-1 concentration for the different polymerization procedures. It is evident from Figure 4 that rate depression occurred with all of the 4-MP-1 concentrations used for the copolymerizations and that the extent of the rate depression increased with an increase in the concentration of the 4-MP-1 used. When the ethene/4-MP-1 copolymerizations were performed with the prepolymerized catalyst, there was either rate enhancement or rate depression depending on the concentration of 4-MP-1 used.

The more gradual decrease in the ethene polymerization rate with an increase in the amount of 4-MP-1 used for the prepolymerization is particularly revealing as will be discussed later.

From Table 1, it is clear that when the silica-supported catalyst system was used in these polymerizations, the kinetic parameters did not vary very much with the polymerization procedure. It is also evident that both catalyst prepolymerization and copolymerization resulted in rate enhancement when the supported catalyst was employed.

The results in Table 1 and in Figure 4 suggest that the Cp_2ZrCl_2/MAO catalyst system becomes more stable after prepolymerization, heterogenisation or both. Figures 5 and 6 reveal a higher level of decay in the rate-time profiles for polymerizations carried out using the homogeneous catalyst system (especially in the last 30 min of the polymerization) when compared to the profiles for runs in which the silica-supported catalyst was used. This further highlights the higher level of stability of the supported catalyst than the homogeneous analogue.

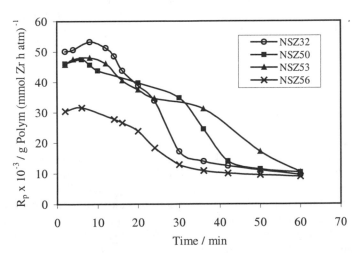

Fig. 5. Ethene Homopolymerization and Ethene/4-MP-1 Copolymerization Using the Homogeneous Cp_2ZrCl_2 / MAO Catalyst System: Comparison of Rate-time Profiles for Runs with and without Catalyst Prepolymerization. <u>Experimental Conditions</u>: Same as in Table 1. NSZ32: Ethene homopolymerization; NSZ56: Ethene/4-MP-1 copolymerization; NSZ50: Prepolymerization; NSZ53: Prepolymerization + copolymerization.

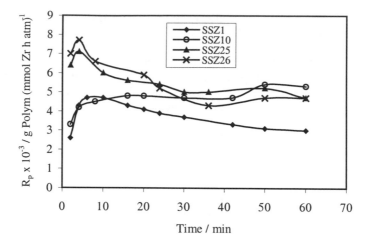

Fig. 6. Ethene Homopolymerization and Ethene/4-MP-1 Copolymerization Using the Silica-supported Cp_2ZrCl_2 / MAO Catalyst System: Comparison of Rate-time Profiles for Runs with and without Catalyst Prepolymerization. <u>Experimental Conditions</u>: Same as in Table 1. SSZ1: Ethene homopolymerization; SSZ10: Ethene/ 4-MP-1 copolymerization; SSZ25: Prepolymerization; SSZ26: Prepolymerization + copolymerization.

3.2 Active Centre Studies

Table 2 shows the variation of the initial active centre concentration, C_o^*, and the apparent propagation rate coefficient, k_p, (herein after simply referred to as the rate coefficient) with polymerization procedure. 4-MP-1 was used as the comonomer in these prepolymerizations and / or copolymerizations. $R_{p(av)}$ and CEF values are also included in Table 2 for the purpose of comparison. It is clear from the table that the C_o^* values for polymerizations carried out using the homogeneous Cp_2ZrCl_2/MAO catalyst system are higher than those obtained when the supported catalyst was used. It is also evident from the table that the C_o^* values associated with prepolymerization and / or copolymerization are lower than the C_o^* value obtained in ethene homopolymerization when the homogeneous catalyst system was employed for these polymerizations. However, when the supported catalyst system was used for the polymerizations, the C_o^* values varied only slightly with the polymerization procedure. The C_o^* values obtained for the different polymerization procedures do not bear any direct relationship with the corresponding CEF values. On the other hand, there appears to be some general correlation between the CEF values for the different polymerization procedures and the corresponding evaluated k_p values. The last statement suggests that the rate coefficient plays an important role in the "comonomer effect".

Table 2. Comparison of C_o^*, $R_{p(av)}$, k_p and CEF values for the Homogeneous and Silica-supported Cp_2ZrCl_2 / MAO Catalyst Systems

Polymerization Procedure	$R_{p(av)}$ x 10^{-3} / g Polym (mmol Zr h atm)$^{-1}$	C_o^* (As % of initial [Zirconocene])	k_p x 10^{4}/ dm^3(mol s)$^{-1}$	CEF
	Homogeneous Catalyst System:			
Homopol	25.5	85.0	3.7	1.00
Copol	16.6	60.3	3.4	0.65
Prepol	27.5	29.8	11.4	1.08
Prepol + Copol	30.7	46.1	8.3	1.20
	Silica-supported Catalyst System:			
Homopol	3.6	20.6	2.2	1.00
Copol	4.8	17.5	3.4	1.33
Prepol	5.3	13.7	4.8	1.47
Prepol + Copol	5.1	16.7	3.8	1.42

Polymerization Conditions: Same as in Table 1

4. Discussions

The kinetic results obtained in this study show that the homogeneous and silica-supported Cp_2ZrCl_2 / MAO catalyst systems behave differently during olefin polymerizations. The comonomer effect factor was generally higher for polymerizations carried out using the silica-supported catalyst system than those performed with the homogeneous catalyst, and was dependent on the nature and concentration of the comonomer used in both the prepolymerizations and copolymerizations. These observations are in general agreement with those reported in the literature [5,9–11].

The influence of the comonomer on the prepolymerization and copolymerization kinetics can, in most cases, be rationalised in terms of the formation, stability and behaviour of the active centres in the presence of the comonomer. The active species in metallocene catalyst systems has been shown to be a cationic metallocenium ion which is stabilised by a counterion through the formation of separated ion pairs [4,7]. The "comonomer effect" in prepolymerizations and / or copolymerizations involving zirconocene catalysts may be related to perturbations of the ion pairs at the active centres [7]. It has also been postulated that α-olefins can function as ligands at the active centres [17]. By coordinating to the active centre, an α-olefin alters the charge density on the cationic zirconocenium ion as well as the extent of interaction of the cationic zirconium centre with the aluminoxane-surrounded counterion. It is felt that the coordination of the α-olefins to the active centres and the resultant perturbation of these centres modify the centres and hence bring about the comonomer effect. The extent to which the comonomer modifies the catalytic centre will depend on the type of catalyst system, the type and concentration of the comonomer and the conditions employed in the polymerization.

A decrease in the CEF value with an increase in the alkyl chain length of the comonomer, as observed in the present study, can be attributed to steric interference between alkyl groups of the α-olefin monomer inserted into the growing polymer chain and the incoming ethene molecule. The steric interference will slow down the insertion of ethene into the growing polymer chain and hence decrease the ethene polymerization rate. Although 1-hexene and 4-MP-1 are isomeric, the CEF value obtained in the presence of 1-hexene was higher than that obtained with 4-MP-1 as the comonomer. This may be due to the fact that the branching in 4-MP-1 results in an increase in the extent of steric interference between the 4-MP-1 and ethene during the insertion process.

When compared to the performance of the homogeneous catalyst system, the polymerization rates for runs carried out using the supported catalyst decreased by a factor of 7 in homopolymerization and by a factor of 3.5 in ethene / 4-MP-1 copolymerization in which a 4-MP-1 concentration of 126.2 mmol dm^{-3} was used. In an earlier publication [11], it was reported that at a 4-MP-1 concentration of 252.4 mmol dm^{-3}, the $R_{p(av)}$ values for ethene/4-MP-1 copolymerizations employing both the homogeneous and silica-supported Cp$_2$ZrCl$_2$/MAO catalyst systems were virtually the same. The rate of copolymerization decreases with an increase in the concentration of 4-MP-1 when the homogeneous catalyst system is employed for the copolymerization (Fig. 4) but is not very much affected when the supported catalyst is used (Fig. 3). Thus the results of the relative performance of the homogeneous and the supported catalyst systems during the copolymerization of ethene with different concentrations of 4-MP-1 demonstrate that the supported catalyst is more stable under the polymerization conditions than the homogeneous analogue. The higher level of stability of the silica-supported Cp$_2$ZrCl$_2$/MAO catalyst system compared to the homogeneous one can be attributed to a reduction in the deactivation of the catalytic centres. In this respect, it can be considered that the silica acts as a "spacer", thus reducing the extent of bimolecular deactivation of the active catalytic centres.

The gradual decrease of the ethene polymerization rate with an increase in the amount of 4-MP-1 used in the prepolymerization step (Fig. 4) is an indication of an increase in the stability of the catalyst system after the prepolymerization step. Since α-olefins are bulkier than ethene, it is expected that the α-olefins would polymerize more slowly than ethene during the prepolymerization step and hence would insert into the active centre more slowly than ethene. It is felt that in the slow insertion process, the α-olefin modifies the electronic properties of the active centre in such a way as to stabilise the catalyst system in the subsequent ethene homopolymerization.

The active centre results obtained reveal that whereas the rate depression observed during copolymerization with the homogeneous catalyst system is attributable to a decrease in the active centre concentration, the CEF values ≥ 1 obtained during copolymerization with the supported catalyst are due to an increase in the apparent rate coefficient, k_p. The results also show that the rate activation obtained after catalyst prepolymerization is due to an increase in the rate coefficient. The changes in the k_p values suggest that both the immobilisation of the catalyst onto the support and catalyst prepolymerization modify the ligand environment of the active centre, thus creating the right environment for the insertion of ethene into the active centre and the growing polymer chain. This is because the k_p value is a measure of the intrinsic reactivity of the active centres. Thus the immobilisation of the catalyst onto a support and catalyst prepolymerization both result in an increase in the intrinsic reactivity of the catalyst.

5. Conclusions

- Immobilization of Cp_2ZrCl_2 on a support results in a less active but more stable catalyst for olefin polymerization
- The CEF in ethene homopolymerization following catalyst prepolymerization and in ethene/α-olefin copolymerization (with or without prepolymerization) is dependent on the nature and concentration of the α-olefin used
- Higher CEF values are obtained in prepolymerizations and copolymerizations carried out using the silica-supported Cp_2ZrCl_2 / MAO catalyst system than when the homogeneous catalyst is used
- Prepolymerizing the homogeneous or silica-supported Cp_2ZrCl_2 / MAO catalyst system prior to ethene homopolymerization or ethene / α-olefin copolymerization reduces the active centre concentration but increases the propagation rate coefficient
- The propagation rate coefficient contributes significantly to the comonomer effect
- Modification of the active centres by coordinated comonomer and perturbation of the electronic environment of the active centres may be responsible for the comonomer effect after catalyst prepolymerization and during copolymerization.

Acknowledgement

One of us (JAMA) wishes to acknowledge with gratitude, a studentship from the Government of Ghana.

References

1. Tait, P.J.T., Downs, G.W. and Akimbami, A.A., in *"Transition Metal Catalysed Polymerizations, Ziegler-Natta and Metathesis Polymerizations"*, Quirk, R.P., Hoff, R.E., Klingensmith, G.B., Tait, P.J.T. and Goodall B.L. (eds), Cambridge University Press, Cambridge, **1988**, 835.
2. Tait, P.J.T. and Berry, I.G., in *"Catalyst Design for Tailor-made Polyolefins"*, Soga, K. and Terano, M. (eds), Elsevier-Kodansha, Tokyo, **1994**, 55.
3. Soares, J.B.P, and Hamielec, A.E., *Polymer*, **1996**, *37(20)*, 4599.
4. Tait, P.J.T., Abozeid, A.I. and Paghaleh, A.S., *Metallocenes '95*, April 26-27, Brussels, Belgium, **1995**.
5. Seppälä, J.V., Koivumäki, J. and Liu, X., *J. Polym. Sci., Polym. Chem. Ed.*, **1993**, *31*, 3447.
6. Tsutsui, T. and Kashiwa, N., *Polym. Commun.*, **1988**, *29*, 180.
7. Karol, F.J., Kao, S-C., Wasserman, E.P. and Brady, R.C., *New J. Chem.*, **1997**, *21*, 797.
8. Uozumi, T. and Soga, K., *Makromol. Chem.*, **1992**, *193*, 823
9. Koivumäki, J. and Seppälä, J.V., *Macromolecules*, **1993**, *26 (21)*, 5535
10. Koivumäki, J., Fink, G. and Seppälä, J.V., *Macromolecules*, **1994**, *27*, 6254
11. Tait, P.J.T. and Awudza, J.A.M., in *"Progress and Development of Catalytic Olefin Polymerization"*, Sano, T., Uozumi, T., Nakatani, H. and Terano, M. (eds), Technology and Education Publishers, Tokyo, **2000**, 214.
12. Chien, J.C.W. and Nozaki, T., *J. Polym. Sci., Polym. Chem. Ed.*, **1993**, *31*, 227.
13. Soga, K., Yanagihara, H. and Lee, D., *Makromol Chem.*, **1989**, *190*, 995.
14. Tait, P.J.T., Monteiro, M.G.K., Yang, M. and Richardson, J.L., *MetCon '96,* June 12-13, Houston, TX, USA , **1996**.
15. Burfield, D.R. and Tait, P.J.T., *Polymer*, **1972**, *13*, 315.
16. Awudza, J.A.M., *Ph. D Thesis*, UMIST, Manchester, **2000**.
17. Karol, F.J., Kao, S-C. and Cann, K.J., *J. Polym. Sci., Polym. Chem. Ed.* **1993**, *31*, 2541.

Unique Flowability Behavior of Ethylene Copolymers Produced by a Catalyst System Comprising Ethylenebis(indenyl)hafnium Dichloride and Aluminoxane

Jun-Ichi Imuta, Toshiyuki Tsutsui, Ken Yoshitugu, Tomoaki Matsugi and Norio Kashiwa

Organo-metal Complexes Catalization Lab. MITSUI CHEMICALS, INC. 580-32, Nagaura, Sodegaura city, Chiba, 299-0265, Japan,
E-mail: Junichi.Imuta @mitsui-chemco.jp

Abstract. This paper relates to novel ethylene copolymers and a process for preparing the same and more particularly to novel ethylene copolymers excellent in flowability in spite of the fact that they are narrow in molecular weight distribution (M_w/M_n) in comparison with conventionally known ethylene copolymers, and to a process for preparing the same. In another aspect, this paper relates to olefin polymerization catalysts capable of polymerizing olefins with excellent comonomer incorporation and capable of giving olefin polymers having high molecular weights.

Introduction

Olefin polymerization catalysts composed generally of titanium compounds or vanadium compounds and organoaluminum compounds have heretofore been used for preparing ethylene copolymers. In recent years, however, Kaminsky and coworkers found that a soluble catalyst system comprising a bis(cyclopentadienyl) zirconium compound and methylaluminoxane(MAO) showed very high activity for ethylene polymerization in 1980[1].

On the basis of this catalyst performance, the alkyl substituted metallocenes were prepared, and the effects of substituents on polymerization activities and molecular weights were examined[2]. After this work, many types of metallocenes have been developed with the aim of new performances (high stereospecificity, high reactivity). Brintzinger[3] and Ewen[4] have succeeded to prepare Zr *ansa*-metallocenes and produced highly isotactic and syndiotactic polypropylene respectively. Recently we have summarized the differences of catalytic behavior of these metallocene catalysts[5].

In the case of polyethylene or ethylene copolymers, many researches have been done with incorporation of various comonomer as well as high molecular weight and high polymerization activity. Another important factor for use in commercial scale is processabilities of copolymers. When molded into articles such as film, copolymers of ethylene and alpha-olefins of 3 to 20 carbon atoms are desired to have excellent mechanical strength such as tensile strength, tear strength or impact strength and also excellent heat resistance, stress crack resistance, optical characteristics and heat-sealing properties in comparison with conventional high-pressure low density polyethylenes, and are known as materials particularly useful for the preparation of inflation film or the like.

If the ethylene copolymers having such excellent characteristics be come to narrower in molecular weight distribution represented by the ratio (Mw/Mn) of weight average molecular weight (Mw) to number average molecular weight (Mn), the molded articles obtained therefrom, such as film, they are found to be less tacky. However, when these ethylene copolymers having a narrow molecular weight distribution are melted, there were such drawbacks that their flowability represented by the ratio (MFR_{10} /MFR_2) of MFR_{10} under a load of 10 kg to MFR_2 under a load of 2.16 kg as measured at 190 °C. is low, with the result that they become poor in moldability.

Therefore, if ethylene copolymers which are low in value of Mw/Mn and narrow in molecular weight distribution and, moreover, large in value of MFR_{10} /MFR_2 and excellent in flowability come to be obtained, such ethylene copolymers are certainly of great commercial value. We have studied under such circumstances as mentioned above, and as a result we have found that by the use of a metallocene **1** wherein hafnium ion is bound to the ethylene bis-indenyl group, an olefin (co) polymer of high molecular weight can be prepared with high in value of MFR_{10} /MFR_2 and an olefin copolymer of much higher comonomer content can be obtained[6].

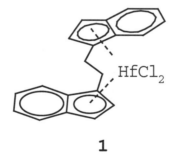

1

Results and Discussion

Catalyst Candidate

Brintzinger[3] and Ewen[4] have succeeded to prepare *ansa*-metallocenes and produced highly isotactic and syndiotactic polypropylene respectively by use of Zr catalyst, as we mentioned in the part of introduction. After this work, a high molecular weight isotactic polypropylene was obtained by polymerization of propylene in the presence of a catalyst system comprising ethylenebis(indenyl)hafnium dichloride **1** and aluminoxane, said isotactic polypropylene having a narrow molecular weight (Mw/Mn) of 2.1-2.4[7]. These results clearly show that filled 4f orbitals in Hf catalyst provides a useful advantage in a olefin polymerization. We applied this catalyst system to the production of ethylene copolymers aiming at obtaining high molecular weight copolymers with high polymerization activities. We examined ethylene-propylene and ethylene-octene copolymerization at 40-90 °C under atmospheric pressure.

The results of polymerization obtained with this catalyst system comprising ethylenebis(indenyl)hafnium dichloride **1** and aluminoxane is summarized in Table 1. The indenyl derivative **1** produced higher molecular weight and low density copolymers (Table1). These preliminary experimental results suggest that metallocene **1** and aluminoxane is one of the catalyst candidates for ethylene copolymers.

Table 1. Polymerization of ethylene and comonomer in toluene under atmospheric pressure.

Run	Met	Comonomer		total volume (ml)	D (g/cm^3)	[η] (dl/g)	MFR$_2$ (g/10 min)	MFR$_{10}$ /MFR$_2$	M$_w$/M$_n$
		Species	Amount (ml)						
1	Hf	Octene	28	1000	0.868	1.80	0.87	-	2.27
2	Hf	Octene	5	500	0.901	1.45	1.80	9.2	2.35
3	Hf	Octene	3	500	0.915	1.67	0.70	11.8	2.53
4	Hf	Octene	4	500	0.907	1.49	1.48	8.7	2.22
5	Hf	Propene	-*	1000	0.887	1.50	0.80	12.7	2.50

*see experimental section.

Our first attempt to produce high molecular weight copolymers was gratifyingly successful. To explain producing such high molecular weight copolymers, we suggest that shorter ligand-metal distances result in lower Lewis acidities for hafnium metallocene relative to zirconium analogues, which will suppress the chain transfer by β-hydride elimination.

Another interesting feature of metallocene **1** is having the ability of relatively high comonomer conversion as exemplified by producing lower density copolymers. It is suggested that the filled 4f orbitals provide a useful advantage in a catalytic reaction.

Unique Flowability Behavior

It is also required for the ethylene copolymers which are small in value of M$_w$/M$_n$ and, moreover, large in value of MFR$_{10}$ /MFR$_2$. We have obtained ethylene co-polymers with various densities using the catalyst system of ethylenebis-(indenyl)hafnium dichloride **1** and aluminoxane (Table 1).

This catalyst system is good for high MFR$_{10}$ /MFR$_2$ with small value of M$_w$/M$_n$ compared to conventional catalyst composed of VOCl$_3$ and aluminum ethyl sesquichloride (Fig.1).

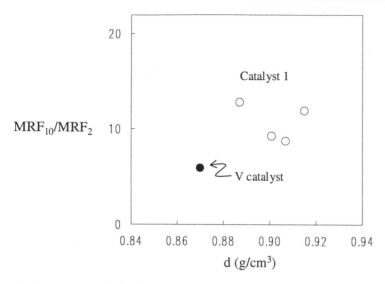

Fig. 1. Differences of flowability between catalysts 1 and V catalyst

The success of this finding of new catalyst system producing ethylene copolymers which have high value of MFR_{10} /MFR_2 has encouraged us to examine the relationship between flow and viscosity of these copolymers. Especially, we are very interested in the reason why our new catalyst system and conventional V catalyst system behaves differently. Indeed, the relationship between MFR_2 and $[\eta]$ of these copolymers by our new catalyst system was quite different from the conventional catalyst composed of $VOCl_3$ and aluminum ethyl sesquichloride (Fig.2).

To explain different behavior in flow and viscosity of these two catalysts, we suggest that a long chain branching is easily formed with a new catalyst system by the insertion of macromonomer having terminal vinyl group which is formed by β-hydride elimination or chain transfer to monomer. In order to demonstrate this hypothesis, the best way is to determine the chemical structure of ethylene copolymers produced by the catalyst system comprising ethylenebis-(indenyl)hafnium dichloride **1** and aluminoxane using ^{13}C-NMR technique. However, it is difficult to discriminate between long chain branching and six carbons branching at this stage. The possible structure of these copolymers is now under evaluation.

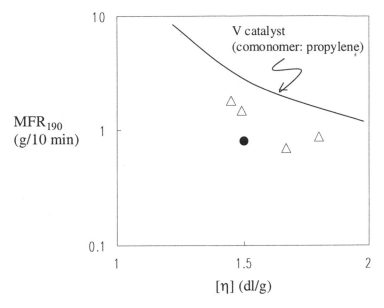

Fig. 2. Relationship between flow and viscosity of ethylene copolymers (• comonomer: propylene; Δ comonomer: octene)

Experimental

Analysis of ethylene copolymers

Intrinsic viscosity[η]: measured in decalin at 135 °C. The molecular weight distribution (M_w/M_n): gel permeation chromatography (GPC). In this connection, a value of (M_w/M_n) obtained in the invention was determined by the following procedure in accordance with Takeuchi, "Gel Permeation Chromatography," Maruzen, Tokyo. *Preparation of methylaluminoxane* Methylaluminoxane was prepared in accordance with the procedure described in Polymer Commun., 29, 180 (1988).

Synthesis of ethylenebis(indenyl)hafnium dichloride)

A nitrogen-purged 200 ml glass flask was charged with 5.4 g of bis(indenyl)ethane [synthesized on the basis of Bull Soc. Chim., 2954 (1967)] and 50 ml of THF, and the flask was cooled with stirring between −30 °C and -40 °C. To the flask was added dropwise 31.5 ml of n-Bu Li (1.6M solution), stirred successively at −30 °C for 1 hours, and the temperature was elevated spontaneously to room temperature, thereby anionizing the bis(indenyl)ethane. Separately, a nitrogen-purged 200 ml glass flask was charged with 60 ml of THF, and the flask was cooled to below –60 °C, followed by gradual addition of 6.7 g of HfCl$_4$ (contained 0.78% by weight of zirconium atoms as contaminants). Thereafter, the flask was heated to 60 °C and stirred for 1 hour. To the flask was added dropwise the anionized ligand, and stirred at 60 °C for 2 hours, followed by filtration with a glass filter. The filtrate was concentrated at room temperature to about 1/5 of the original volume. By this operation, the solids were separated. The separated solids were filtered with a glass filter, followed by washing with hexane/ethyl ether and vacuum drying to obtain ethylenebis(indenyl)hafnium dichloride. The hafnium compound thus obtained contained 0.40% by weight of zirconium atoms.

Polymerization

Run 1: A thoroughly nitrogen-purged 2 liter glass flask was charged with 950 ml of toluene and 28 ml of 1-octene, and ethylene gas was passed at a rate of 160 l/hr. The temperature in the system was elevated to 65 °C, and 1.88 mmoles in terms of aluminum atom of methylaluminoxane and 7.5 x 10^{-3} mmole of ethylenebis-(indenyl)hafnium dichloride were added to the system to initiate polymerization. The polymerization was carried out at 70 °C for 10 minutes under atmospheric pressure while continuously feeding ethylene gas. The polymerization was stopped by the addition of small amounts of methanol, and the polymerization solution obtained was poured in large amounts of methanol to separate the polymer. The separated polymer was dried at 130 °C for 12 hours under reduced pressure to obtain 17.36 g of a polymer having a density of 0.866 g/cm^{-3}, [η] of 1.71 dl/g, M$_w$/M$_n$ of 2.27, MFR$_2$ of 0.87g/10 min.

Run 5: A thoroughly nitrogen-purged 2 liter glass flask was charged with 1 liter of toluene, and a mixed gas of ethylene and propylene (140 l/hr and 40 l/hr respectively) was passed The temperature in the system was elevated to 75°C, and 1.88 mmoles in terms of aluminum atom of methylaluminoxane and 7.5.times.10^{-3} mmol of ethylenebis(indenyl)hafnium dichloride were added to the system to initiate polymerization. The polymerization was carried out at 80°C for 10 minutes under atmospheric pressure while continuously feeding the above-mentioned mixed gas to the system. Thereafter, the operation was conducted in the same manner as in Exp.1 to obtain 17.5 g of a polymer having a density of 0.887 g/cm^{-3}, the ethylene content of 84.0 mol%, [η] of 1.50 dl/g, M$_w$/M$_n$ of 2.50, MFR$_2$ of 0.80 g/10 min and MFR$_{10}$/ /MFR$_2$ ratio of 12.7.

V catalyst polymer: A copolymer of ethylene and propylene (prepared by using a catalyst composed of VOCl$_3$ and aluminum ethyl sesquichloride) having a density of 0.87 g/cm^3, MFR$_2$ of 2.9 g/10 min and M$_w$/M$_n$ of 2.16 was found to have MFR$_{10}$ /MFR$_2$ ratio of 5.90.

References

1. H.Sinn, W.Kaminsky, H.J.Vollmer, and R.Woldt, Angew.Chem., Int.Ed.Engl., 19, 390 (1980)
2. J.A.Ewen; Catalytic Polymerization of Olefins(eds.T.keii and K.soga, Kodansha), p271, Tokyo(1986)
3. F.R.W.P.Wild, M.Wasincionek, G.Huttner, and H.H.Brintzinger, J.Organomet. Chem., 288, 63 (1985)
4. J.A.Ewen, J.Am.Chem.Soc., 106, 6355 (1984)
5. Jun-ichi Imuta and Norio Kashiwa; Handbook of Polyolefins (edited by Cornelia Vasile, Marcel Dekker, Inc.), P71(2000)
6. T.Tsutsui , A.Toyota; EP 685498
7. J.A.Ewen, and Luc Haspeslagh , J. Am. Chem. Soc., 109, 6544 (1987)

Diffusion Measurements in Porous Polymer Particles

John Georg Seland and Bjørn Hafskjold

Department of Chemistry, Norwegian University of Science and Technology,
N-7491 Trondheim, Norway
E-mails: johnsel@chembio.ntnu.no; bhaf@chembio.ntnu.no

Abstract. During a heterogeneous polymerization of polyolefins the original catalyst particles fracture, and are encapsulated in solid polymer. Porous polymer particles are made, and they grow as the polymerization proceeds, giving particle microreactor effects [1-5]. The diffusion resistance to monomers during a polymerization to polyolefin particles may have a significant effect on the observed activity of the catalytic system. This diffusion resistance depends on the diffusivity of the monomers in the different areas of the particles, and this diffusivity is, in general, unknown [1-5]. In this study we have used Pulsed Field Gradient Nuclear Magnetic Resonance (PFG-NMR) [6] to measure diffusion of small organic molecules in porous Polyethylene (PE) particles produced in a heterogeneous metallocene polymerization. When an organic liquid is added to the PE particles, it will occupy different areas of the particles. The two main areas will be the semi crystalline polymer and the cavities inside the porous particle. In addition there is a broad distribution of pore sizes having different influences on the measured diffusivity. In order to study the effect of different pore sizes, diffusion measurements performed in the PE particles are compared with diffusion measurements done in a model systems consisting of mono disperse porous polystyrene particles with more narrow pore size distributions. From the diffusion experiments we have determined the tortuosity values of different liquids in different areas of the PE particles. The diffusivity of a liquid in a PE particle can be divided into three distinctive areas; diffusion in semi crystalline PE, diffusion in small cavities (<1 μm), and diffusion in large cavities (10-60 μm). The tortuosity values may be used as an input to a intraparticle mass transfer model.

Introduction

Polyolefins are often made using heterogeneous catalysis. During such a polymerization the original catalyst particle fracture, and are encapsulated in solid polymer. Porous polymer particles are made, and they grow as the polymerization proceeds [1-5]. The diffusion resistance to monomers during polymerization to polyolefin particles may have a significant effect on the observed activity. This diffusion resistance depends on the effective diffusivity of the monomers in the different areas of the particles.

One of the commercially important polyolefin processes is the slurry process. Here, the heterogeneous catalyst and growing polymer particles form a suspension in an inert hydrocarbon dilutent such as hexane or heptane. In this process, it is the diffusion of the monomer in the pores of the particles that is usually considered to be the potential rate-limiting step, while diffusion in semi crystalline PE is not considered to be important [1-5].

Different models are suggested in order to describe the intraparticle mass transfer during the polymerization of polyolefins. The most common one is the multi grain model (MGM) [1-3], where the polymer particle consists of small aggregated grains of the size of 1-10 μm. Fragments of the catalyst are incapsulated in these small grains. A wide range of values for the effective diffusivity are suggested and used in the modeling of intraparticle mass transfer [1-5], but reliable values for the effective diffusivity in such a system are scarce.

In this study we have used Pulsed Field Gradient (PFG) NMR to measure diffusion of liquid added to polyethylene (PE) particles. The PE particles have a complex morphology and geometry. When a liquid is added to this system, it will occupy different domains of the particles. One can expect that the two main domains will be the semi crystalline phase of the polymer, and in the cavities inside the porous particle. In addition there is a distribution of pore sizes, which will, because of restricted diffusion, have different influences on the measured diffusivity.

It is the goal of this study to determine self diffusion coefficients for the different areas of the PE particles, and to suggest how to use these measurements to estimate parameters which can be used in a intraparticle mass transfer model.

Theory

PFG-NMR is a well-established method for the measurement of self-diffusion coefficients [6]. In a PFG-NMR experiment, the echo attenuation for an ensemble of molecules undergoing free Brownian motion is in general given by [6]

$$\ln \frac{I}{I_0} = -\gamma^2 \delta^2 g_e^{\ 2} D t_d \tag{1}$$

where γ is the gyromagnetic ratio, g_e is the effective strength of the applied magnetic field gradient pulse, and δ is the effective length of this pulse. D is the self diffusion coefficient, and t_d is the effective observation time. The parameters g_e and t_d will depend on the type of PFG sequence used. By measuring the echo attenuation as a function of $g_e^{\ 2}$, keeping the other parameters constant, the self diffusion coefficient can be determined.

When working with heterogeneous media it is convenient to define the so-called reciprocal lattice wave vector, $q = \gamma g_e \delta / 2\pi$, which describes the reciprocal space of the restricting geometry in the system.

When a heterogeneous sample is placed in a static magnetic field, internal magnetic field gradients may be induced, and this may cause a erroneous interpretation of the diffusion measurement [7-8]. This error can be suppressed by the introduction of applied bipolar gradients in the pulse sequence [9-13], but in previous works we have experienced that even when this method is applied, one has to analyze the results carefully at long observation times [14].

It is also important to take into consideration the effects of domains having different diffusivities, effects of restricted diffusion, and the influence these effects have on the measurements performed.

In the limit of slow exchange between the different domains, the process is described by

$$\frac{I}{I_0} = \sum_{i=1}^{n} p_i e^{-4\pi^2 q^2 D_i t_d} \tag{2}$$

where D_i is the diffusion coefficient describing the diffusion of the fractions of molecules, p_i in domain i of the system, and the other symbols are as described in the previous equations. Eq. 2 does not take into account relaxation effects. If the relaxation times are significantly different for the different domains, each fraction, p_i will be a function of the different time intervals in the pulse sequence used in the experiment.

In the limit of fast exchange the attenuation is described by an average diffusion coefficient

$$D_{av} = \sum_{i=1}^{n} p_i D_i \tag{3}$$

The attenuation is then linear, and is given by

$$\ln \frac{I}{I_0} = -4\pi^2 q^2 D_{av} t_d \tag{4}$$

Between these two limits mentioned above one has to consider different degrees of exchange between the various domains during the observation time.

Restricted diffusion has different influences on the measurements. As the diffusion time increases, more molecules will be influenced by the restricting boundaries, giving a lower value of the measured diffusion coefficient. Thus, the measured diffusion coefficient will be time dependent.

The diffusion behavior in the limit of long (infinite) observation time, when the heterogeneity of a porous system is probed, can generally be described by the tortuosity, τ defined as

$$D_\infty = \frac{D_0 \varepsilon}{\tau} \tag{5}$$

where D_0 is the value of the diffusion coefficient at t=0, i.e. the value for the bulk diffusion of the liquid, and D_∞ is the corresponding value at $t = \infty$. ε is the porosity of the system. This expression is often used in order to estimate the effective diffusion coefficient in a porous system, like porous catalysts and also in porous polymer particles [3-4].

In a PFG-NMR experiment the effect of the porosity is already taken into account, and one has the following relationship

$$\frac{D_\infty}{D_0} = \frac{1}{\tau} \tag{6}$$

Thus, by measuring the diffusivity at long observation time and in bulk liquid using PFG-NMR, one can determine the tortuosity in the system of interest.

Experimental

The PE particles were delivered from Borealis A/S, where they were produced in a full scale slurry phase reactor. The grain sizes of the particles studied were 250-500 μm in diameter.

Pore size distribution of the particles were determined by use of mercury intrusion at a Pascal 140 instrument, and the porosity was determined by use of a Micrometric AccuPyc 1330 Helium-pycnometer, and at the Pascal 140 instrument. Scanning Electron Microscopy (SEM) pictures of both the outer surface of the particles, and of sliced particles were obtained. A razor blade was used for slicing of the particles.

Toluene and methanol were separately added to the PE particles. The purpose was to have liquids with different solubility in the amorphous phase of PE, and thus making it possible to vary the ratio between the amount of liquid in the semi crystalline phase of PE, and in the cavities inside the particles.

NMR samples of the PE particles were prepared by filling the particles in a 5 mm NMR tube. Toluene or methanol was then added in excess. The samples were dried at 40°C until it was observed visually that the inter particle liquid had evaporated; the remaining liquid are then to be found inside the particles. The tubes were then sealed off.

In addition to the samples of PE particles, some samples of liquid added to mono sized polystyrene particles were investigated. Three different types of polystyrene particles were studied. One system consisted of totally compact spheres with a mean diameter of 20 μm (sample PS1), and was delivered from Duke Scientific (U.S.A). The other two systems were porous particles with a mean diameter of 15 μm, and with different interior pore size distributions (sample PS2 and PS3). These particles were produced at SINTEF Applied Chemistry.

Polystyrene particles were filled into the 5 mm NMR tubes, and were then totally immersed in liquid, so that both the internal cavities and the cavities between the particles were occupied by liquid. Distilled water was added to sample PS1, while toluene was added to samples PS2 and PS3.

Diffusion experiments were performed on a Bruker Avance DMX200 instrument (Magnetic field strength = 4.7 Tesla, resonance frequency for protons = 200.13 MHz.) using a commercial diffusion probe from Bruker (PH MIC 200 WB 1H SAT 5/10). An applied gradient strength in the range 0-600 Gauss/cm was used. Unless stated otherwise, the experiments were performed at 25.0 ± 0.5°C.

In each diffusion experiment the different time intervals were kept constant, while the strength of the gradients was varied.

In the systems of polystyrene particles, the internal magnetic field gradients were found to be insignificant, and the monopolar stimulated echo pulse sequence was applied [6]. However, in the systems of PE particles, the internal gradients were significant, and pulse sequences applying bipolar gradients were used [12-13].

Results and Discussion

SEM pictures of the PE particles are shown in Fig. 1. a). The particles have a very irregular surface, indicating a rather complex structure and geometry.

The interior of the particles consists of rather large cavities, with a diameter of 10-60 μm, and the interior surface of these cavities seems to be made of smaller grains. From the look of the surface of the larger cavities one may assume that the interior of the particles consists of smaller grains, according to the MGM model.

The pore size distributions of the polyethylene particles and the porous polystyrene particles are shown in Fig. 1. b). The size distribution of the pores to be found inside the PE particles are very broad. The mean pore radius is around 0.5-1.0 μm, but there are pores varying from about 100 nm up to a few μm's. The interior porosity of the particles were measured to be 8 %. The mercury intrusion measurements did not detect pores/cavities larger than around 10 μm. and in addition isolated pores are not detected with this method. On can therefore assume that the porosity is higher than the value of 8 % measured by mercury intrusion porosimetry. The interior of the particles can be divided into two main domains, the large cavities, 10-60 μm in diameter visualized by SEM, and the areas detected by mercury intrusion, where smaller pores, ranging from a few μm down to about 100 nm, are found.

On the other hand, the polystyrene particles have narrow pore size distributions. In sample PS2 the mean pore radius of the internal pores is around 0.2-0.5 μm, while for the PS3 sample it is around 0.02-0.03 μm. The radius of the cavities between the particles are 2-3 μm in both samples.

By performing PFG-NMR diffusion experiments of a liquid in the polystyrene particles one can have a better understanding of how to interpretate the results obtained in the PE particles.

The diffusion experiments in the system of polystyrene particles showed that because of fast exchange, one can not separate diffusion of a liquid found inside the particles (small cavities) and between the particles (large cavities). A mean value has to be used. The same approach has to be used in the interpretation of the PFG-NMR results obtained for diffusion of liquids in the cavities of the PE particles, which have a broad size distribution.

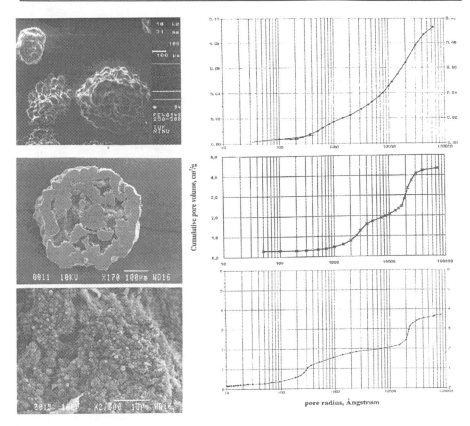

(a) Scanning Electron Microscopy (SEM) pictures of the PE particles. Top: unsliced particles, middle: sliced particle, bottom: a magnified area of the interior surface of a cavity.

(b) Pore size distributions. Top: PE particles, middle: polystyrene particles PS2, bottom: polystyrene particles PS3.

Fig. 1. SEM pictures and pore size distributions of the particles studied.

In the system of compact polystyrene spheres immersed in distilled water (sample PS1), the obtained tortuosity value, $\tau=1.45$, is as expected in a system of random loose packing of mono sized spheres [11]. Sample PS2 has slightly higher tortuosity value than sample PS1, and longer observation time is necessary in order to reach the tortuosity limit. This is as expected, because of the presence of the rather large interior cavities in sample PS2, giving an additional diffusion path through the particles. In sample PS3, on the other hand, the situation is different. Here the internal cavities are so small that they will give a strong restriction on the diffusion, and the tortuosity limit is higher in this sample, $\tau=1.64$.

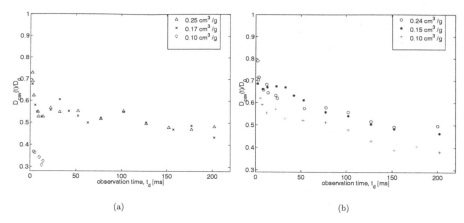

(a)   (b)

Fig. 2. Diffusion measurements in the samples where toluene (a) and methanol (b) was added separately to the PE particles. The normalized diffusion coefficient in the cavities, $D_{cav}(t)/D_0$, is plotted as a function of the observation time. The bulk values, D_0 for the liquids were $2.28 \cdot 10^{-9}$ m^2 s^{-1}, and $2.42 \cdot 10^{-9}$ m^2 s^{-1} for toluene and methanol, respectively.

The diffusion experiments performed in the PE particles were analyzed according to Eq. 2, but by also taking into account effects of restricted diffusion [15]. Two components were assumed, representing diffusion of liquid in semi crystalline polymer, and in the cavities of the particles. The diffusion in the cavities was treated according to Eq. 4, assuming fast exchange between small and large cavities, giving an average diffusion coefficient. Details regarding the analysis are given in [15].

The diffusion coefficients representing diffusion in the cavities (D_{cav}) vary with both type and amount of liquid added, and with the observation time. Because the rather low amount of liquid found in the semi crystalline PE, the signal-to-noise ratio was rather poor for the part of the experimental attenuation curve representing this diffusivity. The diffusion coefficient representing diffusion of liquid in the semi crystalline PE (D_{sc}), varied randomly in the range 1-4 10^{-11} m^2 s^{-1}, a value which corresponds with similar results found in the literature [16-17].

Though the variation in the obtained D_{sc}'s was random, there was a tendency of obtaining higher values at short observation time, indicating effects of restricted diffusion. Since the main interest in this study is the diffusivity in the cavities, the values from the semi crystalline domains are not presented here.

From the calculated fractions of liquid dissolved in semi crystalline PE, using Eq. 2, it is possible to estimate the amount of liquid to be found in semi crystalline PE, and the amount to be found in the cavities, for each sample. When these amounts are compared with the pore size distribution in Fig. 1 b), it is possible to approximately estimate the size of the pores which are filled in each sample. All the important parameters determined in this manner are summed up in Table 1. The relative amount of liquid dissolved in semi crystalline PE is approximately constant in each type of sample with different filling degree, but clearly, the solubility of methanol is lower than toluene.

The results obtained for the time dependent diffusivity in the cavities of the PE particles in the different samples studied are given in Fig. 2.

For the samples with toluene added, there is no significant difference for the results obtained in the two samples having the highest filling, and the tortuosity limit obtained is the same. For the sample having the lowest filling (0.10 cm^3/g) there is a significant difference. Already at the shortest observation time, the tortuosity limit is reached, and the tortuosity value is significantly higher than in the other samples. As it is indicated by the estimated amount of liquid found in the cavities in this sample, only the smallest cavities are filled up. In the other two samples the small cavities are filled, but in addition a certain fraction of the larger cavities are filled.

Table 1. Diffusion experiments in the samples of PE particles. The total amounts of toluene and methanol added (cm^3/g)$_{tot}$, are given. p_{sc} is the fraction of liquid found in the semicrystalline PE. The estimated amounts of liquid in semicrystalline PE (cm^3/g)$_{sc}$, and in the cavities (cm^3/g)$_{cav}$, is given for each sample. The tortuosity values τ, determined from the diffusivity limit at long observation time, is given with an uncertainity of $\pm$ 0.15

	Toluene added			Methanol added		
(cm^3/g)$_{tot}$	0.25	0.17	0.10	0.24	0.15	0.10
p_{sc}	0.2	0.4	0.7	0.05	0.15	0.2
(cm^3/g)$_{sc}$	0.05	0.07	0.07	0.01	0.02	0.02
(cm^3/g)$_{cav}$	0.20	0.10	0.03	0.23	0.13	0.08
τ	2.1	2.1	3.0	2.0	2.0	2.5

In the samples with methanol added, the two samples having the highest filling have about the same tortuosity limit as the corresponding samples with toluene, and we may assume that here both the small and large pores are more or less filled up, which is also indicated by the estimated amount of liquid found in the cavities given in Table 1. In the sample with lowest filling the tortuosity value is higher, but not as high as in the corresponding sample with toluene. When the same amount of toluene and methanol is added separately to the PE particles, a higher fraction of toluene dissolves in semi crystalline PE, while in the sample with methanol a certain fraction of the larger cavities will be filled up in addition to the small cavities. This explains the difference in the time dependent diffusivity for these two samples, and the difference in tortuosity value obtained.

The results presented in Fig. 3 show that the estimated tortuosities vary with the filling degree of the different cavity sizes. When only pores smaller than 1 μm is filled with liquid, the tortuosity value is around 3. This represents diffusion in the areas having the smallest pores. As larger pores are filled with liquid, the tortuosity value decreases to a plateau value of around 2.

These tortuosity values are lower than the ones normally used in the modeling of mass transfer in polymerizations of polyolefins [3-4], especially the values for the larger cavities. This suggest that mass transfer in the larger cavities is not significant, and that these cavities are in direct contact with the bulk liquid outside the particle, and may function as intraparticle reservoirs for monomers, as suggested by McKenna et al. [5].

With additional information about the porosity, these tortuosity values may be used to determine the effective diffusivity to be used in an intraparticle mass transfer model using Eq. 5.

The results presented here show how the geometry of the different areas in the particles influences the effective diffusivity. The PE particles can be divided into three different mass transfer areas; diffusion in large cavities (10-60 μm), diffusion in small cavities (≤ 1 μm), and diffusion in semi crystalline PE.

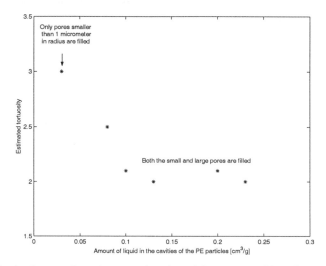

Fig. 3. The obtained tortuosity values as a function of the amount of liquid in the cavities of the PE particles, $(cm^3/g)_{cav.}$

Conclusions

Diffusion measurements performed in samples of randomly packed mono sized porous and non porous polystyrene particles show that because of fast exchange, one cannot separate diffusion of liquid in small and large pores. This is also the situation for the diffusion process in the porous PE particles.

By varying the amount of liquid added to the PE particles, cavities of different size are occupied, and the tortuosity value, τ, in the different areas of the particles can be determined.

Clearly, the diffusion coefficients, and thus the tortuosity, vary with the filling degree of the cavities in the PE particles. The tortuosity values estimated for the different areas of the particles may be used as an input to a intraparticle mass transfer model.

Acknowledgment

We gratefully acknowledge financial support from the Research Council of Norway (NFR) through the Polymer Science Program.

References

1   Floyd, S, Choi, K. Y., Taylor, T. W., and Ray, W. H., *J. Appl. Polym. Sci.* **1986**, *32*, 2935.
2   Hutchinson, R. A., and Ray, W. H., *J. Appl. Polym. Sci.* **1987**, *34*, 657.
3   Hutchinson, R. A., Chen, C. M., and Ray, W. H., *J. Appl. Polym. Sci.* **1992**, *44*, 1389.
4   McKenna, T. F., Dupuy, J., and Spitz, R., *J. Appl. Polym. Sci.* **1997**, *63*, 315-322.
5   McKenna, T. F., Cokljat, D., Spitz, R., Schweich, D., *Catalysis Today*, **1999**, *48*, 101.
6   Price, W. S., *Concepts in Magn. Reson.* **1998**, *10*, 197.
7   Zhong, J., Kennan, R. P., and Gore, J. C., *J. Magn. Reson.* **1991**, *95*, 267.
8   Hurliman, M. D., *J. Magn. Reson.* **1998**, *131*, 232.
9   Karlicek, R. F., and Lowe, I. J., *J. Magn. Reson.* **1980**, *37*, 75.
10 Cotts, R. M., Hoch, M. J. R., Sun, T., and Markert, J. T., *J. Magn. Reson.* **1989**, *83*, 252.
11 Latour, L. L., Li, L., and Sotak, C. H., *J. Magn. Reson. B.* **1993**, *101*, 72.
12 Sørland, G. H., Hafskjold, B., and Herstad, O., *J. Magn. Reson.* **1997**, *124*, 172.
13 Sørland, G. H., Aksnes, A., and Gjerdåker, L., *J. Magn. Reson.* **1999**, *137*, 397.
14 Seland, J. G., Sørland, G. H., Zick, K., and Hafskjold, B., *J. Magn. Reson.* **2000**, 146, 14.
15 Seland, J. G., Ottaviani, M., and Hafskjold, B., to be submitted to *J. Coll. Int. Sci.*
16 Fleischer, G., *Colloid & Polymer Sci.* **1984**, *262*, 919.
17 Fleischer, G., *Polymer Comm.* **1985**, *26*, 20.

Printing: Saladruck, Berlin
Binding: Buchbinderei Lüderitz & Bauer, Berlin